World Soils Book Series

Series editor

Prof. Alfred E. Hartemink
Department of Soil Science, FD Hole Soils Laboratory
University of Wisconsin–Madison
Madison
USA

For further volumes:
http://www.springer.com/series/8915

Aims and Scope

The *World Soils Book Series* brings together soil information and soil knowledge of a particular country in a concise and reader-friendly way. The books include sections on soil research history, geomorphology, major soil types, soil maps, soil properties, soil classification, soil fertility, land use and vegetation, soil management, and soils and humans.

International Union of Soil Sciences

Toma Shishkov • Nikola Kolev

The Soils of Bulgaria

Springer

Toma Shishkov
Institute of Soil Science,
Agrotechnologies and Plant Protection,
Nikola Poushkarov
Sofia
Bulgaria

Nikola Kolev
Agricultural Academy
Sofia
Bulgaria

ISSN 2211-1255 ISSN 2211-1263 (electronic)
ISBN 978-94-007-7783-5 ISBN 978-94-007-7784-2 (eBook)
DOI 10.1007/978-94-007-7784-2
Springer Dordrecht Heidelberg New York London

Library of Congress Control Number: 2013958218

Printed on acid-free paper

Springer is part of Springer Science+Business Media (www.springer.com)

Preface

Imagine a country of diversity and magnitude, that extends along the Balkan mountains in the North to the Rhodope mountains in the South, and to the Black sea in the East. Not unlike the land, each of its soils are a unique representation of the region's rich history and the natural processes that have come to be defined as characteristic of Bulgarian soils' formation. To exemplify the variety of soils one can observe just how the main type of soil in Bulgaria, the chernozem—a well—structured soil by definition, essentially calcareous and susceptible to humus accumulation – alone is sub-divided further into four types: the calcareous chernozems, the typical chernozems, the leached chernozems, and the degraded chernozems. The chernozem, however, is one type, among the multiplicity of soils present in Bulgaria. This structure of soils being highly complex and distinctive is thus a main theme in the discussion of soil research in Bulgaria.

In this perspective, this book provides the reader an insight into soil concepts that are particularly representative of Bulgaria. It presents contemporary scientific interpretation of soil processes based on the newest references for soil classification (USDA Soil Taxonomy, World Reference Base for Soil Resources and FAO's Revised Legend of the World Soils).

The book contains 16 chapters which deal in-depth with various aspects of soil. The first two chapters delve into the history of Bulgarian soils and discuss the soil science progress in Bulgaria, as well as environmental conditions responsible for soil diversity in the country. Furthermore, they approach soils as having a key role as a source of nutrients and water. Chapters 3–12 analyze the basic features and soil forming processes of some of the major groups of soils in the country, applying a pedological approach. The last two chapters focus on land use in the country and the ecology of soil formation.

Acknowledgments

I would like to thank everybody who has helped in the preparation of this book, but I am especially grateful to:

Dr. Kercheva Milena, Chief of the Physic department at the Institute of Soil Science, Agrotechnologies and Plant Protection Nikola Poushkarov for assistance with the interpretation of information and statistics;

Dr. Jokova Mariyka, for comments on soil chemistry and extractable forms of iron, manganese and aluminium;

Nedyalkov Simeon from the Agency for Soil Resources (former) at the Ministry of Agriculture and Food for permission to use the latest data regarding soil classification and soil salinezation;

Prof. Rousseva Svetla, Head of the Soil Science Direction at the Institute of Soil Science, Agrotechnologies and Plant Protection Nikola Poushkarov;

Dr. Dimitrov Emil at the Physic department at the Institute of Soil Science, Agrotechnologies and Plant Protection Nikola Poushkarov for support in the visual aids.

The suggestions and information they provided were valuable contributions to this book.

Contents

Introduction

Historical

Soil is the most important phenomena for the existence of life on Earth.

In this book, the history of mankind is divided into two periods: the prehistoric and the historic; and the development of pedology is also divided into two stages: one going back many centuries from which time it collected facts and empiric knowledge, followed by the period when soils became the subject of academic study.

Throughout the history of mankind, human beings have consciously looked at how earth can be used in various ways. Soils promote the prosperity of a society. Fertility is the most significant feature of the earth and is a synonym for soil. Soil fertility is a result of soil regimes. From ancient times, man has evaluated soil in terms of crop production. That is why much was known about soil fertility long before pedology appeared.

As a result of the practices of earth cultivation and land evaluation, pedology emerged as a natural science during the second half of the nineteenth century. As a natural science, pedology adopted the methods of more advanced sciences such as chemistry, physics, physical chemistry, mineralogy, geology, biology, and others. With its origins in regard to how to obtain higher yields, pedology rapidly became a contemporary science with its own set of principles, concepts, and methodology which applied advanced methods to find solutions to problems.

Pedology (or soil science) is concerned with gaining a scientific understanding of the natural object "soil." The word taxonomy originates from the Greece "taxis" which means an order, or from the Latin "takso" which means to evaluate and "nomos" which means a law (Kovda and Rozanov 1988). The taxonomy units are consequently subordinated categories which reflect the existent soils as an object in nature. Knowledge of the world's soil diversity and the links between and within the different soil groups (taxons) is the main task of pedolody.

This means that the object "soil" should be described in such a way that its interpretation reveals characteristics based on the defined diagnostic criteria, features, and properties in compliance with contemporary soil science.

Nowadays, the main aim of soil science is the understanding of soil regulatory function maintaining the exchange of energy and matter and thus supporting life development.

Some of its essential and ongoing trends are the compilation of georeferenced soil information, development of effective spatial models, assessment of the implication of space technologies for observation, adequate prediction of threats, soil conservation, and fertility management based on the integrated indexes.

Purpose of the Book

While there are a number of books that cover different aspects of soils in Bulgaria, there is a need to collect and clarify the data that have accumulated over the last four decades about the methods of nomenclature and classification of soils. It must be mentioned that this book intends to present a fundamental level of information about concepts of soils of Bulgaria to readers who specialize in branches other than pedology. Specific detailed local information is gained from reference to the soil survey reports or data from authorized centers of information.

The book describes and discusses the composition of the major soils that occur in the terrain of Bulgaria as space continuous natural bodies. It deals with basic soil processes. The authors take responsibility for the accurate presentation of the facts.

Reference

Kovda VA, Rozanov BG (1988) Soil Science. Part I The soil and soil development. High school Press, Moscow, p 400 (In Russian)

1 General Facts of Soils in Bulgaria

1.1 Soil Research in Bulgaria

The development of soil science in Bulgaria came about due to the urgent need to increase the economic prosperity of the country and find effective ways to raise crop yields.

A brief historic review of soil science in Bulgaria outlines four periods (Garbouchev 1972).

The first period from, 1911 to 1948, was characterized by the collection of empirical data, the training of specialists and skilled workers, the commencement of the Soil map of Bulgaria at scale 1:200,000 (Tanov and Koynov 1956) and the publication of the monograph "Soils of Bulgaria" in 1948 (Antipov-Karataev and Gerassimov 1948).

The period from 1948 to 1960 was characterized by the growth of Bulgarian pedology based on the genetic conception of soil evolution which was introduced after two Bulgarian-Soviet expeditions in 1947 and 1954 and as a result of scientific work on the monograph "Soils in Bulgaria" (Antipov-Karataev et al. 1960).

The period from 1960 to 2000 was characterized by active close cooperation at the international level, the strengthening of relations between world organizations and the adoption of advanced international methodologies for further research. At the end of 1974, a soil survey at scale 1:25,000 covered the whole territory of the country and in 1975 a soil survey was undertaken at scale 1:10,000, 1:5,000 for the cultivated lands in the country (Dimitrov 1989).

The period after 2000 is marked by the association of Bulgaria as a member state of the European Union. The Strategy of soil threats (2002) in the European Union is widely applied.

The first paper describing Bulgarian soil classification based on the content of sand, clay, lime, and humus was published in Belgrade in 1853.

The first period of soil science development in Bulgaria started in 1911 when a small department of agro-geology was set up at the former Central experience farm station in Sofia headed by the accomplished scientist Nikola Pouskarov (Fig. 1.1). Later, in 1924, the second center for soil study was established at the Department of common agriculture at the faculty of Agrarian-forestry in the Sofia National University. Its chair was the academician Ivan Stransky who wrote the first textbook on soil science in 1929 (Fig. 1.2). Being a Russian alumnus, he had a great influence on the adoption and popularization of the ideas and methods of Dokuchaev scientific school.

At that time, for many reasons, the Bulgarian government did not give attention or financial support to pedology or seek to stimulate an increase in the number of specialists. Research results were mainly based on the efforts of the specialists. In evaluating this period it should be mentioned that despite the difficulties, Bulgarian soil scientists did undertake an important study related to the origin, nature, features, and geographic distribution of the few new unknown soils in Bulgaria. Nikola Poushkarov (1913) presented the first soil map of the region of Sofia at a scale of 1:126,000. The main achievement during that period was the creation of the first version of Soil map of Bulgaria at scale 1:500,000 edited by Nikola Poushkarov (1931). This map showed the geographic distribution of the main soil units. In 1940, Nikola Poushkarov was authorized by the Ministry of Agriculture to create a Soil map of Bulgaria at scale 1:250,000.

Nikola Pouskarov was dismissed from his job in 1941 for his democratic ideas, and died in February 1943. During this whole first period, Bulgarian soil science had a goal to define soil resources in the country.

The second period started in 1946 with the creation of a soil map at scale 1:250,000. For this purpose the first Bulgarian-Soviet expedition, led by Academician Gerasimov and Professor Antipov-Karataev, was undertaken in 1947, followed by a second one in 1954. On the Bulgarian side Academician Ivan Stransky and many pedologists.

The expedition solved few problematic aspects. As a result of this a soil map at scale 1:1,000,000 was edited by the Academician Gerasimov along with a book of the classification of soils in Bulgaria (Antipov-Karataev and Gerassimov 1948).

T. Shishkov and N. Kolev, *The Soils of Bulgaria*, World Soils Book Series, DOI: 10.1007/978-94-007-7784-2_1,

Fig. 1.1 Nikola Poushkarov (1874–1943)

Fig. 1.2 Ivan Stranski (1886–1959)

In 1947 Academician Ivan Stransky initiated the setting up of the Soil Science Institute authorized by the Bulgarian Academy of Science. The establishment of the Soil museum was of great service to soil science.

In 1948, the Ministry of Agriculture set up a new Institute of Soil Science named in memory of Nikola Poushkarov.

As a result of the efforts of soil scientists a Soil map of Bulgaria at scale 1:200,000 was completed in 1956, authored by Tanov and Koynov (who elaborated Thracian lowland).

The monograph "Soils in Bulgaria" was published in 1960 (Antipov-Karataev et al. 1960), a result of collaboration between Bulgarian and Soviet pedologists. The monograph contained all the available data on morphological, physical, chemical, and physicochemical characteristics of the main soil units in Bulgaria.

The third period of Bulgarian soil science was brought about by the demands for more detailed data concerning soil resources in the country. For this reason, after 1957 huge efforts were made to undertake a large systematic soil survey of the country.

At the end of 1968 a soil survey at scale 1:25,000 was completed for about 60 % of the entire territory and approximately 75 % of the arable lands. In December 1974 a soil survey at scale 1:25,000 covered almost the whole territory of the country (Garbouchev 1972). That survey contributed to the detailed (according to the scale) classification of soil varieties, presented agrarian status of land resources, and determined soil features responsible for the effective maintenance of soil fertility.

State farms have had soil maps with land management regulations which provided a scientific basis for the most effective use of soils. A large soil map at scale 1:25,000 contained data of land cover diversity in the country, revealed soil genetic individuality, and characterized soil associations.

The vast amount of material pertaining to geographic distribution and soil characteristics was a basis for the creation of a new soil map of Bulgaria at scale 1:400,000 with enclosed description (Koynov et al. 1968). This map generalized the objective information on the geographic distribution of soil units based on the survey 1:25,000, and information on parent rocks influence, relief specifics, and shallow soils was also included. The map identified 67 soil units at the levels of type and subtype, soil texture class, and degree of erosion.

As a result soils were divided into groups according to their productivity, as follows: calcareous and typical chernozems—780,000 ha; slightly, medium, and strongly leached chernozems—700,000 ha; degraded chernozems, dark gray and gray forest soils—200,000 ha; smolnitsa—370,000 ha; cinnamonic forest soils—1 000,000 ha, alluvial and colluvial soils—780,000 ha, and the rest 200,000 ha were brown forest soils, rendzinas, boggy, and sandy soils. Most soils were favorable for cultivation and crop production. The main limitations were shortage of nitrogen, phosphorous, and deficiency of soil moisture.

After the generalization of the map at 1:400,000 scale, the new soil map at the scale of 1:1,000,000 was compiled and issued in the Geographic atlas of Bulgaria (Koynov 1973a, b). It distinguished 45 soil units. The Geographic atlas of Bulgaria also included maps of the geographic distribution of the soils according to their texture and soil reaction (both at 1:3,000,000 scale), maps of the soil geographic regions and soil erosion regions (both at 1:2,000,000 scale), and the distribution of the soil resources within the administrative districts of the country (Koynov 1973a, b).

Professor Tzvetan Staykov at the Institute of Soil Science contributed greatly toward the development of the analytical studies. From 1956 to 1962 many analyses of soil nutrition components were carried out, including mobile forms of nitrogen, phosphorous, potassium, and soil reaction.

A new cycle of agrochemical analysis of the county lands based on new improved methodology started in 1965. At the end of 1968 the study of mobile forms of the main soil nutrients covered 2,200,000 ha of the cultivated lands.

During the second period basic studies of physical properties, sorption capacity, clay minerals, erosion control, reclamation, and microbiology contributed to a better understanding of soils in Bulgaria.

Forestry pedology was established as a new branch of soil science in Bulgaria in 1950 when a Department of forestry pedology was set up at the Institute of Forestry, and regular teaching of the discipline of forestry pedology began in 1954 (The Institute of Forestry is now called the Forestry University).

Those 25 years of soil science development in Bulgaria were actually a tangible basis for future soil research in the country.

The third period of the development of Bulgarian soil science began in 1960 when two institutes of soil science (the first belonging to the Bulgarian Academy of Science and the second belonging to the Ministry of Agriculture) merged into the Institute of Soil Science Nikola Poushkarov. The Institute was founded with the purpose to coordinate scientific research within the country with that at the international level and to solve issues of soil fertility improvement.

A large scale soil survey department and agro-chemical service was founded at the Institute of Soil Science Nikola Poushkarov in 1965. In 1975 a campaign was started to carry out a large soil survey at scale 1:10,000, 1:5,000 and set up a systematic study of nutrients balance in Bulgarian soils. The main achievements were obtaining qualitative and quantitative indices of soil processes as well as soil productivity maintenance.

An inventory of soil resources was included for the purpose of cadastre; it reflected soil classification harmonization, and the improvement of practical activities with regard to erosion control and soil melioration.

The main problematic aspects of investigations during that time were: the chemistry of total and available forms of nutrient elements (N, P, K, Ca, Mg, Na and trace elements) based on the characteristics of the sorption, clay minerals, organic matter, and soil reaction in soil varieties (Raykov and Ganev 1971; Ganev 1989); the regulation of physical conditions and development of indices linked to fertility of soils, obtaining to rich experimental material of soil regimes state (Kercheva and Dilkova 2005a, b); the study of microbiological activity in different soils, crop rotations, and ecological conditions.

The distribution of clay minerals for Bulgarian soils has been carried out with regard to ecological conditions (Boneva 2012).

Aiming at the further improvement of fertilization efficiency the temporary geographic net of well equipped experimental bases related to the particular soil units was established. Agro-chemistry research and practical works have been carried out to further conserve the nutrient status of cultivated Bulgarian soils based on the balance models of plants uptake.

It was established that erosion affected 60 % of the arable lands in the country; and a scheme of the regions at scale 1:600,000 with the intensity rates of erosion was created. Effective erosion prevention practices were elaborated in different soil and ecological conditions. The necessity of organizing systematic erosion control measures led to the establishment of permanent watershed nets of observation. The four experimental ground fields in different regions were a model for systematic control against erosion in the territory of the entire country.

For the time being, field plot experiments have been carried out at nine experimental sites for soil erosion research across Bulgaria, representing diverse soil and climate conditions, in order to study soil protection, the reduction of surface runoff, and the economic efficiency of different practices for soil erosion control (Lazarov et al. 2002). The data from the field plot studies have been used to validate widely used erosion prediction models, such as USLE and WEQ, to be integrated with GIS for soil erosion risk assessments and optimization of measures for soil erosion control (Rousseva 2002; Rousseva and Stefanova 2006).

Melioration investigations were carried out with regard to groups of soil with unfavorable features presented at 70,000 ha of saline and swampy soils, 11,000 ha of sandy soils, 460,000 ha acidic water-logged light gray pseudo-podzolic and cinnamonic podzolic pseudopodzolic strongly gleyic soils, and 110,000 ha slightly gleyic soils.

Bulgaria is situated in a zone where seasonable agricultural drought is a regular annual occurence that lasts a long time and can impact crop production. It is a growing concern that the best practises of irrigation and water management are scientifically maintained (Popova 2012).

Two of FAO's international meetings held in May 1979 and June 1981 were hosted in Sofia by the UNEP and FAO with the aim to produce a widely acceptable world legend of soil units. For the first time, a new approach was applied in soil nomenclature.

For 25 years the soil monitoring grid 16 × 16 km with 256 plots in the forest ecosystems has been carried out by means of systematic analysis of soil parameters guided by the Institute of Forestry at the Bulgarian Academy of Science and Forestry University.

According to Annex V (1994) the goal of the United Nation Convention to Combat Desertification (UNCCD) is to prevent and limit land degradation in Europe. The UNCCD stipulates soil monitoring in article 10(a), which states that early warning systems be established.

Since 2004 the National Land and Soil Monitoring System (NLSMS) has been a subsystem of the National System for Monitoring of the Environment (NSME), conducted by the Executive Environment Agency, and the Ministry of Environment and Water. The systematic determination of soil variables to record their temporal changes was at grid 16 × 16 km with 407 sites for cultivated and grazing lands. A series of pedogenic studies have provided update specificity of soil variability (Shishkov et al. 2009).

The European Community Treaty, article 174, states that one of the objectives of the European Union is to maintain a high level of environmental protection. The decision No 1600/2002/EU determines the 6th Community Environment Action Program with the aim to protect the environment and promote sustainable land use.

In a Communication COM/2002/179 of the European Commission entitled: Towards a Thematic Strategy on Soil Protection, published on 16 April 2002, eight soil threats were recognized for the soils of the European Union.

In 2006 the National Action Program for Sustainable Land Management and to Combat Desertification in Bulgaria promoted the role of soil protection.

The national Soil law came into force on 6 November 2007 with the help of the previously outlined activities of soil scientists. Improvement of access to the spatial soil data was implemented by Directive INSPIRE—Infrastructure for Spatial Information in the Directive (2007/2/EC).

1.2 Classification of Soils in Bulgaria

Soil systematic or classification is a scientific conception of the real distribution of soils in nature. The main purpose of soil systematic is the establishment of the full classification of soils in the world. It is based on the realistic soil diagnostic and comparison of similarities which can be analyzed systematically.

Nowadays the three most widespread classification systems that determine the framework of the contemporary world soil nomenclature are: FAO/UNESCO, USDA Soil Taxonomy, and the World Reference Base for Soil Resources (recently proposed as an official soil classification for the European Union).

Each classification is based on its principles of soil diagnostic and systematic (Table 1.1). FAO/UNESCO is the international soil classification, the result of international cooperation of the national experts; USDA Soil Taxonomy is based on the diagnostic horizons; and WRB is a tool for correlation of the national systematic systems.

While conducting surveys in Russia, Dokuchaev discovered that the nature and properties of the soils vary according to environmental factors, particularly climate and vegetation. He introduced this common principle into the scientific terminology by means of the adoption of a simple vocabulary and recognized for the first time the sequence of horizons.

Russian soil nomenclature has a great influence in the world and particularly in Bulgaria. In Russian tradition, terms are formulated by means of using the base for the word "zem" which corresponds to the English "sol". At the international level, for example, "Chernozem" or "Mollisol" are widely accepted.

Bulgarian soil classification adopted the Russian tradition of giving names to soils; however, there was also the influence of the national linguistic specific. Due to the historical partnership between the soil scientists from both countries, the Bulgarian soil classification was actually set up after the Russian school of pedology and is based on Doukuchaev's principles of soil type (Antipov-Karataev and Gerasimov 1948). Soil type is a basic unit in the Bulgarian classification and it corresponds to the great soil group in the Soil Taxonomy and Reference soil group in WRB. The classification was the basis for the Legend of the Soil map of Bulgaria at scale 1:200,000 that was developed in 1956 (Tanov and Koynov 1956).

Soil type refers to a group of soils where development in similar environmental conditions of climate, hydrology and biological species has resulted in a specific type of soil formation. It comprises similar trends in organic matter decomposition and humic substances formation, migration and accumulation of the solutions, transformation and synthesis of mineral parts, and formation of soil horizons in the solum.

Correspondence between the different systematic units is shown in Table 1.2.

In outlining this fact, we pay tribute to the previous generation of soil scientists in Bulgaria who advanced the system whereby the name of a soil holds the meaning of display of soil feature that belongs to the sequent lower level of soil systematic. For example, according to the Bulgarian classification there is clear relation of leaching

Table 1.1 General principles applied in the Bulgarian soil classification, Soil Taxonomy, and World reference base for soil resources

Extended systematic list, 1976, (Bulgarian soil classification)	USDA Soil Taxonomy system (2010)	WRB classification system, 2006
• Similar genesis	• Criteria are observable or measurable, not inferred	• Refer to present day characteristics and functions
• Sequence of genetic horizons	• Define all soils in soil survey, with natural boundaries in property space	• Significance for management purposes
• Similar relief occupation	• Classify soils as they occur, not concepts	• Soil forming processes are not used as criteria itself
• Similar air, water and temperature regimes	• Soil moisture and temperature 'regimes' are extremely important for interpretations	• Soil correlation amongst national classification systems

Table 1.2 Correspondence between different systematic units

Soil Taxonomy	WRB classification system	Bulgarian classification
ORDER—soil forming processes, determining of presence or lack of the diagnostic horizons	*REFERENCE SOIL GROUP*—primary pedogenetic process produced characteristic soil features	*TYPE*—unity of soil origin and similarity of the soil forming processes
SUBORDER—similarity of conditions of soil temperature and moisture, parent materials, vegetation	*Lower categoric level, combination of Reference Soil Group with QUALIFIERS* predominant secondary soil forming process significantly affected the primary soil features	*SUBTYPE*—Manifested extent of basic processes
GREAT GROUP—similarity in sequence and performing of diagnostic horizons, saturation, temperature, and moisture		*GENUSES*—qualitative aspects connected with the composition of parent material and groundwaters
SUBGROUP—central meaning in great group		*UNIT*—manifested extent of basic process specific to a given soil type leaching, humification, salinization
FAMILY—practical peculiarities, texture, mineralogy, management		*SUBUNIT*—differences of features related to humus horizon thickness, humus content, salinity, etc.
SERIES—characteristic and sequence of horizons, structure, color, chemical features, place names		*VARIETY*—within the species differ by the surface horizon texture

pronounced in the chernozems profile with the classification at the subtype level.

Consequently, the full name in the Bulgarian classification scheme contains all names in different taxonomic levels, starting from soil type, followed by the scheme, and ending according to the scale. An example is slightly leached (*subtype*) chernozem (*type*), noneroded and slightly eroded (*genuses*), low thick (*unit*), loam (*variety*).

The process of defining the soil type includes sequence of genetic horizons in the soil profile; type of landscape and position; geo-referencing of the area of soil distribution; specifics of soil forming processes; and specifics of solutions migration and accumulation.

In 1964 the Bulgarian Soil Classification (Koynov et al.) was developed in compliance with the soil survey of the country at scale 1:25,000.

The campaign to carry out a systematic large soil survey at scale 1:10,000 was commenced in 1975. As a result of this, in 1976 the updated version of classification named "The Extended Systematic List" was created comprising full nomenclature of occurred mapping units (Yolevski and Hadzhiyanakiev 1976).

There were a few more classifications in 1983 (Yolevski et al.), and 1992 (Penkov et al.); the latter were compiled for the purpose of land parcel fragmentation, but not used in practice.

In 1993, based on the 1976 classification, the improved version was prepared for the purpose of the National Soil Agency and the Ministry of Agriculture and Food. Since 1993 it has been applied in the national GIS and digital information by the National Soil Agency.

A contemporary correlation deals with the Bulgarian classification (Yolevski and Hadzhiyanakiev 1976) and uses the latest soil classifications found in USDA Key to Soil Taxonomy (edition 2010), World Reference Base for Soil Resources (edition 2006), and FAO's Revised Legend of the World Soils (FAO-UNESCO-ISRIC 1990) (Shishkov 2011).

1.3 Conditions and Factors of Soil Formation

1.3.1 General Conditions

Bulgaria is situated on the Balkan peninsula and occupies part of southeastern Europe. It is located at latitude 43° north of the Equator and longitude 25° east of Greenwich. To the east Bulgaria borders with Romania as most of the border runs along the Danube (471.5 km). The eastern Bulgarian border runs along the coast of the Black Sea (378 km). South Bulgaria borders Greece (493 km) and the southeast of Turkey (259 km); to the west it borders with Serbia and Macedonia (506 km). Bulgaria's total territory is 110,994 km^2, and its surface is very diverse (1982). Lowlands to an altitude of 200 m occupy 31.5 % of the country's territory, hilly lands with altitude 200–600 m—41 %, low mountainous lands with altitude 600–1,000 m occupy 15 % and mountain lands over 1,000 m—12.5 % (Fig. 1.3). Lands of an altitude of 600 m are of great importance, representing 72.5 % of the total area.

Geomorphology of Bulgaria

The territory of the country is dominated by hilly mountainous relief, approximately 66 % of lands occupy slopes over 3°, lands with slopes exceeding 6° are 35.2 % of the country's territory (Fig. 1.4).

Detailed information on the Global and National Soil and Terrain Digital Databases (SOTER) methodology with special emphasis on small-scale physiographic mapping is given in the SOTER Manual (van Engelen and Wen 1995). The slope classes compared to those used for major landforms (simple) are as follows:

W 0–2 % flat, G 2–5 % gently undulating, U 5–8 % undulating, R 8–15 % rolling, S 15–30 % moderately steep, T 30–60 % steep, and V ≥60 % very steep.

The topology of Bulgaria in combination with hypsometric grouping and degree of dissection has delineated several large physiographic units which were formed in parallel direction and have a deep geological disposition (Geography of Bulgaria 1982). One or more typical combinations of surface forms alternate from north to south with extension from west to east and can be subdivided into terrain combinations as follows: hilly plains (occupy 41 % of the territory), mountains with intermontane valleys, lowlands and highlands (occupy 2.5 % of the territory). These relief forms are determined from specific geological structures formed by different tectonic movements which have occured in the geological history of Bulgaria.

Bulgaria is a part of the tectonic zone between two lithospheric plates—Eurasian and African, which determines numerous tectonic movements occurring on the territory.

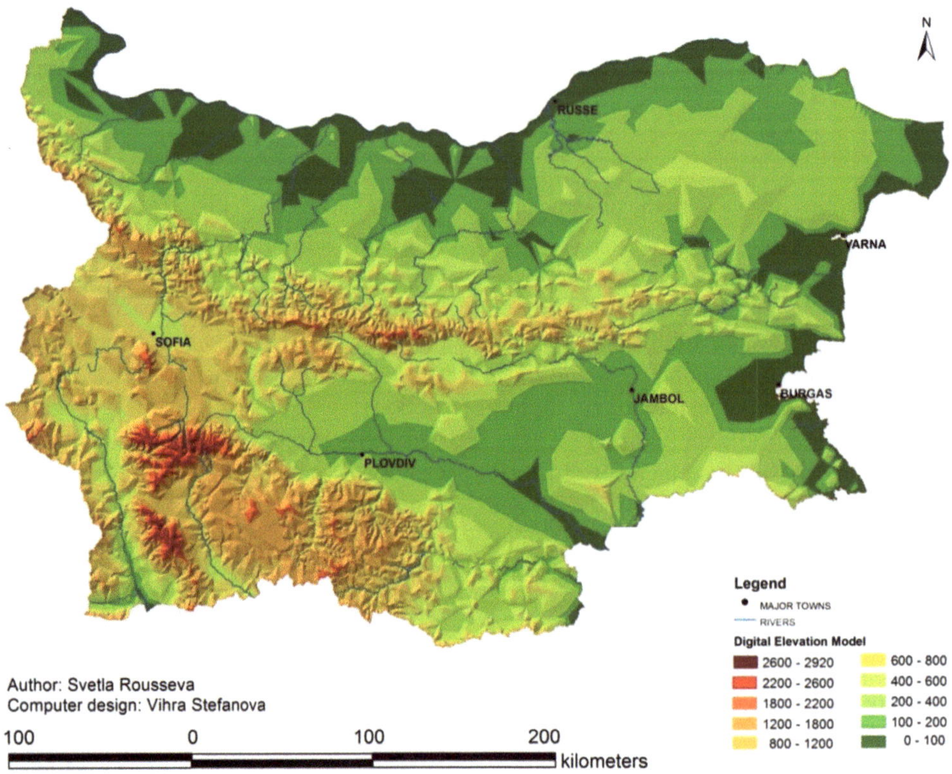

Fig. 1.3 Digital hypsometric model grouping of the major landforms on the territory of Bulgaria

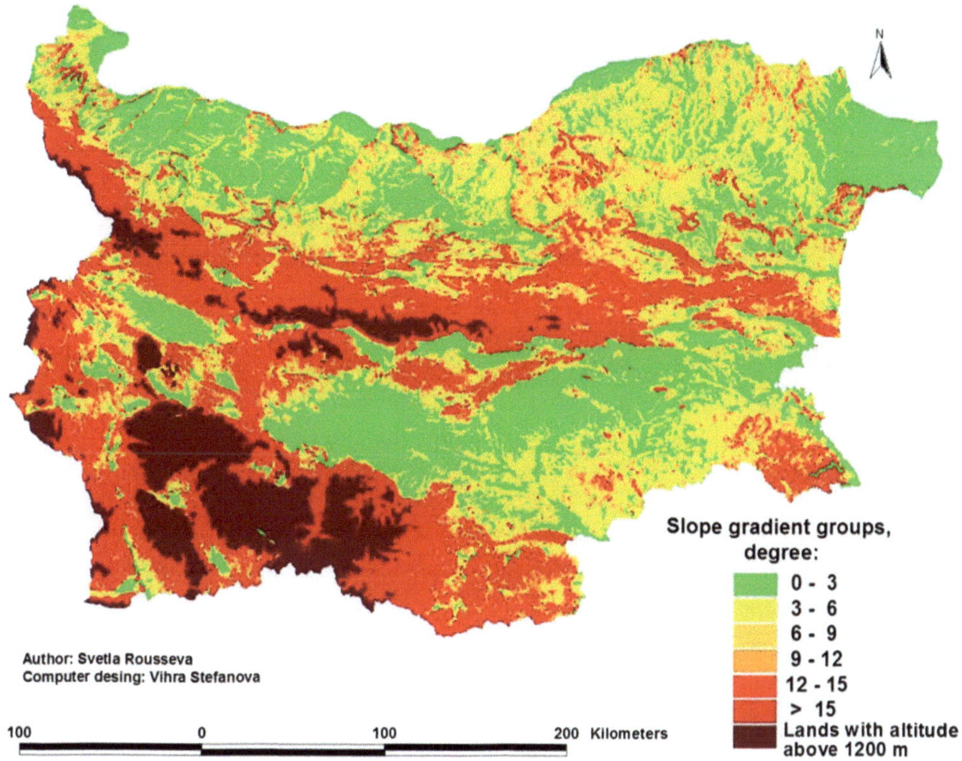

Fig. 1.4 Slope gradient groups on the territory of Bulgaria

The typical surface form in the northern part of the country is the hilly Danube Plain developed on the Moesian platform; in the central part is the Balkan Mountains chain system; the Fore Balkan is transitionally situated between the Danube Plain and the Balkans; in the western part is located the so-called transitory mountain and valley floors province; the Upper Thracian and Mid Tundzha Lowland stretches in a west–east direction throughout the country and in the southern part is the Rila-Pirin-Rhodope Mountains province.

The hilly Danube Plain is delineated to the north of the Danube River bordering with Romania, to the south by the Fore Balkan, to the west by the river Timok bordering with Serbia, and to the east reaches the coast of the Black Sea.

The hilly Danube Plain has a weak internal difference in the segmentation of relief characterized by lowlands (0–100 m altitude), valleys (100–200 m), and rolling plateau topography (200–450 m altitude) (Fig. 1.5). The total area of the Danube Plain is 31,522.6 km^2 or 28 % of the country's territory.

The northwestern part is dominated by large watershed areas with an altitude of 100–250 m with monocline northeast, and is dissected by the valleys of tributaries of the Danube River, namely the Archar, Lom, Tsibritsa, Ogosta, Skat, and Vit rivers. Lowlands with an altitude of 100–200 m are 84.7 % of the territory and the remaining 15.3 % of the terrain is rolling. Along the Danube River alluvial valleys are spread as follows: Bregovska, Vidinska, Archarska, Orsoyska, Kozloduyska, Ostrovska, and Chernopolska.

The middle part of the Danube Plain has developed large dissected flat ridges between large valleys of meandering river beds of the Danube tributaries—the Vit, Osam, and Yantra rivers. Tributaries of the Danube River have irregular flow in canyon-like valleys, and between these are located flat watersheds with an altitude 100–200 m. Flat watersheds gradually rise from west to east, from 250 to 400 m altitudes. Lowlands with an altitude of 100–200 m are 82.2 % of the territory and the remaining 17.8 % of the terrain is rolling. Along the Danube River alluvial valleys are spread as follows: Koynarska, Osamska, Belene, and Vardimska.

In the eastern part of the Danube Plain the relief is rolling plateau with a maximum altitude of 502 m on Shumen plateau. Flat watersheds occupy vast areas on Ludogorsko plateau. The relief of Dobrudzha is typical plateau with an altitude of 100–200 m; the Provadiisko and Shumen plateau are highly dissected. Lowlands with an altitude of 100–200 m are 60.4 % of the territory and the remaining 39.6 % of the terrain is rolling. Along the Danube River alluvial valleys spread as follows: Pobrezhie, Aydemirska, and Popino-Garvanska.

Despite the intensive development of agriculture in the northeastern Dobrudzha region, there are places that remain untouched by anthropogenic activities. One of the main migratory routes of birds, the *Via Pontica*, passes above it. There are coastal wet areas of global importance, a relict steppe with more than 400 plant varieties, of which 40 are

Fig. 1.5 The hilly Danube plain

rare and endemic. The reserve Horn Kaliakra (beautiful horn) is of 687,5 ha including the sea shelf. The reserve Baltata is of 197,7 ha with the most northerly situated longoze (archaic dense) forest, the protected area of "Yailata" (meaning pasture), is of 45,3 ha. All reserves are situated on the coast of the Black Sea.

The Fore Balkan is transitionally situated between the Danube Plain and the Balkan Mountains and is characterized by hilly ridge and partly mountainous relief. The Fore Balkan is a system parallel to the Main Chain of Balkans consisting of high and low mountains and hills, divided by valleys and gorges. To the west the Fore Balkan borders with Serbia and to the east with the Black Sea coast. Vast structural changes have greatly diversified the landscape of the western part of the Fore Balkan where relief is low mountainous and hilly. The monoclinic highs are the peaks Babin nos (1,108 m), Belogradchishki Venets (904 m), Shiroka Planina (941 m), Veslets (781 m), and Milin Kamak (460 m). The great presence of limestone determines extensive karst relief. Vast denuded karst formation has led to the creation of complex forms of abysses, stalactite caves, whirlpools, and cavities.

The middle part of the Fore Balkan is the widest and is characterized by anticlinal ridges, namely Vasilov peak (1,490 m), Elenski highs (902 m), Dragoitsa (967 m), Strazhata (788 m), Krushevski ridge (582 m), and other ridges with an average height of 420 m.

The eastern part of the Fore Balkan is dominated by hilly terrain with an average height of 296 m, the highest point being Lisa mountain (peak Golyam Sakar is 1,054 m), followed by Preslavska mountain and Anton highs.

Differences between the relief of the Main Chain and Fore Balkans are as follows: lowlands and hills with an altitude below 600 m occupy 41 % of the terrain in the Main Chain and 90 % in the Fore Balkan; low mountains and highlands occupy 55 % of the terrain in the Main Chain and 10 % in the Fore Balkan.

Morphographically the Balkan Mountains are divided into the Main Chain spreads with a total area of 11,596.4 km^2 and the (Fore Balkan) foothills which lie to the north and cover about 14,389.8 km^2 of the total area.

The Balkan Mountains are located from the western to the eastern border of the country. Slopes are covered with deciduous and coniferous woods and grasslands. The Balkan Mountains serve as a climatic border because they stop the cold air that invades from the north.

The Balkans chain is delineated by two parallel faults to the south and north (Fig. 1.7).

The width of the various parts of the Main Chain varies between 15 and 40 km. The most extensive is the western part consisting of Chiprovska mountain (peak Mindzhur 2,168 m), Koznitsa, Ponor, Etropolska mountain, Rzhana, and Murgash (1,687 m).

The central part is the National Park and is the tallest consisting of the highest Balkan Mountain ridge of Tetevenska mountain (peak Vezhen 2,198 m), Kalopherska mountain (peak Botev 2,376 m, Kupena 2,169 m, Levski 2,166 m, Triglav 2,276 m), peak Ispolin 1,524 m, peak Chumerna 1,536 m, and peak Paskal (Fig. 1.6). The eastern part of the Main Chain of Balkans consists of Slivenska mountain (peak Bulgarka 1,181 m), the low mountains Kotlenska, Emino-Kamchiyska, Aytoska, and Karnobatska with an altitude of about 600 m and hills with an average height of 365 m.

The western part of the transitory mountain and valley floors province include the mountains Konyavska (peak Viden 1,487 m), Golo brdo, Strazha, Vitosha (peak Cherni Vrukh 2,290 m), Lyulin (1,256 m), Plana (1,338 m), Veskyar (1,077 m), and Kraishte (peak Golemi Vrukh 1,481 m) that are actually faults which are separated from each other by elongated valley floors to the west—Kyustendilska, Radomirska, Pernishka, Sofiyska (550 m), Trunska, Ihtimanska, and Zlatishka-Pirdopska (750 m) (Fig. 1.7). Vitosha mountain has been a National Park since 1934 and was the first on the Balkan peninsula with a total area of 26,606 ha (Fig. 1.8).

In the central part of the province is situated the Sredna Gora mountain (peak Bogdan 1,604 m), Ychtimanska Gora, Sarnena Gora (peak Bratan 1,236 m), Hisarska Gora, and the eastern Sub-Balkan valley floors—Karlovo and Kazanlak (350 m), which are picturesquely situated between the Balkan and Sredna Gora mountains (Fig. 1.9). Characteristic forms of relief are river valleys and gorges. Being young, the valleys have well-expressed fault structures and are characterized by high seismicity accompanied by thermal waters. The plane relief combined with a relatively warm climate and fertile soils make favorable conditions for growing the Bulgarian Rose. Further east is located the Slivenska valley (150 m altitude) and Karnobatska (100 m).

The Upper Thracian and Mid Tundzha lowlands occupy flat and hilly terrains on both sides of the middle stream of the Maritza and Tundzha Rivers and spread east to the Burgas valley. The lowlands occupy the alluvial plain. Structural lowering of lowlands is indicated by the presence of seismicity and thermal waters. To the north the area is adjacent to the southern slopes of the Sredna Gora Mountain, to the south to the fault northern slopes of the Rhodopes and Sakar Mountain (peak Viishegrad 856 m), and to the east to the Strandzha Mountain (peak Gradishte 710 m on the Bulgarian territory) (Fig. 1.10).

Fig. 1.6 National Park, the central Balkans

Fig. 1.7 The foot of the Balkan Mountains and Zlatishka-Pirdopska valley floors

Fig. 1.8 National Park—the Vitosha Mountain and Sofia valley floor

Fig. 1.9 The Sredna Gora Mountains

Fig. 1.10 The southeastern region of Black Sea and Natural Park of the Strandzha Mountain

The western part is occupied by Pazardzhic and Plovdiv lowland (100–300 m altitude), which is drained by the Maritsa River and has an average altitude of 168 m. The central part is occupied by Stara Zagora lowland (with 150–170 m altitude); and the eastern part includes Yambol and Elhovo lowland (with 80–100 m altitude), drained by the Tundzha River. Smooth relief, combined with transitional climate between temperate continental and transitional Mediterranean and alluvial soils, provides excellent conditions for intensive agriculture.

Rila-Pirin-Rhodope province occupies the southern part of the country. The region includes the mountain chains of Osogovo (peak Ruen 2,251 m), Vlahina, Ograzhden, Belasitsa (peak Radomir 2,029 m), and Malashevska (peak Yilyo 1,803 m), with a total area of 2,828.1 km^2. There are several deeply incised valleys drained by the Struma River—Blagoevgradska, Simitliyska, Sandanska, and Petrichka. The Belasitsa mountain is a Natural Park.

The mountains are Rila (peak Mussala 2,925 m, Irechk 2,852 m), Pirin (peak Vichren 2,914 m, Kutelo 2,908 m), and Slavyanka (peak Gotsev 2,212 m), with a total area of 5,206 km^2 separated by the big faults of the rivers Struma and Mesta (Fig. 1.11). The National Parks are the Rila and Pirin mountains.

The Rhodopes are a Natural Park, with a total area of 12,233 km^2 on the Bulgarian territory being divided into eastern and western parts (Fig. 1.12). The average height is about 785 m. In the western Rhodopes is located Mursalitsa mountain (peak Big Perelic 2,191 m) and Suytka (peak Suytka 2,186 m). A large part of the western Rhodopes has middle mountainous relief characterized by high rounded surfaces covered with forests and grasslands, and deep gorges carved between the ridges. Karst develops in the areas that consist of marble. The eastern Rhodopes are characterized by low mountainous and hilly relief (peak Veykata 1,463 m, peak Alada 1,241 m) and extensive erosion depressions in the basin along the Arda River.

Igneous rocks form the landscape of the Vitosha, Sredna Gora, Rila, Pirin, Phodope, and Strandzha mountains. Sedimentary rocks cover the Danube Plain, Upper Thracian Lowland, Burgas valley, Fore Balkan, Eastern Balkan, and Eastern Rhodope mountains. Metamorphic rocks are distributed in the mountain regions of southern Bulgaria in Pirin, Rila, Rhodopes, Osogovska, and Belasitsa mountains.

The geology of Bulgarian lands was shaped over a period of 3.5 billion years.

The country's lands have undergone several stages of geochronological development, namely Precambrian, Paleozoic (when granite plutons were formed in the Rhodopes), Mesozoic (when during the Alpine-Himalayan crimp the Sredna Gora mountain was formed), Cenozoic (when the Balkans were finally shaped), Paleogene (when the main tectonic structures were formed), and Neogene. The main stages in the development of Bulgaria's landscape

Fig. 1.11 The southwestern region of the Struma river valley and the National Park in the Rila Mountain

Fig. 1.12 The Natural Park in the Rhodope Mountains (peak Snezhanka 1926 m)

are Preneogene, Neogene, and Quaternary. The Neogene stage is characterized by strong uplift which today covers the western and central parts of the country; in the late Neogene stage the separate blocks of the Balkans, Vitosha, Pirin, and Rila mountains were formed; and the lakes that covered graben cuts of the Upper Thracian Lowland and of Struma and Mesta valleys dried up. During the Quaternary stage the modern landscape of Bulgaria was shaped, including aspects such as the river network, terraces, and denudation in the mountains. During the Pleistocene (in Quaternary) stage the highest parts of Rila and Pirin mountains became iced and loess cover was formed in the Danube Plain.

The northeast Black Sea coastal territory of contemporary wave abrasion is 42 % of the total affected coastal territory of the country. It is an inherited relict consequence of an ancient slides formation. Formation of the terraced relief is related to the end of the Pleistocene stage, when the ocean level decreased during the glacial period. At the beginning of Holocene the sea level was 90–100 m lower than it is today. After Würm glaciation during the warm period 10,000 years BP, the sea flooded waste coastal valleys and transformed them into Holocene terraces.

The Black Sea (*Pont Euxin*) was formed 7,500 years ago when the waters of the Mediterranean Sea invaded after the rupture of the Bosphorus.

Climate

The influence of the four large soil provinces of Europe can be traced on the territory of the country: the Steppe and Forest-Steppe Eastern European, the Mediterranean South-European, the Forest-Atlantic West European, and the Eastern Black Sea regions (Koynov et al. 1974).

Based on the climate type three zones can be singled out on the territory of Bulgaria, comprising 7 sub-zones and belts, 28 provinces, and 63 regions in all (Fig. 1.13).

The northern Bulgaria Forest-Steppe zone includes the Danube Plain and the Fore Balkan and covers the territory with altitude between 100 and 800 m. With respect to climate it belongs to the temperate continental climatic region, which is a succession of the Middle European temperate continental and partly of the south Russian steppe climatic region. In this zone, in a north–south direction, changes in the relief, climate, vegetation, parent rock, etc., are observed. This zone is divided into three sub-zones: Danube sub-zone of chernozems, Danube plain and hilly Fore Balkan sub-zone of gray forest soils, and Fore Balkan sub-zone of light gray (pseudopodzolic) forest soils.

The southern Bulgaria xerothermal zone covers the territory with altitude up to 800 m, and is characterized by both semi-continental climate type and transitional continental with the Mediterranean climate influence. The zone is divided into two sub-zones: middle Bulgarian sub-zone of

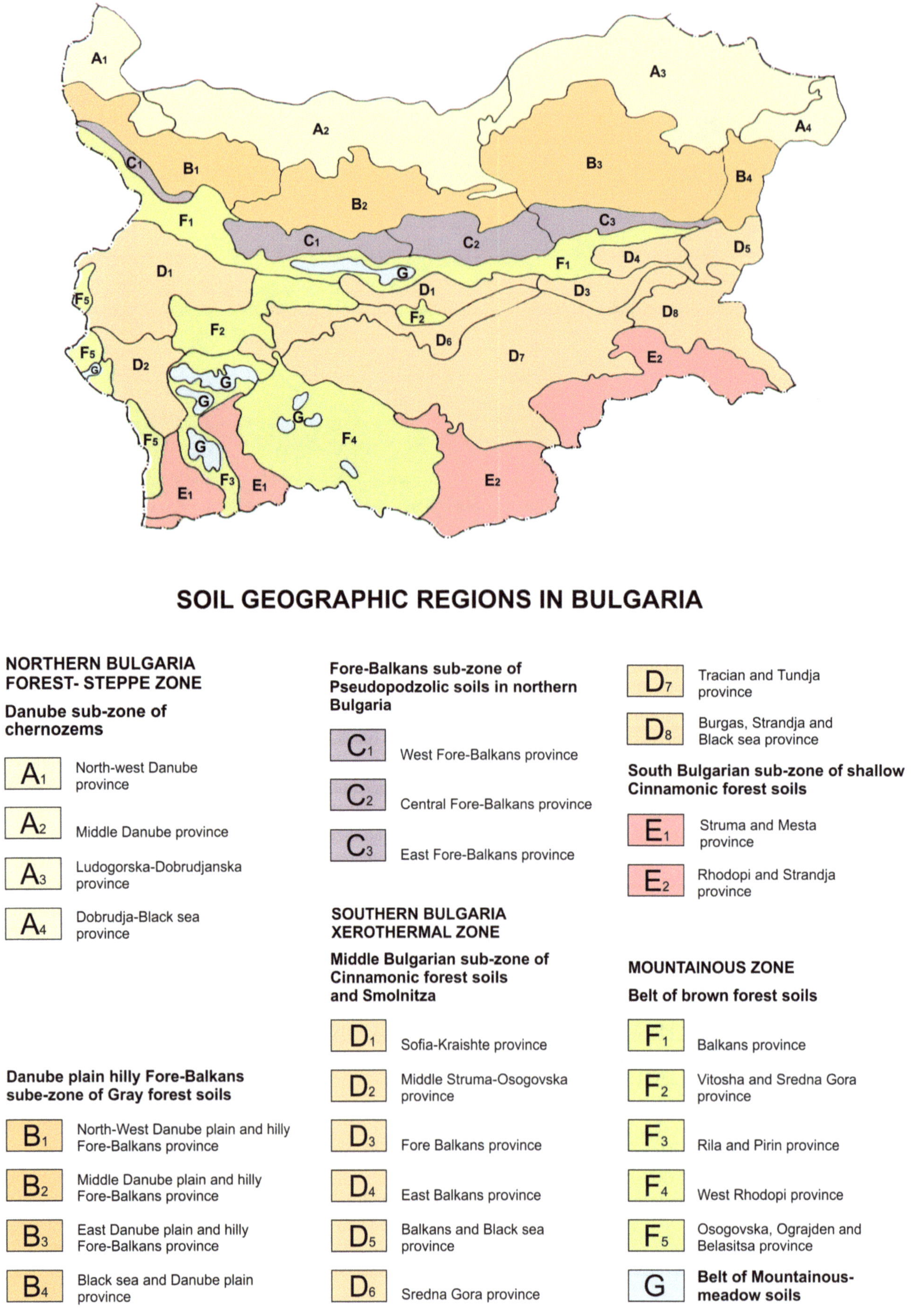

Fig. 1.13 Scheme of the soil geographic regions in Bulgaria (Koynov et al. 1998)

cinnamonic forest soil and smolnitsa and south Bulgarian sub-zone of cinnamonic soils.

The mountainous zone comprises mountains in the country with altitude over 700–800 m. Two belts are distinguished: the belt of brown forest soils and that of mountainous meadow soils at the highlands.

The general sequence of soil distribution in Bulgaria can be seen in Table 1.3.

The spatial soil distribution on the territory of Bulgaria is shown on the soil map at scale 1:1,000,000 (Fig. 1.14).

Soil units according to (FAO-UNESCO 1990)	Abbreviation
Eutric Vertisoils	Vre
Haplic Solonetzs	SNh
Eutric Planosols	Ple
Luvic Phaeozems	PHI
Chromic Luvisols	LVx
Haplic Luvisols	LVh
Lithic Leptosols	LPq
Calcic Chernozems	CHk
Eutric Fluvisols	Fle
Rendzic Leptosols	LPk
Humic Cambisols	Cmu
Eutric Cambisols	Cme
Dystric Cambisols	CMd
Haplic Chernozems	CHh
Haplic Acrisols	Ach

According to the 1:400,000 Soil map of Bulgaria (Koynov et al. 1968) the most widely spread soil group is Chromic Luvisol (21.91 % of the country's territory), followed by Chernozems (20.23 %), Cambisols (15.58 %), Haplic Luvisols (10.24 %), Pseudopodsolic soils (9.75 %), Fluvisols (8.97 %), and Vertisols (5.37 %) (Fig. 1.15). Rendzic Leptosol (2.74 %) and Leptosols (1.55 %) have a limited spread. The least spread soil groups are Solonetz and Solonchaks (0.22 %). The remaining groups (Regosols, Arenosols, Calcisols, and Anthrosols) can be found in association with the other soil groups.

Intensity of relief, parent rocks, provincial bio-climatic conditions, and other factors benefit various soil covers. Deep soils in the plains and plateaus occupy 6,500,000 ha or 58.6 % of the entire territory of the country, and the remaining 41.4 % of the territory, about 4,500,000 ha, is shallow soils in the mountain regions.

Lands with high potential risk of erosion comprise 65,000 km^2 or 59 % of the entire territory. The most affected terrains are in the regions near to the towns of Blagoevgrad, Kardzhali, Smolyan, Xaskovo, Kustendil, and

Table 1.3 Bulgarian soil resources, based on the inventory of materials of the soil survey at scale 1:25,000 and generalized soil mapping units shown on the soil map of Bulgaria at scale 1:400,000 (Koynov et al. 1972)

Soil varieties	Total area (ha)	(%) of the entire territory
Zonal soils		
Calcareous and typical chernozems	831943.4	7.52
Slightly and medium leached chernozems	1075099.6	9.72
Strongly leached chernozems	36830.7	0.33
Strongly leached, podzolized chernozems, and dark gray forest soils	311024.4	2.81
Chernozems-karasolutcy and dark gray forest soils	58486.2	0.53
Smolnitsa, calcareous, typical, and leached	591023.3	5.34
Dark gray forest soils	198788.9	1.80
Gray forest soils	1163555.2	10.5
Light gray pseudopodzolic soils	571273.7	5.16
Typical and leached cinnamonic forest soils	2106473.5	19.1
Strongly leached slightly podzolized cinnamonic forest soils	453832.8	4.10
Cinnamonic podzolic (pseudopodzolic soils)	358217.3	3.24
Cinnamonic podzolic (pseudopodzolic soils) accumulated	88187.6	0.80
Jeltozem podzolic soils	26946.9	0.24
Soils of vertical zones		
Brown forest soils	1631006.5	14.7
Dark mountainous soils	110113.4	0.99
Meadow mountainous soils	168201.3	1.52
Azonal soils		
Rendzinas	265841.0	2.40
Alluvial and alluvial meadow soils	585037.3	5.29
Colluvial and colluvial meadow soils	140308.9	1.27
Meadow cinnamonic	54585.0	0.49
Meadow smolnitsa	12854.4	0.23
Meadow cernozems	150116.6	2.9
Meadow boggy and boggy alluvial meadow soils	14464.2	0.28
Peat boggy soils	7355.8	0.14
Saline soils	21052.3	0.41
Sands	11307.9	0.22
Ponds, lakes, peats	16603.8	0.15
	11060591.9	

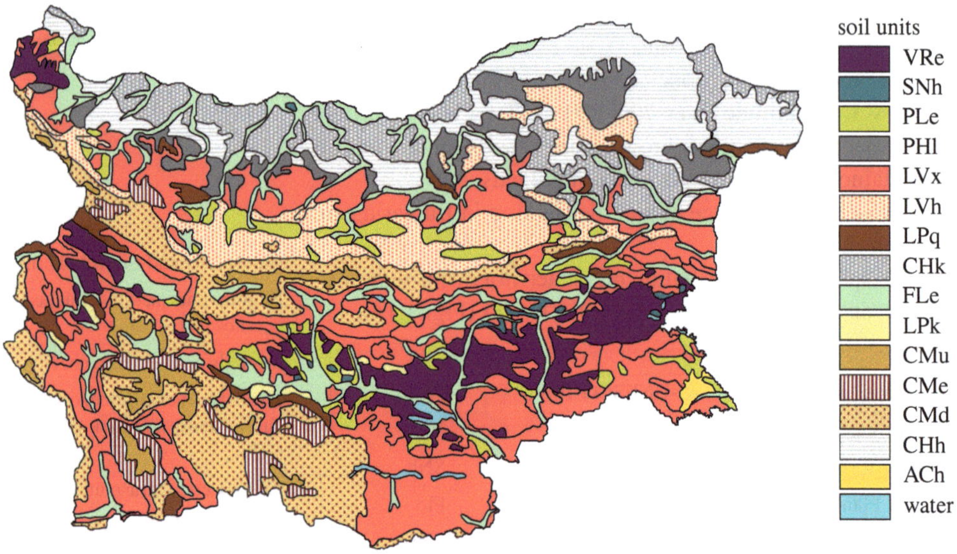

Fig. 1.14 Soil map of Bulgaria based on the digitized soil map at scale 1:1 million (Kolchakov et al. 1994) in compliance with the Revised Legend of the Soil Map of the World (FAO-UNESCO-ISRIC 1990)

Razgrad towns. About 11.8 % or 1,300,000 ha of the deep soils are affected by moderate and severe erosion. Yearly soil loss for the country is 40,000,000 m^3.

Soils in the different regions of the country possess a number of provincial features related to the type of parent rocks and duration of weathering. Some of the soils may gradually completely change into other soils laterally in space, whereas others continue unaltered and have great spatial distribution. The rate in change in horizons is an important soil characteristic by variability in color and in physical or chemical properties.

There is sequence in the spatial relation of the soils in the country. Chernozems, dark gray-brown, and gray-brown forest soils prevail in north Bulgaria. There is a belt of calcareous and typical chernozems along the Danube River; in central north Bulgaria they gradually turn into leached

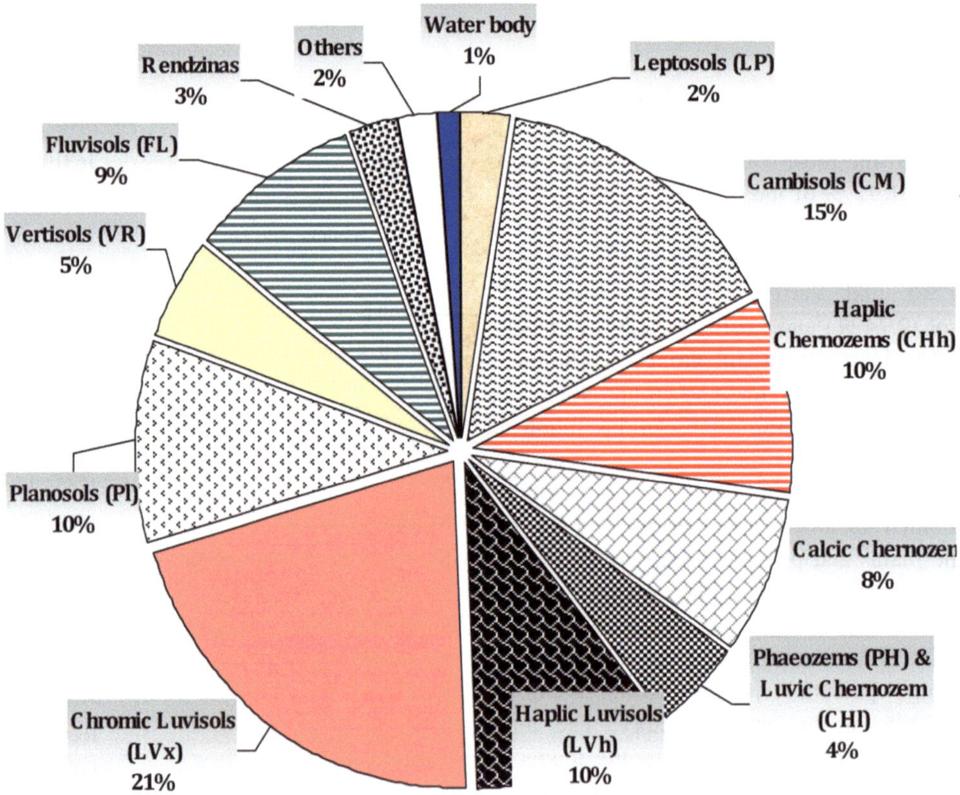

Fig. 1.15 Distribution of soil resources in Bulgaria according to the digital map 1:400 000 (Koynov et al. 1968)

and degraded (podzolized) chernozems and finally into gray-brown forest soils to the south of the hilly Danube Plain (Koynov 1964). A broad belt of the Fore Balkan is occupied by pseudopodzolic soils. The situation in south Bulgaria is as follows: cinnamonic forest soils are widespread in the hilly areas on various parent materials; smolnitsa occupy plains with old weathered materials of Pliocene, Pleistocene and Quaternary origin. The mountain areas above 900 m altitude are covered by acid non-podzolized brown forest soils. The alluvial meadow soils are at the river valleys and in the depressions where Quaternary depositions prevail. Young soils in means of duration of evolution like alluvial, alluvial meadow, meadow cinnamonic, meadow smolnitsa, and salinized varieties of soil occupy low topography terrains and depression areas.

Farming is developed mainly on terrain with the varieties of chernozems, smolnitsa, gray-brown forest, and cinnamonic forest soils. The most fertile, with the best properties for growing the main cereal crops are leached and degraded chernozems, dark gray-brown forest soils, smolnitsa, and alluvial meadow soils; light gray pseudopodzolic forest, and cinnamonic podzolic (pseudopodzolic) soils are less fertile and have unfavorable properties. Seasonally, smolnitsa, gray-brown forest, and cinnamonic forest soils have poor technologic tillage properties, are suitable for ploughing for a limited period, and result in higher expenses.

The ground experiments with nitrogen isotope confirmed data that the mineral regime in Bulgarian soils was poor and also that the nutrient cycle took place over shorter periods. About 40–50 % of the mineral nitrogen applied as a fertilizer was used by plants and the rest was lost in the form of gas.

It was established that in the different regions of the country the biological fixation of atmospheric nitrogen had as a limiting factor an unfavorable ratio of nitrogen and phosphorous. It was also established that all soils in Bulgaria were poorly supplied with available forms of phosphorous for plants.

Bulgarian soils had a comparatively high supply of potassium for plants; however, the main resource is at inextricable form. Balance studies of potassium in different climatic conditions revealed relatively high intensity in the nutrient supply for plants compared to the other soils in Europe.

1.3.2 Factors of Soil Formation

Parent materials

Mineral material is the most widespread type of parent material consisting of a number of rock minerals. The original source bedrock is not considered soil parent material. When bedrock occurs at land surface the weathering and formation of parent materials occur.

The consolidated mineral material includes rocks like granite, basalt, and conglomerate; and the unconsolidated materials are loess and various deposits.

The igneous intrusive rocks are granite, diorite, granodiorite, syenite, gabbro, etc.; and the igneous effusive rocks are rhyolite, andesite, latite, basalt, diabases, volcanic tuffs etc., which are characterized by crystallized structural links, of high mechanical strength, are water permeable through the cracks.

Metamorphic schistose rocks are gneisses, phyllites, mica, hornblende, etc., which are characterized by crystallized and amorphous solid structural links, of high mechanical strength, are water permeable through the cracks, and are weakened by wetting.

Metamorphic not schistose rocks are marbles, which are resistant to wetting.

Semi-rocks are the sea lithological type of rocks such as conglomerates, breccias, sandstones, limestones, dolomites, marls, shales, etc., which are characterized by crystallized and water-colloidal structural links, of medium mechanical strength, and are greatly weakened by wetting.

Continental sedimentary rocks and sediments are conglomerates, sandstones, limestones, marls, shales; loess (sandy, typical, clayey, clay-like); lacustrine, alluvial, and deluvial (clays, sands, boulders, and gravels).

Usually Bulgarian soils developed from hard rocks occur in the mountains and are commonly shallow. Colluvial sediments occur at the base of slopes in mountainous areas where gravity is the dominant force. During the Pleistocene period wind activity was widespread in areas peripheral to glaciers. The wind picked up and transported fine materials over long distances and then deposited them as loess. Today, large areas are covered by loess or old redeposited materials as a result of geomorphologic activities. Alluvial sediments occur as an inclusion or are the dominant parent materials in valleys.

The physical weathering of rocks and minerals results in a wide ranging size of products from stones, to gravel, to sand, to silt, and clay. The proportion of sand, silt, and clay in a soil determines the soil's texture. Since 1926, based on the agreement of the International Soil Science Society, it has been accepted that soil elements have spherical form and diameter (mm) as follows: very coarse and coarse sand 2.0–0.2, medium and fine sand 0.2–0.02, very fine sand and silt 0.02–0.002, and mineral colloids or clay below 0.002 (mm).

Silt 0.05–0.002 (mm) (according to USDA) in many soils is dominated by quartz and feldspar; mineral colloids of 2.0–0.2 microns (μm) consist of quartz, illite and kaolinite; 0.2–0.08 (μm) consist of illite and kaolinite; and below 0.08 (μm) is montmorillonite.

Quartz is a common soil mineral with a density of 2.65 g/cm^3, granite from 2.50 to 2.66 g/cm^3, diorite

2.55–2.82 g/cm^3, rhyolite 2.10–2.59 g/cm^3, andesite 2.15–2.90 g/cm^3, basalt 3.01 g/cm^3, marble 2.60–2.82 g/cm^3, and feldspar has densities ranging from 2.56 to 2.76 g/cm^3. The average particle density for mineral soils is usually 2.65 g/cm^3.

Chemical weathering of the minerals is due to the silicates hydrolysis by H_2O. Usually the amount of silicate and/or non-silicate (free) iron indicates the advance of rock weathering. A very low amount of the non-silicate (free) iron, released due to the chemical weathering of biotite, pyrocsene, amphibole, and other iron containing silicates, is included in the newest clay mineral formation, which is contra to the aluminum (Scheffer and Schachtschabel 1979). The largest amount of iron remains at the weathering point or after translocation accumulates as Fe^{3+} oxides.

Weathering solutions remove downwards substances much deeper than at soil depth.

Clay formation results mainly from chemical weathering; however, the real picture of clay formation is not clear in the soil. Resistance to weathering varies; the time taken for 1 % clay to form in rock parent materials ranges from 500 to 10,000 years. The nature of the influence of parent rocks is prominent.

The most important properties of clay minerals are first, their cation exchange capacity and capacity to absorb water. The clay fraction usually has a net negative charge and absorbs nutrient cations Ca^{2+}, Mg^{2+} and K^+. Second, is their ability to be aggregated or dispersed depending upon the nature of cations in the soil.

The important objective is to gauge the specific surface area (surface area per gram) of particles in parent materials which directly influence the advance of soil formation.

Microelements, needed by plants occur in a number of minerals.

Climate

Average annual temperature varies from 13.6 °C in the warmest places of the Struma valley to −2.9 °C at peak Mussala on Rila Mountain. The quantity of rainfall varies from 450 to 500 mm in northeastern Bulgaria and the western part of the Thracian Lowland to more than 1,000 mm in the regions over 1,600 m altitude.

Contemporary climatic conditions impose five climatic regions on the territory of Bulgaria. The temperate continental climatic region is characterized by cold and prolonged winters, and the mean annual precipitation from 500 to 750 mm typically has summer maximum and winter minimum. The transitional continental region is characterized by mild winters, cold springs, and hot summers. The mean annual precipitation is near 600 mm, with maximum (June and November) and minimum (February and August). The transitional Mediterranean region is characterized by droughts during summer. The mean annual precipitation is from 600 to 995 mm, with maximum (June and December) and minimum (March and August). The region with Black Sea climatic influence is characterized by a mild climate and low annual temperature amplitude. The mountainous climatic region is characterized by cold conditions, low annual temperature amplitude, and about 1,250 mm mean annual precipitation, generally of snow.

Precipitation varies considerably from place to place. As altitude increases so does the amount of precipitation.

The most important indicators of solar energy reaching the earth are changes in atmospheric and soil temperature (Fig. 1.16). The amount of radiation that reaches the earth's surface is determined by season and altitude. Atmospheric conditions, such as humidity and temperature, play a major role in determining how quickly water is lost through transpiration and evaporation.

Bulgaria is a country with a cycle of contrasting seasonal climates. Soil has an annual temperature cycle, becoming warm during the summer and then cool to freezing during the winter. The effect on soil is determined by the rates of chemical and biological reactions and evaporation.

Variations in the soil's capacity to absorb moisture depend on structure, permeability, and clay content.

The territory of the country can be divided into 40 agricultural and 10 forest ecological regions. This classification has been made on the basis of major qualitative and quantitative criteria including the prevailing type of soil, water regime during the vegetative period, temperature, some extreme climatic changes, altitude, and relief (Yolevski et al. 1989). These regions form seven large groups which correspond to a very great extent to the major soil and climatic zones and sub-zones of the country (Fig. 1.17, Table 1.4).

Organisms

Carbon in the atmosphere is absorbed by plants to compose organic matter. Residues further decompose, emitting carbon back into the atmosphere or fixing as humus. The constant carbon and nitrogen cycle is from the soil to the plants and then back to the soil.

Trees are the major producers of organic matter, contributing leaves and wood to the surface of the soil. Grass and roots also bring a large amount of organic compounds back to the soil.The diversity of vegetation cover in the country is changed by man and his activity. Natural, mostly forest vegetation has been disturbed in the construction of communities, and shrubs, grasses, and crops have replaced it in over 50 % of the territory.

The vegetation of Bulgaria consists of 3,500 plant species: 31.5 % of them with Euro-Asian distribution, 21.34 % European distribution, 9.25 % Circumpolar, 5.43 % Cosmopolitic, 4.0 % Mediterranean, 12 % distributed on the Balkan peninsula (of which 4.17 % are endemic), and a few

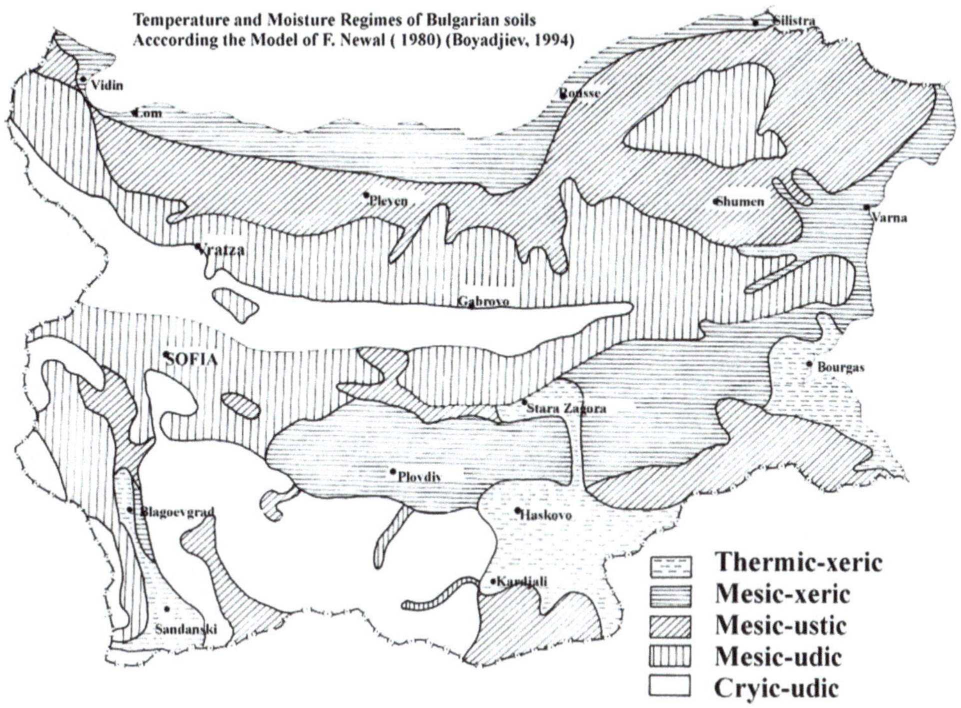

Fig. 1.16 Map-scheme of the general temperature and moisture regimes of Bulgarian soils according to the model of Newal F. 1980 (Boyadzhiev 1994)

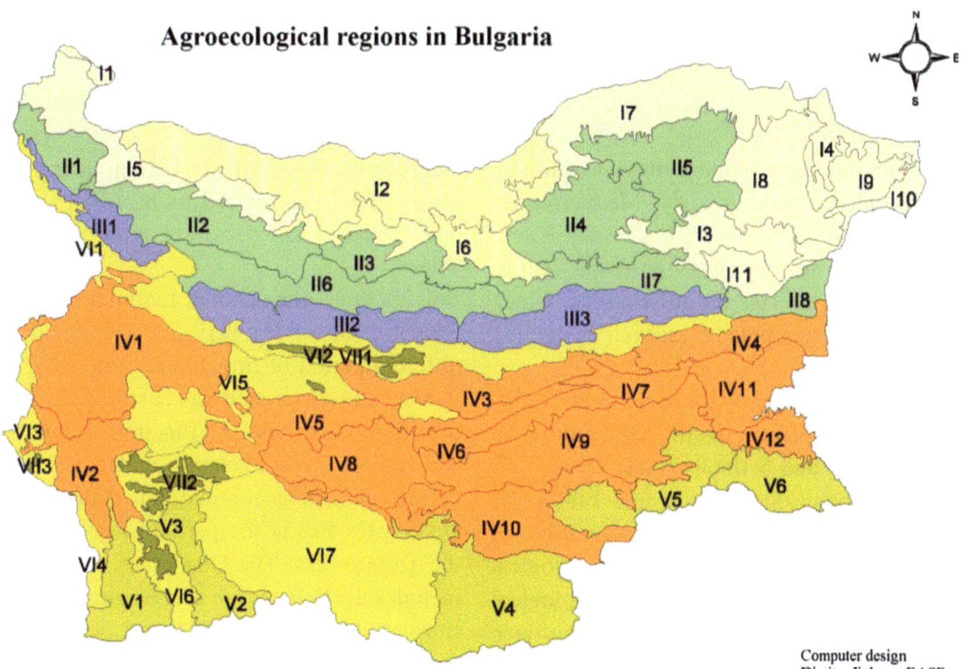

Fig. 1.17 Map of the agroecological regions of Bulgaria based on the original map at scale 1:600 000 (Yolevski et al. 1980)

other species with another distribution. More of the species are hemicryptophytes (near the soil surface) 55.27 %, therophytes (desert plants) 27.0 %, phanerophytes (normally woody perennials) and chamaephytes (woody perennials 25 cm height above the soil surface) 11.61 %, and geophytes (resting on dry land) 6.0 %. Phanerpphytes

Table 1.4 Agroecological regions on the territory of Bulgaria (Yolevski et al. 1980)

Agroecological regions of Chernozems		*Agroecological regions of gray-brown pseudopodzolic forest soils (Planosols)*		*Agroecological regions of Cinnamonic forest soils (Chromic luvisols)*	
I1	Zlatenrogski-Novoselski				
I2	Lomsko-Svishtovski	III1	Berkovski	V1	Petrichko-Sandanski
I3	Novopazarski	III2	Botevgradsko-Gabrovski	V2	Gotzedelchevski
I4	Kardamsko-Durankulashki	III3	Elensko-Smyadovski	V3	Razlojki
				V4	Karjaliiski
I5	Vidinsko-Beloslatinski	*Agroecolgical regions of Vertisols and Cinnamonic forest soils (Vertisols and Chromic Luvisols)*		V5	Topolovgradsko-Fakiiski
I6	Plevensko-Pavlikenski			V6	Malkotarnovski
I7	Rusensko-Silistrenski			*Agroecological regions of Brown forest soils (Cambisols)*	
I8	Tervelski				
I9	Dobrichki	IV1	Sofiisko-Pernishki		
I10	Balchishki	IV2	Kiustendilski	VI1	Petrohanski
I11	Provadiiski	IV3	Karlovsko-Kazanlashki	VI2	Srednostaroplaninski
Agroecological regions of gray-brown forest soils (Haplic luvisols)		IV4	Sungurkarski	VI3	Transko-Osogovski
		IV5	Srednogorski and	VI4	Ograzhdenski
			Rodopski lowlands	VI5	Samokovsko-
II1	Kulsko-Belogradchishki	IV6	Novozagosrki		Srednogorski
II2	Montansko-Lukovitski	IV7	Slivensko-Straldjalski	VI6	Rilo-Pirinski
II3	Loveshki	IV8	Pazardjisjko-Plovdivski	VI7	Rodopski
II4	Popovsko-Razgradski	IV9	Chirpansko-Yablanski	*Agroecological regions of Mountain-meadow soils*	
II5	Isperihski	IV10	Haskovski		
II6	Mezdrensko-Sevlievski	IV11	Karnobatski-Bourgaski		
II7	Velikotarnovski-Preslavski	IV12	Grudovsko-Sozopolski	VII1	Visokoplaninski
				VII2	Visoko-Rilo-Pirinski
II8	Avrenski			VII3	Visoko-Osogovski

form the forest communities and build natural vegetation. Therophytes and geophytes have mainly Mediterranean distribution (Bondev 1991).

Mesofauna consist of invertebrates—protozoa, earthworms, mites in acid soils, insects, and their larvae, which produce a large amount of fecal pellets—as well as digging vertebrates.

Microorganisms play a part in making nutrients available for another cycle of plant growth. The groups of organisms (bacteria, fungi, and actinomycetes, algae, etc.) are involved in the decomposition of organic substances and synthesise of resistant to further decomposition of humus. Fine roots form the space for the microbiological activity of the rhizosphere. The most important biological process is humification of organic substances and nitrogen transformation. The formation of humus is a little understood process where organisms play a part in plant decomposition.

Mineralization is the conversion of an element from an organic into an inorganic form as a result of microbial activity. Nitrification is the process during which organisms decompose organic residues and produce nitrogen compounds. Most plants obtain their nitrogen supply as nitrate (NO^{3-}) produced by bacterial activity.

Time

Soil is a three-dimensional continuum in space and time. The formation of soil is a very slow process and commenced in a different period of time (Gerassimov 1972). The change of climate and vegetation has complicated horizon differentiation. Soil properties are a function of time. The surface of the soil is more prone to change with time.

The general geochronological data are as follows: the Quaternary Era is the period of geological time following the Tertiary Era. The Quaternary Era (2,500,000 years BP) includes the Pleistocene and Holocene periods. The Tertiary Era lasted from 75,000,000 to 2,500,000 years BP. During the Pliocene period (12,000,000 years BP) the climate was cooling and during that time the creature we call man developed.

The Holocene (post glacial) stage lasted from 10,000 years BP, and the Pleistocene from 2,500,000 years BP (with 4–5 glacial and interglacial stages). Historically, currently is sub-Atlantic phase that commenced 2,500 years BP.

The process of continuous leaching has taken place since 2,500–5,000 years BP and peat areas were dominant between 5,000 and 7,500 years BP.

The oldest soils have been exposed to weathering since the Tertiary period.

The rock weathering that took place during the Holocene period was negligible compared to that which took place during the Tertiary period. Some of the features in soils are relict. For example, the feature that is found as a blue-black coating on the surface of aggregates or within them and bounded with iron is manganese dioxide which usually occurs at places where water-logging periodically occurs.

Topography

The earth's surface is characterized by its topography or synonymous relief. This influences soils in many ways according to elevation, aspect, slope, and moisture. The position of the landscape plays an important role in affecting the depth of the water table and soil drainage.

In Bulgaria, the great amplitude of altitude determines the topography of the territory (Koynov 1964): valleys are located at the altitude of 0–200 m and occupy 31.5 % of the entire territory of the country; hilly regions are at the altitude of 200–600 m and occupy 35.5 % of the entire territory; hilly-mountains are at the altitude of 600–1,000 m and occupy 20.5 % of the entire territory; mountains are over 1,000 m and occupy 12.5 % of the entire territory.

As the angle of slope increases, erosion hazard takes place resulting in shallow soils on strongly sloping terrain.

In Bulgaria nearly 3,730,000 ha or 65 % of land use is affected by water erosion, and nearly, 1,350,000 ha or 24 % of land use is affected by wind erosion (Rousseva et al. 2006). In sloping areas erosion is common due to the accelerated run-off. About 66 % of the land has a slope over 3°, and 35.2 % of the land has a slope over 6°. The value of the index of soil susceptibility to water erosion (USLE-K) varies from 0.003 to 0.055 t ha h/ha MJ mm; and susceptibility to wind erosion (WEQ-I) is from 100 to 375 t/ha y. Soils in Bulgaria are highly susceptible to water erosion which affects (>0.03 t ha h/ha MJ mm) or over 60.1 % of the country's territory; and >150 t/ha y or over 26.1 % of the territory is susceptible to wind erosion (Rousseva et al. 2010a, b).

Many soils on sloping land are in equilibrium in terms of erosion rates and rate of soil formation. Soils on steep mountains slopes are thin and often stony. Gently sloping surface areas accumulate the distributed fine materials, particularly clays.

Human beings

The use of land for agriculture, forestry, grazing, and urbanization has produced changes in soils in terms of soil erosion, drainage, salinity, depletion of organic matter, addition of nutrients, compaction, flooding, soil sealing, and landslides. Some aspects of soil degradation are slowly reversible (e.g., declines in organic matter) while others are irreversible (e.g., erosion).

In the past, wooded lands were typical in Bulgaria. *Ultimos minor, Fraxinus oxycarpa,* and *Quercus pedunculiflora* grew in the valleys; and *Q. cerris, Q. frainetto, Q. pubescens,* and *Q. Virgiliana* grew on hilly lands.

Intensive agriculture has a serious impact on the environment and thus on soils. Cropping, together with the removal of food and the nutrients it contains, results in reduced soil fertility.

For an agronomic point of view, the effects of increasing organic matter are mainly positive, related to improvement of the structural stability of soils, in levels of nutrients, and water reserves. However, these effects are difficult or even impossible to quantify.

Soils differ widely in terms of their capacity to support environmental, productive, and other functions. The ability of soils to recover from damage also varies. Recognition of this diversity is essential for effective soil protection policy.

Policies dealing with soil quality improvement commonly provide a range of approaches, including investment to promote conservation practices, and advice on soil management. Guidelines for Good Agricultural Practice (GAP) or Good Farming Practice (GFP) define sustainable practices but give limited evidence of their environmental performance. In 2006 the National Action Program for Sustainable Land Management and to Combat Desertification in Bulgaria stipulated soil protection as essential for community prosperity.

1.3.3 Soil Forming Processes

Major soil forming processes in different soils are as follows:

(a) Physical processes are related to aggregation, fragmentation of the rocks, solution, and masses movement due to the gravity, freezing, and thawing.
  - The process of aggregation is when smaller particles are brought together forming large, differently shaped units. Type and degree of aggregation is referred to as soil structure.
  - Wetting and drying are related to the moisture regimes. Wetting initiates soil development. The wetting and drying cycles results in cracking, the formation of wedge shaped aggregates, large columns in clay content soils.
  - Freezing and thawing help to better structure shape formation, orient the stones to the surface of soil, and help in the physical weathering of rock fragments.
  - Leaching and lessivage are related to the movement of solution and masse downward due to the force of

gravity. Leaching takes place through the soil to the underground waters and finally to the river system. The easily soluble salts including nitrates, chlorites, and sulphates are removed, whereas carbonates may be at various positions in the soil.

- Lessivage is a process during which colloidal materials and fine particles can be removed from the upper horizon down to form cutans deeper. This movement can be within a horizon or to affect deeper horizons.

In soils on slopes there is some lateral movement through the soil and down the slope.

- Weathering. Hydrolysis of the primary silicates and loss of some elements in solution is the basis for parent rocks alteration. Iron is presented in a reduced (ferrous) state in the primary minerals. As a result the resistant residue appears in a form of quartz and zircon; altered compounds are hydrous oxides, silica, clay minerals, and solution.

(b) Main chemical processes are hydration, hydrolysis, and solution.

- Hydration is a process during which water is absorbed by compounds and slightly alters them.
- Hydrolysis is a process of alteration of minerals or leads to complete decomposition, whereby hydrogen ions are exchanged for cations.
- Oxidation is an energy-releasing process of removal of electrons from a substance; in biological systems by the removal of hydrogen (or sometimes by the addition of oxygen).
- Reduction is a process of an element or compound accepting an electron during a chemical reaction.

References

Antipov-Karataev IN, Gerassimov IP (1948) The soils of Bulgaria. Zemsnab Press, Sofia (In Bulgarian)

Antipov-Karataev IN, Galeva V, Gerassimov IP et al (1960) The soils in Bulgaria. Zemizdat Press, Sofia (In Bulgarian)

Bondev I (1991) Vegetation in Bulgaria, map at scale 1:600 000 with explanatory text. St. Kliment Oxridski University Press, Sofia (In Bulgarian)

Boneva K (2012) Study on soil characteristics related to calibration and use of "crop-water-yield" simulation model. In: Popova Z (ed) Risk assessment of drought in agriculture and irrigation management through simulation models. Institute of Soil Science N Poushkarov, Agricultural Academy, pp 141–165 (In English)

Boyadzhiev T (1994) Soil map of Bulgaria according to the soil taxonomy explanatory notes. Soil science. Agrochem Ecol J XXIX(4–6):43–51 (In Bulgarian)

Communication COM/2002/179 of 16 April 2002 from the commission to the council, the European parliament, the economic and social committee and the committee of the regions: towards a thematic strategy for soil protection, COM 2002, pp 179

Dimitrov D (1989) Equipment, state, organization and methods of soil mapping in Bulgaria. Lectures on soil science. FAO project TCP/BUL/4502 (T), Institute of Soil Science Nikola Poushkarov, Sofia, pp 140–156 (In English)

Directive 2007/2/EC of the European parliament and of the council of 14 March 2007 establishing an infrastructure for spatial information in the European community (INSPIRE). Official Journal of the European Union L 108, 50

Ganev S (1989) Fundamental chemical indexes necessary for soil surveys. Lectures on soil science. FAO project TCP/BUL/4502(T), Institute of Soil Science Nikola Poushkarov, Sofia, pp 166–191 (In English)

Garbouchev I (1972) Development and forthcoming tasks of the soil science in Bulgaria. First national soil science congress. Bulgarian Academy of Sciences Press, Sofia, 23–25 Sept 1969 (In Bulgarian)

Geography of Bulgaria (1982) Physical geography. Natural conditions and resources. Bulgarian Academy of Sciences, Sofia, p 514 (In Bulgarian)

Gerassimov IP (1972) Absolute and relative age of the soils determined by radio-active Carbon. First national soil science congress. Bulgarian Academy of Sciences Press, Sofia, pp 85–100, 23–25 Sept 1969 (In Bulgarian)

Kercheva M, Dilkova R (2005a) Bulk density as indicator of soil aeration conditions. In: Dilkova et al. (eds) Proceedings of the conference "management, use and protection of soil resources", pp 264–270, 15–19 May 2005

Kercheva M, Dilkova R (2005b) Assessment of soil physical conditions via productivity index. In: Dilkova et al. (eds) Proceedings of the conference "management, use and protection of soil resources", pp 271–275, 15–19 May 2005

Kolchakov I et al. (1994) Project CESD-Communataree and Nikola Poushkarov Institute of Soil Science and Agroecology, N 392 EHL/SW

Koynov V (1964) Peculiarities of the main Bulgarian soil types and their basic soil-genetic processes. In: Gerasimov IP et al. (eds) Soils of south–eastern Europe. Proceedings of the international symposium of soil science, June 1963, Sofia. Bulgarian Academy of Sciences Press, Sofia: 15–62 (In Russian)

Koynov V (1973a) Soil map of Bulgaria at scale 1:1000 000. In: Bezhkov (ed) Atlas of Bulgaria. Bulgarian Academy of Sciences, General Office of Geography and Mapping, Sofia (In Bulgarian)

Koynov V (1973b) Map of the soil geographical regions. In: Bezhkov (ed) Atlas of Bulgaria. Bulgarian Academy of Sciences, General Office of Geography and Mapping, Sofia (In Bulgarian)

Koynov V, Gyurov G, Fotakieva E (1974) Soil geographic zoning of Bulgaria. Institute of Soil Science Nikola Poushkarov, Agricultural Academy, Bulgarian Academy of Sciences Press, Sofia (In Bulgarian)

Koynov V et al (1964) Classification and systematic of soils in Bulgaria. In: Gerasimov IP et al. (eds) Soils of south–eastern Europe. Proceedings of the international symposium of soil science, June 1963, Sofia. Bulgarian Academy of Sciences Press, Sofia: 231–246 (In Russian)

Koynov V, Trashliev Ch, Yolevski M et al (1968) Soil map of Bulgaria at scale 1:400 000 with description, Institute of soil science Nikola Poushkarov. Bulgarian Academy of Sciences Press, General Office of Geography and Mapping, Sofia (In Bulgarian and English)

Koynov V, Trashliev Ch, Yolevski M et al (1972) Soil resources of Bulgaria and their use. First national soil science congress. Bulgarian Academy of Sciences Press, Sofia, 23–25 Sept 1969 (In Bulgarian)

Koynov V, Kabakchiev I, Boneva K (1998) Atlas of the soils in Bulgaria. Zemizdat Press, Sofia (In Bulgarian)

Lazarov A, Rousseva SS, Stefanova V et al (2002) Geographic database and evaluation of different soil erosion prediction models for the purposes of the soil information system. Final report of research project contract No 1108-2556. Ministry of Environment and Water, Sofia (In Bulgarian)

Penkov M et al (1992) Soil classification and diagnostic in Bulgaria in relation to the land reform fragmentation. Zemizat Press, Sofia (In Bulgarian)

Popova Z (2012) Risk assessment of drought in agriculture and irrigation management through simulation models. Institute of Soil Science N Poushkarov, Agricultural Academy, p 342 (In English)

Poushkarov N (1913) Soil-geological review of Sofia's valley. State Agricultural Station Press, Sofia (In Bulgarian)

Poushkarov N (1931) Soil map of Bulgaria at scale 1:500 000. Sofia

Raykov L, Ganev S (1971) Chemical composition of soils in Bulgaria, Institute of Soil Science Nikola Poushkarov, Agricultural Academy. Bulgarian Academy of Sciences Press, Sofia (In Bulgarian)

Rousseva SS (2002) Information bases for developing a geographic database for soil erosion risk assessments. Monograph. Nikola Poushkarov Institute of Soil Science, Sofia (In Bulgarian)

Rousseva S, Stefanova V (2006) Assessment and mapping of soil erodibility and rainfall erosivity in Bulgaria. In: Morel M et al. (eds) BALWOIS 2006 conference on water observation and information system for decision support, Ohrid, Republic of Macedonia. Book of abstracts (ISBN 9989-9594-1-2), pp 28–29. http://balwois.com/balwois/administration/full_paper/ffp-607.pdf. Accessed 23–26 May 2006

Rousseva S, Lazarov A, Stefanova V et al (2006) Soil erosion assessments in Bulgaria. In: Morel M et al. (eds) BALWOIS 2006 conference on water observation and information system for decision support, Ohrid, Republic of Macedonia. Book of abstracts (ISBN 9989-9594-1-2), pp 29–30. http://balwois.com/balwois/administration/full_paper/ffp-608.pdf. Accessed 23–26 May 2006

Rousseva S et al (2010a) Soil erosion risk in Bulgaria and recommendations for soil protective use of agricultural lands. Part I Northern Bulgaria, Agricultural Academy, Institute of Soil Science Nikola Poushkarov. PublishSaySet-Eco, Sofia (In Bulgarian)

Rousseva S et al (2010b) Soil erosion risk in Bulgaria and recommendations for soil protective use of agricultural lands. Part II Southern Bulgaria, Agricultural Academy, Institute of Soil Science Nikola Poushkarov. PublishSaySet-Eco, Sofia (In Bulgarian)

Scheffer F, Schachtschabel P (1979) Lehrbuch der Bodenkunde. Ferdinand Enke Verlag, Stuttgart

Shishkov T, Georgieva G, Kolev N (2009) Soil monitoring and framework of Bulgarian policy for environment sustainability. J Balkan Ecol 12(2):127–133

Shishkov T (2011) Implementation of World Reference Base and Soil Taxonomy within the framework of Bulgarian soil classification. In: International conference "100 Years Bulgarian Soil Science", 16–20 May 2011, Sofia, 1: 86–102

Tanov E, Koynov V (1956) Soil map of Bulgaria at scale 1:200 000 (Koynov elaborated thracian lowland). General Office of Geography and Mapping, Sofia

United Nations Convention to Combat Desertification (1994) In those countries experiencing serious drought and/or desertification, particularly in Africa. Annex V. Regional implementation annex for Central and Eastern Europe

Van Engelen VWP, Wen TT (1995) Detailed information on the global and national Soil and Terrain Digital Databases (SOTER), procedures manual. International Soil Reference and Information Center (ISRIC)

Yolevski M, Hadzhiyanakiev A (1976) Agro-grouping of soils in Bulgaria, Extended systematic list. Code of groups and soil varieties. Institute of Soil Science Nikola Poushkarov, Sofia (In Bulgarian)

Yolevski M, Hadzhiyanakiev As, Georgieva Y et al (1980) Map of agroecological regions of Bulgaria at scale 1:600 000, Sofia (In Bulgarian)

Yolevski M, Ninov N, Krastanov S et al (1983) Brief diagnostic and criterion of the main taxonomic soil divisions in Bulgaria. Precise classification scheme, Institute of Soil Science Nikola Poushkarov, Sofia (In Bulgarian)

Yolevski M, Hadzhiyanakiev As, Georgieva Y et al (1989) Agroecological regioning of territory of Bulgaria. Proceedings of the IV national soil conference, Sofia, pp 136–142 (In Bulgarian)

2 Major Soil Types in Bulgaria

2.1 Chernozems

In this chapter, information is given regarding those soils that have been accorded Bulgarian classification, such classification being made official by use during the large soil survey of the country as follows: Koynov et al. (1964), Yolevski and Hadzhiyanakiev 1976.

Calculation of the mechanical composition (texture) from the entire soil mass was carried out according to the method of Kachinskiy (revised), where carbonates were destroyed and particles were peptized with alkali (Kachinskiy 1965). Fractions larger than 1 mm were separated by means of stirring and sieving.

Soil texture data were transferred from the classification scheme of Kachinskiy to the official scheme used by the USDA, based on the methodology developed by Rousseva (1997).

Methods of laboratory analysis are based on pedological traditions. The analysis of organic matter in mineral soils is conducted according to the Turin method based on wet combustion (Kononova 1963). Chemical determination is based on oxidation of organic matter by an excessive amount of $K_2Cr_2O_7$.

Equivalent $CaCO_3$ is based on destruction of calcareous minerals and emission of carbon dioxide gas (CO_2).

Derived from the Russian words *cherniy* meaning black and *zemlya* meaning earth; and from the Latin word *mollis* meaning soft.

A group of comparatively deep, dark colored, well structured soils developed on the calcareous loess-like materials and characterized by mull type of humus accumulation.

2.2 Concept of the Type

Main concepts

1. Chernozems have thick dark mineral A horizon with high rate of accumulation of organic substances;
2. Characterized by high base saturation (>85 %);
3. Have a remarkable structure consisting entirely of valuable water stable aggregates.

The main processes that take place in chernozems are humification, leaching, and seasonal migration of solutions.

2.3 Ecology of Soil Formation

2.3.1 Distribution

Chernozems are widely distributed on the larger part of the hilly Danube Plain (Fig. 2.1), partly on the North-Western plateau, and on the Dobrudzha and Ludogorie plateaus to the south. Chernozems occupy about 2,250,000 ha or 20 % of the territory of the entire country.

Calcareous chernozems are distributed on the northern part of the hilly Danube Plain in a belt that follows the shoreline of the Danube River (Fig. 2.2). This variety occupies approximately 5.89 % of the entire territory. Calcareous chernozems have been found in the central region and in the southeastern part of the Ludogorie plateau.

Typical chernozems are associated with the occurrence of calcareous chernozems. They spread around the calcareous chernozems on the Ludogorie and Dobrudzha plateaus and continue into Romania; they occupy approximately 1.6 % of the entire territory.

Leached chernozems occupy about 1,120,000 ha or 10 % of the entire territory (Fig. 2.3). Slightly leached and leached varieties are widely distributed on the south Dobrudzha plateau and the periphery of Ludogorie plateau. Strongly leached varieties have been found on the elevated territory of the hilly Danube Plain and on the North-Western plateau (Fig. 2.4).

Degraded (podzolized) chernozems occupy about 2.8 % of the entire territory. Areas of the occurrence are on the Ludogorie plateau and on the southeast of the hilly Danube Plain.

T. Shishkov and N. Kolev, *The Soils of Bulgaria*, World Soils Book Series, DOI: 10.1007/978-94-007-7784-2_2,

Fig. 2.1 The northeastern region of the Danube Plain

Climatic conditions

Chernozems are formed under temperate continental climatic conditions with seasonal contrast in distribution of temperatures and precipitation. Summer is usually hot, often with rainstorms, and winter is freezing. The tendency is for precipitations to increase in intensity from north to south, and from east to west. The maximum falls are during the summer.

In the northwestern province of the Danube Plain the mean annual temperature is 10.7–11.6 °C. In January the mean annual temperature is −1.5 (−2.4) °C. In July the mean annual temperature is 21.6–23.3 °C. The mean amount of precipitation varies from 501 to 594 mm.

In the central province of the Danube Plain the mean annual temperature varies from 10.6–11.1 °C to 11.6–12.1 °C. In January the mean annual temperature is −1.1 (−3.6) °C. In July the mean annual temperature is 21.5–24.1 °C. The mean amount of precipitation varies from 507 to 611 mm.

In the northeastern province in the region of Ludogorie the mean annual temperature is 9.7–10.8 °C. In January the mean annual temperature is −1.6 (−2.4) °C. In July the mean annual temperature is 20.2–21.4 °C. The mean amount of precipitation varies from 520 to 611 mm.

In the eastern province in the region of Dobrudzha the mean annual temperature is 11.8–12.1 °C. In January the mean annual temperature is 0.6 (−1.2) °C. In July the mean annual temperature is 22.2–22.6 °C. The mean amount of precipitation varies from 423 to 480 mm.

Generally, soils have formed under conditions of moisture deficit. Occasionally in spring, when the amount of precipitation is low, chernozems are wetted at field capacity to a depth of 40–60 cm.

The freezing of uppermost (10–15 cm) is common during winter, but freezing below 25–30 cm is rare. The earliest freezing of the top soil (0–2 cm) occurs in the middle of November and the last freezing is in the middle of March.

Topography

The terrain of soil formation is flat to moderately undulating dissected hilly plain. The typical landscape is gently undulating watersheds dissected by the alluvial valleys and depression areas.

The entire hilly Danube Plain is characterized by a slight gradient from the north to the south. The average rates of elevation vary from 100 to 300 m.

Relief in the eastern part of the hilly Danube Plain and Ludogorie plateau is strongly dissected. The average rates of elevation vary from 40 to 200 m. The altitude at the Ludogorie plateau gradually decreases to the north.

In the central part of the hilly Danube Plain the Shumen plateau is located with an average altitude of 350 m; the highest point is the top of Tarnov Dyal with an altitude of 502 m. Because of geological specificity, there is no running water on the plateau, but there are more than 70 springs at its footstep.

The northeastern part comprises the regions of the hilly Danube Plain (with an altitude of 150–200 m) and the Fore

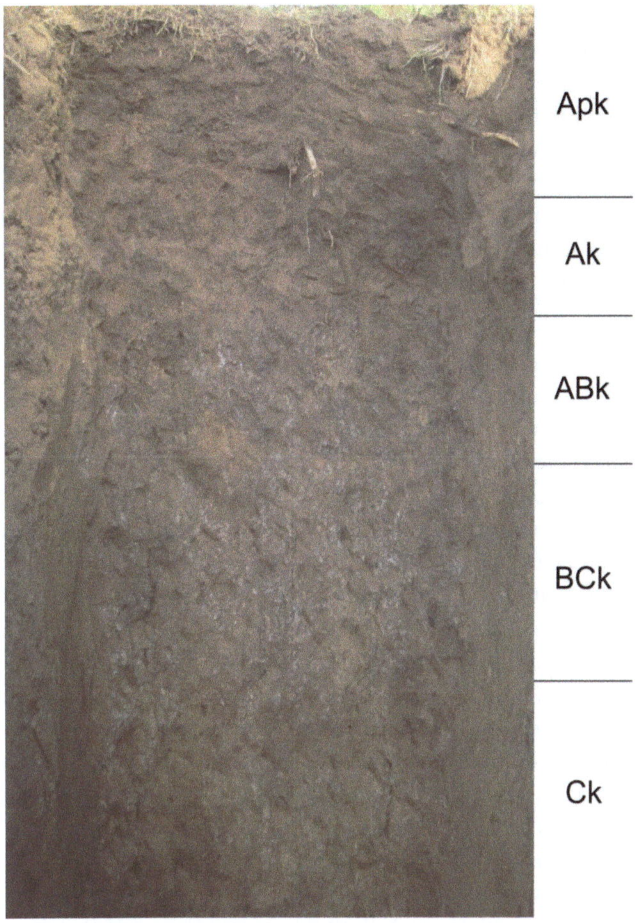

Fig. 2.2 *Calcareous or Typical chernozem*, Bulgarian classification Yolevski and Hadzhiyanakiev (1976). *Epicalcic Chernozem Siltic*, WRB (2006), *Typic Calcixerolls*, Keys to Soil Taxonomy (2010), *Calcic Chernozem*, FAO-UNESCO-ISRIC (1990)

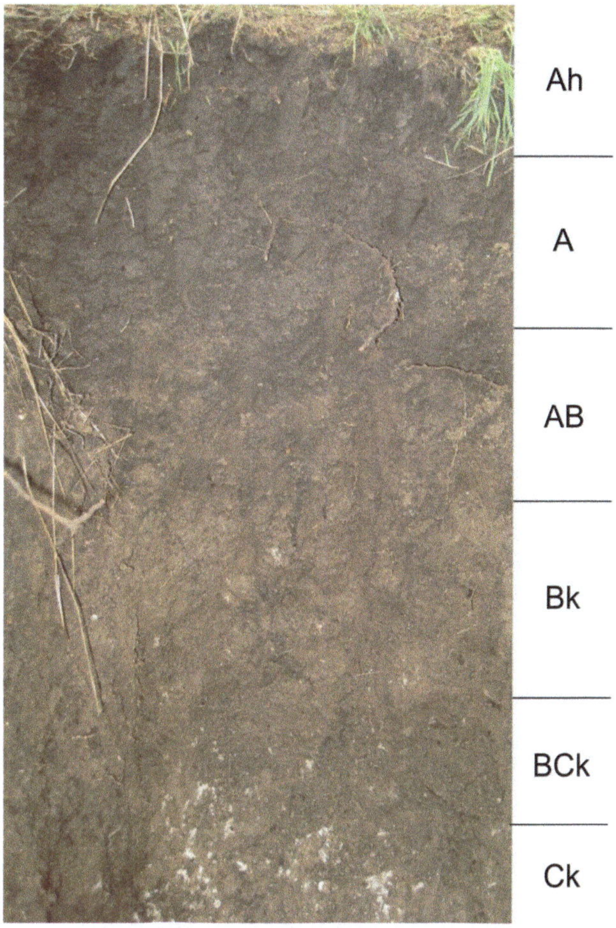

Fig. 2.3 *Leached chernozem*, Bulgarian classification Yolevski and Hadzhiyanakiev (1976). *Endocalcic Chernozem Siltic*, WRB (2006), *Typic Calciustolls*, Keys to Soil Taxonomy (2010), *Haplic Chernozem*, FAO-UNESCO-ISRIC (1990)

Balkan (with an average altitude from 200–800 m). The low plains of the Vidinska, Archaro-Usoiska, Dolnotsibarska and Kozloduy-Oryahovska have an average altitude of 20–30 m.

Parent materials

Chernozems have generally formed on the Quaternary Age eolian sediments (loess, sandy loess, clayey loess) or calcareous parent materials containing easily weatherable minerals (Table 2.1). At the central part of the hilly Danube Plain reddish brown calcareous clays are distributed. At the North-Western plateau and south Dobrudzha plateau the parent materials are loess and loess-like materials (Boykov 1936) (Minkov 1968). In the southeastern part of Ludogorie plateau the parent materials are marl clayey products. Dark non-calcareous clays are located on the North-Western plateau.

Common Quaternary loess is a porous parent material that contributes to the formation of fine structure.

The main peculiarity is the carbonate content in loess materials. Usually, soil texture is like that of the parent materials.

Some provincial differences in soils are determined by the character of these materials.

Thus the calcareous chernozems along the Danube River except on the Quaternary loess have also formed on the Quaternary Age alluvial sediments on the higher terraces.

Leached chernozems in the central part of the hilly Danube Plain except on the Quaternary loess have also formed on the eluvium of the Lower Cretaceous Age (Mesozoic Era) marls and clayey marls with sandstone interbeds and orbitoline sandstones (Aptian).

In the northwestern part of the country leached chernozems karasolutcy have mainly formed on the marine sediments (Fore-Carpathian basin) in North Bulgaria of the Neogene/Miocene (Middle-Upper Sarmatian) Age

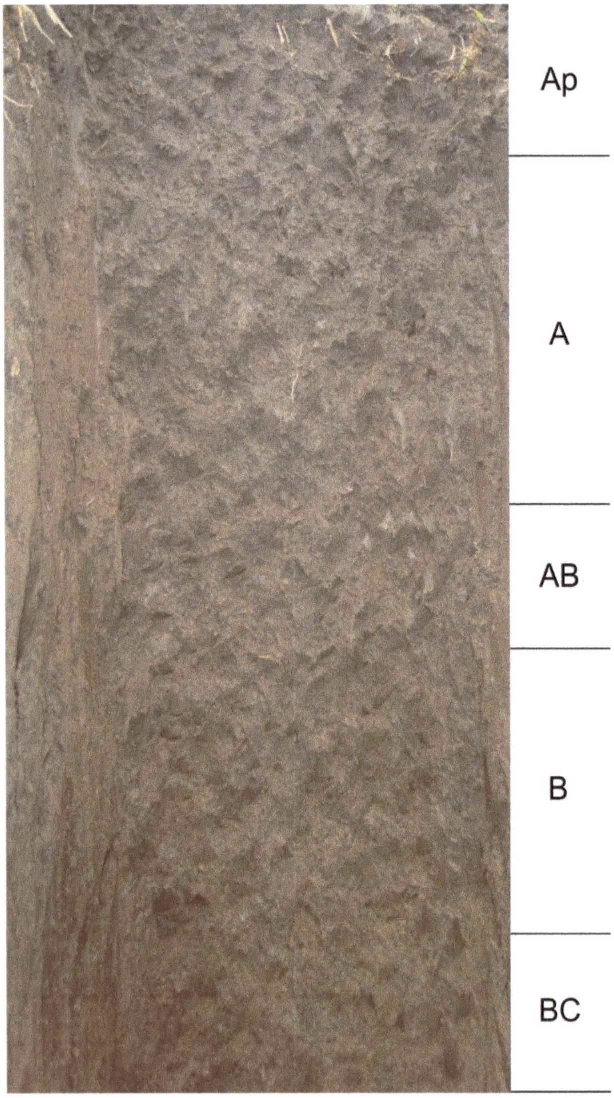

Fig. 2.4 *Strongly Leached chernozem*, Bulgarian classification Yolevski and Hadzhiyanakiev (1976). *Bathycalcic Chernozem Siltic*, WRB (2006), *Typic Haplustolls*, Keys to Soil Taxonomy (2010), *Haplic Chernozem*, FAO-UNESCO-ISRIC (1990)

limestones and banded clays, and on the Lower-Middle Sarmatian Age clays, sands, and limestones.

In the eastern part of the country in the South Dobrudzha region the leached chernozems except on the Quaternary loess have also formed on the marine sediments (Fore-Carpathian basin) in North Bulgaria of the Neogene/Miocene (Middle-Upper Sarmatian) Age on the eluvium of limestones, mactra limestones, banded clays, and carbonates.

In the eastern part in the region of Kamchia River chernozems have formed on the marine sedimentary rocks (Euxino-Kaspian basin) in North Bulgaria on the Neogene/Miocene (Konkian) Age sands with clay interbeds, sandstones, and limestones; on the Paleogene/Paleocene (Middle-Upper Eocene) Age eluvium of marls locally with interbeds of detrital sandstones; and on the Paleogene/Lower Eocene Age clayey sands, sands and sandstones, and nummulitic limestones.

Degraded (podzolized) chernozems on the Ludogorie plateau except on the Quaternary loess have also formed on the eluvium of the Lower Cretaceous Age (Mesozoic Era) limestones, and clayey limestones (Aptian).

Vegetation

Soils from the northern part of the hilly Danube Plain are associated with the southern forest-steppe zone of chernozems distribution.

Chernozems evolution takes place on the grasslands and steppe territories, where grass vegetation has grown mainly in the spring and early summer period and the Oak forest covered uplands of the North-Western, Ludogorie, and south Dobrudzha plateaus. The natural vegetation is tall steppe grasses in the open areas and deciduous woodland dominated by species of pedunculated oak (*Quercus pedunculata*), pubescent oak (*Quercus pubescens*), elm (*Ulmus* sp.), oriental hornbeam (*Carpinus diunensis*), flowering ash (*Fraxinus ornus*), common smoke-tree (*Cotinus coggygria*), Christ's thorn (*Paliurus spina-Christi*) as well as beardgrass (*Andropogon* sp.), bedstraw (*Gallium*), feather grass (*Stipa stenophylla, S. joannis, S. capillata, S. Lessingiana*), fescue (*Festuca* sp.), wild oats (*Avena* sp.), brome grass (*Bromus L.*), crested hair-grass (*Koeleria cristata*), wheat-grass (*Agropyrum pectiniforme*), Bermuda grass (*Cynodon dactylon*), bent grass (*Agrostis capillaries*), sedge (*Carex* sp.), clover (*Trifolium montanum, T. lupinaste*), sage (*Salvia nutans*), and bluebell (*Campanula steveni*) (Bondev 1991).

Following the annual decay, deeply penetrating grass roots directly supply the soil with a considerable amount of nutrient elements. The high intensity of residues transformation occurs directly in the soil where high biological activity has taken place.

Time

Most chernozems are relatively young to moderately old soils formed in a different time during Holocene (10,000–12,000 years BP).

Human influence

Historically, chernozems use has resulted in a dramatic reduction of pristine lands. Human activities over the centuries have resulted in most of the soils being turned into arable lands. Today the natural woodland that used to cover the territories of chernozems occurrence has disappeared.

In the northwestern part of the country the total area of the arable land is 600,000 ha (Koynov et al. 1972), where the cultivated land is 321,000 ha, generally occupied by cereal crops (>67 %).

Table 2.1 Physical properties of loess deposits

Loess deposits	Particle-size distribution (%)				Particle density (g/cm³)	Bulk density (g/cm³)	Pore volume (%)
	>2.0	2.0–0.05	0.05–0.005	<0.005			
Sandy loess	–	46	47	7	2.74	1.45	46
Typical loess	–	27	57	16	2.75	1.38	50
Clayey loess	–	25	54	21	2.74	1.49	46
Clay-like	–	26	40	34	2.72	1.57	41

In the central-northern part the total area of the arable land is 755,000 ha, where the cultivated land is 445,200 ha, generally occupied by cereal crops (>58 %).

In the northeastern part the total area of the arable land is 630,000 ha, where the cultivated land is 371,700 ha, generally occupied by cereal crops (>62 %).

Intensive cultivation causes degradation of structure. The reversal of pristine lands into arable has resulted in an alteration of structure, whereby what was once naturally fine and granular has become loose and non-coherent in plough horizons. Fine soil particles have become detached and are easily prone to erosion hazard. Subsoil has become more vulnerable to compaction. Decreases in pore space have reduced the rates of aeration and water movement through the soil.

Irrigation practices force soil negative processes and cause secondary salinization at some sites. The greater use of fertilizers of fertilizers together with increased leaching reaches the deeper parent materials and ground waters.

Erosion strongly affects all varieties of chernozems.

2.3.2 Genesis

Chernozems have been an object of research since pedology emerged two centuries ago. In 1763, the Academician Lomonosov produced a treatise titled "On the Earth's strata" in which he described for the first time the idea of the organic origin of soil (which could be due to the natural decay of flora and fauna) as well as peat, coal, petroleum, and amber, and thus made a significant contribution to mineralogy.

Dokuchaev formulated his idea in the monograph "Russian chernozem" (1883). This book was the basis for the establishment of the genetic conception in Russian pedology.

The original idea about the dominance of biological process and the significance of microorganisms in producing humus accumulation and favorable soil structure through decomposition of grass roots, came from Kostichev (1886).

Bulgarian chernozems were first presented on the Soil map of Romania created by Enculescu in 1924. In 1927, Bonchev included the chernozems on a Soil map of Bulgaria. In 1931, Poushkarov showed the areas of chernozem distribution on the Soil map at scale 1:500,000.

The most recent popular hypothesis of chernozems origin is the existence of a paleo-hydromorphic phase in past soil development.

Bulgarian chernozems cannot be separated from the soils of the bordering countries and are defined as micellar chernozems from the Danube valley. The specifics of solution migration determines the micellar form of carbonate segregation in soil profile.

The principal variations within the type are thickness, humus content, and the depth of carbonate appearance after treatment with 10 % HCl.

The main processes that take place in chernozems development are decomposition and mineralization of organic remnants and transformation in humic substances, rapid root turnover, high biological activity, accumulation of humus, seasonal leaching of soluble salts, and carbonate oxidation and reduction.

The profile of the chernozem comprises both the humus accumulative profile and the carbonate accumulative profile. Differentiation is on the basis of forms and distribution of carbonates in the profile.

The geographical localization of the Bulgarian chernozems is associated with the most southern occurrence in Europe. This fact determines specific bioclimatic conditions manifested by the enhanced mineralization of organic matter. Bulgarian soils produce relatively small humus content compared to the chernozems in other parts of the world.

High proportion of organic matter in the soil is derived from root decomposition. Grass formation produces a considerable amount of organic substances not only at the surface, but also particularly in the depth of profile. This often indicates the depth of the plant root system and the thickness of the humus horizon. Organic matter is in the form of humus and dark crumbly partially decomposed substances.

Strong granular to crumb aggregated structure affects soil aeration, increasing water-holding capacity and drainage.

The process of carbonate migration occurs as in pristine and arable soils. This process contributes to the development of buffering conditions in a soil. It reflects on the stable calcium bounded humic acids formation, high rates of colloids precipitation with calcium, neutral soil reaction, and formation of calcareous horizon.

Chernozems varieties in Bulgaria are characterized by different moisture regimes. Periodically in spring during the thaw, soil solutions, being enriched with dissolved $Ca(HCO_3)_2$, reach the deeper layers of parent materials. In summer, due to evaporation, the soil solutions, being enriched with dissolved $Ca(HCO_3)_2$, reach back to the upper horizons of the soil profile.

The carbonate pattern distribution is antagonistic to the clay pattern distribution.

In the calcareous, typical, and slightly leached chernozems there is no clay enrichment in the profile. Some clay enrichment can occur down the profile of strongly leached and degraded (podzolized) chernozems varieties due to weathering in situ or lessivage.

Wind and water erosion affects vast areas. Most prone to the erosion hazard are calcareous, typical, and slightly leached chernozem varieties, while strongly leached and degraded (podzolized) varieties are less prone.

2.4 Soil Diagnostic and Classification

2.4.1 Morphology

Chernozem is a deep, dark colored, well structured soil, developed on calcareous loess materials and characterized by humus accumulation of mull type. Penetration of the deep roots steppe grasses is characterized by rapid root turnover, thus ensuring a considerable amount of organic matter at soil depth.

Ah*(*k*)**, (Ap***)—a sod that can be 8 cm thick, naturally occurs in pristine lands, and consists of grasses, roots, mat, or plough. The fine earth is homogeneous, dark grayish brown (moist), and dark brown (dry) colored; between the roots there are structural aggregates of fine granular structure (1–2 mm diameter) with strong pedality grade, consistence is friable (moist) and slightly hard (dry). Abundance of fine roots (0.5–2 mm); beneath there is clear transition to the lower

A*(*k*)**, (Ap***)—a mineral humus accumulative horizon that can be subdivided into prime, second, etc. Commonly, it is 40–80 cm thick, homogeneous, dark grayish brown (moist) and dark brown (dry) colored; characterized by remarkable granular water stable structure (spheroids or polyhedrons), with coarse size (5–10 mm diameter), and strong pedality grade; consistence is firm (moist) and slightly hard (weakly resistant to pressure) dry. Large aggregates comprise many small ones, which may be composed largely of earthworm casts. Secondary calcium carbonate is deposited as pseudomycelium (threads) in case of presence (Fig. 2.5). Porosity is high. Biological features are many like burrows (unspecified) and open large burrows. Many fine roots (0.5–2 mm) gradually grade beneath to

***AB*(*k*)**—a transitional mineral humus-metamorphic horizon, dark brown or dark yellowish brown colored; structure is granular or crumb (spheroids or polyhedrons), with strong pedality grade, consistence is firm (moist), and slightly hard (dry); porosity is high. Secondary calcium carbonate is deposited as pseudomycelium (threads) in case of presence. High activity of the earthworm population is indicated by the abundance of casts present. Evidence of a dense population of burrows mammals is common. A lot of fine (0.5–2 mm) and medium (2–5 mm) roots are in the soil mass; gradually grade beneath to

***B*(*k*)**—an underneath mineral metamorphic B horizon, which is dark brownish colored. A common feature is the accumulation of secondary carbonate like soft concretions and mycelia (threads) on ped faces. Pedality type is subangular blocky (blocks having mixed rounded and plane faces of aggregates), with medium size (10–20 mm), strong pedality grade, consistence is firm (moist) and hard (dry), porosity is high. Many earthworm casts and vertebrate passages are present in the soil mass. Krotovinas are common in the profile like burrows deeply into sols, caused by blind mole rats and susliks. Few roots of medium size (2–5 mm) occur; there is gradual transition to

***BC*(*k*)**—a transitional mineral horizon to the upper part of parent materials, common dark yellowish brown colored. Pedality type is subangular blocky, with medium size (10–20 mm) and very fine (1–5 mm) size, moderate pedality grade, consistence is firm (moist) and hard (dry), lesser pseudo mycelia and rich with carbonate concretions (Fig. 2.6); porosity is medium. Biological features are burrows. Clear transition to

Ck—partly altered by solutions of loess parent materials. Usually this horizon is accepted as parent material with pale brown or yellowish brown color, massive (structureless), weak pedality grade, porosity is low. Evidence of mesofauna activity is like burrows. There is the most prominent accumulation of secondary carbonate mostly like soft spots and concretions (Fig. 2.7).

The depth of carbonate leaching is hardly to be explained by the contemporary rainfall distribution. The exact location of $CaCO_3$ effervescence (of 10 % HCl) in the profile is a basic classification feature.

Thus in the calcareous chernozems strong effervescence of 10 % HCl is fixed at the surface, and accumulation of carbonate-like pseudomycelium starts in the upper part of humus A horizon. In the typical chernozems leaching is at the uppermost (5–10 cm) below the soil surface. In the slightly leached chernozems the carbonates have been removed from the humus A horizon. In the leached chernozems the secondary carbonates are fixed at the middle part

Fig. 2.5 Subangular blocky aggregate with carbonates like pseudomycelium (threads) fine pores, few channels of worms, and worm casts

Fig. 2.6 High porous subangular blocky agregate of BCk horizon saturated with carbonates

of the transitional AB horizon and there is accumulation at the B horizon. In strongly leached chernozems carbonates have been removed to the lower part in C horizon and there is absence of pseudo mycelia carbonate accumulation.

2.4.2 Classification

Despite the fact that chernozems have been well studied, the question of classification is still open to discussion.

Fig. 2.7 Carbonate concretions with different size and shape from Ck horizon

Chernozems were described by Dokuchaev as an "endless variability of chernozem like soils."

In Bulgaria during the years 1936–1940 some of the properties of the soil group named steppe chestnut soil were described. In 1940, Poushkarov undertook a survey in northern Bulgaria over the watershed of the Iskar and Ogosta rivers, the results of which supported the idea of chernozems occurrence in that period. Later in 1947 the type of Bulgarian chernozem was defined by the Bulgarian-Soviet expedition. Soil classifications of 1948 (Antipov-Karataev and Gerassimov) and 1964 (Koynov et al.) subdivided the type into four soil subtypes, namely calcareous, typical, leached, and podzolized chernozems, which were very similar to the Russian names. Soil classification of 1983 (Yolevski et al.) rejected the subtypes of typical and podzolized (degraded) chernozem and renamed the latter degraded chernozem. There was classification in 1992 (Penkov et al.) which at the theoretic level named the subtypes of chernozems as kastanozem, haplic, luvic, vertic, and gleyc; these have not been applied in practice.

The subtype varieties differ with regard to thickness of humus horizon (slightly thick < 40 cm, medium 40–80 cm and thick >80 cm) and on the degree of humification (poor < 1 % humus content, low 1.0–2.5 %, medium 2.5–5.0 %, and rich >5 %).

Classification after WRB (2006) required *Mollic* diagnostic horizon in the reference group of chernozems.

Keys to Soil Taxonomy (2010) required a *Mollic* epipedon for Mollisol order classification. In the Soil Taxonomy there are demands for subsurface horizons allowing *cambic* expressing mottling, gleying, and structure development; *argillic* horizon with high base saturation (>35 %); *calcic or petrocalcic* horizon; *albic, natric or none*.

Mollic epipedon may occur in other orders without Mollisol classification. Thus subsurface horizon is *oxic* in Oxisols, *argillic or kandic* horizon with low (<35 %) base saturation in Ultisols; *vertic* in Vertisols takes precedence over the Mollic horizon.

2.4.3 Chernozems Subtypes in Bulgaria

2.4.3.1 Calcareous Chernozem

Calcareous (*subtype*) **chernozem** (*type*) is named after the Bulgarian classification of Koynov et al. (1964), and Yolevski and Hadzhiyanakiev (1976) (Fig. 2.2).

Epicalcic Chernozem is named according to WRB (2006) and *Typic Calcixerolls* according to Keys to Soil Taxonomy (2010).

Calcic Chernozem or Calcic Kastanozem: Revised Legend of the World Soils (FAO-UNESCO-ISRIC 1990).

Distinguishing features are less pronounced of humus accumulation compared to the other chernozem subtypes, the appearance of secondary carbonates from the surface, and their accumulation like pseudo mycelia deeper in the humus A horizon.

Before the Bulgarian-Soviet expedition in 1947, it was considered that less color intensity was sufficient argument

Table 2.2 Variables of the particle-size distribution in 30 referenced profiles of calcareous chernozems in Bulgaria

Profile count 30	Horizon	Upper depth	Lower depth	Loss HCl	Particle-size distribution (mm), % (according to method of Kachinskiy revised)							
					>1 mm	1–0.25	0.25–0.05	0.05–0.01	0.01–0.005	0.005–0.001	<0.001	<0.01[a]
Mean	Ak′	0	28	8	0.0	0.2	15.2	39.7	8.3	7.7	21.2	37.1
	Ak″	28	50	11	0.0	0.4	13.6	38.8	8.6	7.4	19.8	35.8
	ABk	50	74	18	0.0	0.2	11.5	38.7	7.7	6.9	17.0	31.5
	BCk	74	103	22	0.1	0.0	13.3	37.6	8.2	6.1	13.1	27.4
	Ck	103	126	24	0.1	0.0	12.4	38.1	7.9	6.6	11.4	25.8
Stdev	Ak′	0	2	6.5	0.0	0.8	7.1	6.0	2.9	2.5	5.4	6.4
	Ak″	2	5	6.7	0.0	1.8	5.3	4.5	2.6	2.1	5.2	6.2
	ABk	5	8	6.4	0.0	1.1	5.2	4.2	2.4	2.2	4.8	6.3
	BCk	8	13	3.7	0.3	0.1	5.6	4.4	2.2	2.3	5.9	6.7
	Ck	13	14	3.1	0.2	0.1	4.8	5.0	2.6	2.8	4.9	5.6

[a] Physical clay, the parameter characterized sum of the particle-size distribution, according to the Kachinskiy method (mm) $\sum <0.01 = <0.001 + (0.005 - 0.001) + (0.01 - 0.005)$

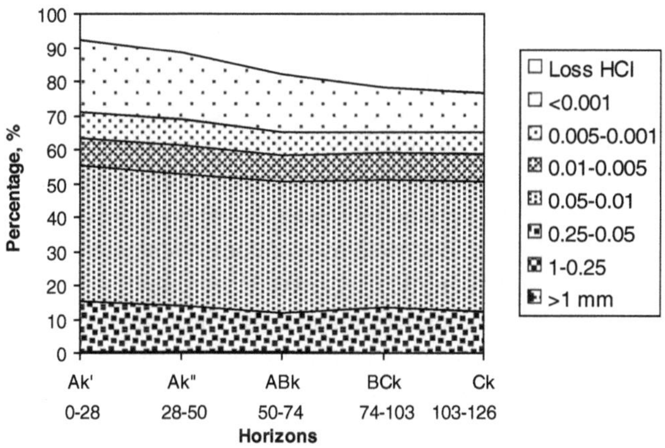

Fig. 2.8 Pattern of the particle-size distribution (mm) in calcareous chernozem (according to revised Kachinskiy method)

for the kastanozem-type classification similar to Russian soil. Subsequently, a number of research studies approved the morphological and analytical identification of calcareous chernozem.

Recently, WRB (2006) has allowed that some soils that have a Mollic horizon can be referred to as Kastanozems in arid climate conditions. There is discussion about whether Kastanozens can be morphologically identified where calcareous chernozems occur (Shishkov 1998).

The pattern of the horizon sequence is: Ak(Apk)–(Ak)–ABk–BCk–Ck. Metamorphic B horizon is not recognized in the profile.

Usually the texture is silt loam. Particle size distribution shows that the coarse silt fraction (0.05–0.01 mm) prevails, followed by the clay (<0.001 mm), (Table 2.2 and Fig. 2.8). Texture is homogeneous at the depth of profile (Table 2.3 and Fig. 2.9).

Water stable aggregates (with size 1–3 mm) are about 80 %. Aggregates are characterized by high internal and external porosity. The average particle density is 2.45–2.6 g/cm.[3] Field water capacity is 64 % of the total porosity and is a result of good aeration. Calcareous chernozems have high water-holding capacity and are characterized by high rates of water percolation.

Development of structure results in the formation of pore spaces, which in turn results in an increase in porosity and a decrease in the bulk density. The size of pore spaces is as important as the total amount of pore space. In the calcareous chernozems 10–15 % of pore space comprises macropores or the pores of aeration (with diameter >300 μm); and over 20 % is mesopores or pores with rapid moisture drainage (>30 μm diameter). Micropore space or water-filled space is about 20 %, comprising slow drainage pores (diameter 0.2–30.0 μm) and 10–15 % of micropores is with none remaining available for roots moisture (diameter <0.2 μm) (Dilkova 1985; Dilkova et al. 1998).

Thickness of humus horizon (A + AB) is 40–80 cm. Content of humus varies from 2.0 to 4.2 % on the upper part of humus horizon (Table 2.3 and Fig. 2.10), with higher values in pristine land. Stock of organic matter is from 250 to 400 tons per hectare to a depth of 100 cm (Filcheva et al. 2002). It stores and supplies from 0.14 to 0.25 % of nitrogen holding ability. The process of transformation of organic residues is accompanied by nitrogen enrichment of humic substances shown by the ratio C/N 10-12 which means high rates. In the humus composition 100 % of humic acids (C_h) are bound with calcium (Petrova and Shishkov 2001).

The profile of the Bulgarian calcareous chernozems is not differentiated in clay and sesquioxides distribution. Content of Al_2O_3 is two times higher than Fe_2O_3, the content of MgO and K_2O is much higher than CaO.

Table 2.3 Variables of humus, pH, carbonates, and particle-size distribution (mm) according to USDA in 30 referenced profiles of calcareous chernozems in Bulgaria

Profile count 30	Horizon	Upper depth	Lower depth	Humus (%)	pH_{KCl}	Carbonates	Bulk density g.cm^3 calc	Clay <0.002	Silt 0.002–0.05	Fine sand 0.05–0.2	Total sand 0.05–2.0
Mean	Ak′	0	28	2.73	7.3	3.83	1.24	29.1	54.8	15.7	16.0
	Ak″	28	50	2.23	7.4	7.75	1.30	29.7	55.4	14.5	14.9
	Abk	50	74	1.52	7.5	14.02	1.24	30.1	56.8	12.9	13.1
	BCk	74	103	1.13	7.6	18.41	1.26	27.8	57.2	15.0	15.1
	Ck	103	126	0.84	7.7	19.24	–	27.4	58.4	14.2	14.3
Stdev	Ak′	0	2	0.45	0.2	4.41	0.08	5.4	5.4	7.0	7.2
	Ak″	2	5	0.51	0.2	5.89	0.06	4.9	4.6	5.1	5.9
	Abk	5	8	0.41	0.2	5.73	0.15	4.5	4.8	5.1	5.5
	BCk	8	13	0.35	0.2	4.70	0.16	5.3	5.4	5.7	5.7
	Ck	13	14	0.32	0.2	2.90	–	4.7	6.0	4.9	4.9

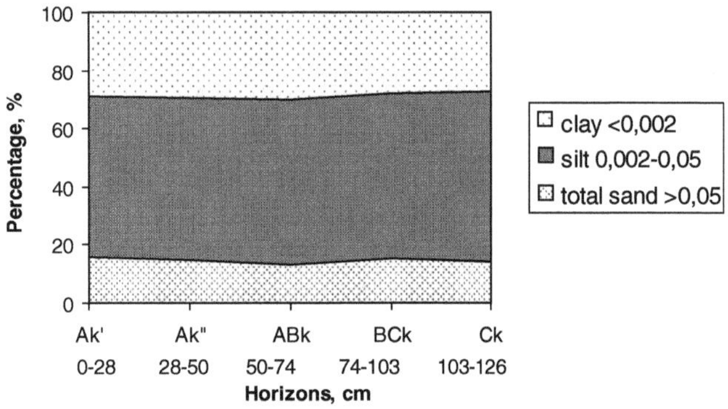

Fig. 2.9 Pattern of the particle-size distribution (mm) in calcareous chernozem (USDA)

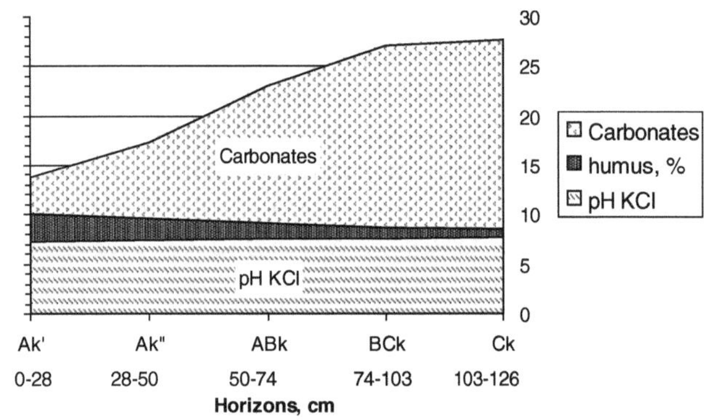

Fig. 2.10 Pattern of pH, humus, and carbonates distribution in the calcareous chernozem

Minerals are the same as in the loess parent materials. Distribution of quartz and feldspars (mainly plagioclase) shows enrichment at the surface and decrease in the loess parent materials. Orthoclase is regularly distributed. The ratio of quartz content to the feldspars decreases down the profile.

The qualitative composition of clay minerals, determined by X-ray diffraction analysis and transitional electronic microscopy, has shown a prevalence of illite at rates of 34–56 % in the clay and about 10 % in the soil at the surface humus (A) horizon, gradually decreasing at the depth, where in the (C) horizon it is about 24–40 % in the clay and 3–4 % in the soil. Followed by montmorillonite, which is at rates of 17–44 % in the clay (<0.001 mm) and 3–12 % in the soil at the surface humus (A) horizon, gradually increasing at the depth, where in the (C) horizon it is about 28–56 % in the clay and 2–9 % in the soil. Kaolinite is about 12–20 % in the clay (<0.001 mm) and 2–4 % in the soil. A significant quantity of quartz is at the surface (A) horizon about 4–11 % in the clay and 1–3 % in the soil, sharply decreasing at the depth (Bechar et al. 1988).

The amount of total iron (Fe_2O_3) (Arinushkina 1970) is about 6 % with uniform distribution in the profile. In all horizons the silicate form of iron (74 % from the total iron Fe_2O_3) greatly dominates.

The non-silicate iron (Mehra and Jackson 1960) is about 26 % (from the total iron Fe_2O_3). All non-silicate iron appears in crystallized form (Hadzhiyanakiev 1989).

The amorphous form of iron (Tamm 1922) is only 2.5–4.0 % (from the total iron).

The free amorphous form of iron in the humus horizon is about 12–14 % (from the total non-silicate amorphous iron) and gradually decreases with depth.

The amorphous form of iron bounded with organic matter (Bascomb 1968) is insignificant and is only 0.6–1.3 % (from the total non-silicate iron).

Carbonates comprise dolomite, calcite, powder calcite in segregations, and needles of lublinite at a depth of 90 cm.

Soil reaction is slightly alkali. Cation exchange capacity is 20–26 meq/100 g soil. Exchangeable cations are mostly Ca^{2+} and Mg^{2+}.

Evidence of high biological activity is found throughout the soil, including burrowing tunnels, holes made by unspecified animals and insects, and especially earthworm casts.

Naturally the spatial relationship of the calcareous chernozems is with typical and leached chernozems as well as with calcareous regosols.

Calcareous chernozem-karasolutcii variety
Within the subtype of calcareous chernozems there is a gender of soils with enhanced vertic features. According to the Bulgarian classification this soil is named *Karasolutcii* or *Epicalcic Vertic chernozem* (after WRB 2006) and *Vertic Calcixerolls* (after Keys to Soil Taxonomy 2010).

Verti-Calcic Chernozem or Verti-Calcic Kastanozem: Revised Legend of the World Soils (FAO-UNESCO-ISRIC 1990).

This variety is characterized by high clay content throughout the profile as well as high content of clay in the parent materials. The profile is not differentiated in clay. Burrowing features and especially the earthworm tunnels are evidence of high biological activity.

Calcareous Regosols soil variety

Calcareous regosols are soil varieties where local factors cause permanent reduction of surface layers; no diagnostic horizons are observed in the profile. Texture is homogeneous at the depth (Tables 2.4, 2.5 and Figs. 2.11, 2.12). The objective origin of calcareous regosol has close links to the bordering soil which is in a state of advanced development. Erosion has seriously affected the soil horizon sequence. This results in a strong enrichment with secondary carbonates, and a low amount of humus and nutrients content (Fig. 2.13). Morphologically, these facts reflect a deterioration.

According to the Bulgarian classification these soils are named *Calcareous Chermozems severely eroded* or *Haplic Regosol* (*Calcaric*) (after WRB 2006), or *Typic Xerorthents* (after Keys to Soil Taxonomy 2010) and *Calcaric Regosol:* Revised Legend of the World Soils (FAO-UNESCO-ISRIC 1990).

Texture is homogeneous at the depth (Tables 2.4, 2.5 and Figs. 2.11, 2.12, 2.13).

2.4.3.2 Typical Chernozems

Typical (*subtype*) **chernozems** (*type*) is named after the Bulgarian classification of Koynov et al. (1964), and Yolevski and Hadzhiyanakiev (1976).

Epicalcic Chernozem is named according to WRB (2006) and *Typic Calcixerolls* according to Keys to Soil Taxonomy (2010).

Calcic Chernozem: Revised Legend of the World Soils (FAO-UNESCO-ISRIC 1990).

Typical chernozems are associated with calcareous chernozems distribution and grade into slightly leached chernozems.

Distinguishing features are similar to the calcareous chernozem except the objective fact that the fixed leaching leaching depth is at the uppermost (5–10 cm) below the surface, always in the upper part of the humus horizon and with pseudomycelium appearance at the depth of 30–40 cm. At the surface the amount of carbonate is about 0.5 %. The exact time of the leaching and the factors that are responsible for it are open to discussion.

The content of humus varies from 2.0 to 4.0 % on the upper part of humus horizon, with higher values in pristine land. The stock of organic matter is from 250 to 450 tons per hectare to a depth of 100 cm. In the humus composition 100 % of humic acids (C_h) are bound with calcium.

The pattern of the horizon sequence is: A(Ap)–(Ak)–ABk–Bk–BCk–Ck.

Humus accumulation and active migration of soil solutions is typical for the chernozems.

The common texture is silt loam and silty clay loam. The profile is not differentiated in terms of clay and sesquioxides distribution. The qualitative composition of clay minerals, determined by X-ray diffraction analysis and transitional electronic microscopy, has shown a prevalence of of motmorillonite at rates of 44 % in the clay and 12 % in the soil at the surface humus (A) horizon, gradually increasing at the depth, where in the (C) horizon it is about 56 % in the clay and 9 % in the soil. This is followed by illite, which is at rates of 34 % in the clay (<0.001 mm) and 10 % in the soil at the surface humus (A) horizon, gradually decreasing at the depth, where in the (C) horizon it is about 24 % in the clay and 4 % in the soil. Kaolinite is about 13–16 % in the clay (<0.001 mm) and 2–4 % in the soil. Quartz is about 3 % in the clay and 1 % in the soil at the surface humus (A) horizon, sharply decreasing at the depth (Bechar et al. 1988).

Biological activity is high, and is especially evident in the occurrence of earthworm casts.

Soil reaction is slightly alkali. Exchangeable cations are mainly Ca^{2+} and Mg^{2+}.

Typical chernozems have high water-holding capacity and are characterized by high rates of water percolation.

Within the subtype of typical chernozems there is a gender of soils with pronounced vertic features. This variety is named *Karasolutcy* and is *Epicalcic Vertic Chernozem* (after WRB 2006) and *Vertic Calcixeroll* (after Keys to Soil Taxonomy 2010).

The karasolutcy variety is characterized by high clay content in the profile as well as high clay content in the parent materials.

Table 2.4 Variables of the particle-size distribution in nine referenced profiles of the calcareous regosols in Bulgaria

Profile count 9	Horizon	Upper depth	Lower depth	Loss HCl	Particle-size distribution (mm), % (according to method of Kachinskiy revised)							
					>1 mm	1–0.25	0.25–0.05	0.05–0.01	0.01–0.005	0.005–0.001	<0.001	<0.01[a]
Mean	ABk	0	27	19.7	0.0	0.0	17.4	35.9	7.8	6.2	13.0	27.1
	BCk	27	53	26.9	0.2	0.0	15.4	37.7	6.8	5.1	8.0	19.9
	Ck1	53	86	26.0	0.4	0.0	16.1	37.2	10.8	4.0	5.9	20.7
Stdev	ABk	0	3	9.7	0.0	0.0	7.0	8.5	2.6	3.0	7.7	7.9
	BCk	3	7	5.9	0.6	0.0	6.6	8.3	2.5	2.8	6.3	8.0
	Ck	7	18	6.9	1.0	0.0	7.0	9.2	6.1	2.1	5.6	7.5

[a] Physical clay, the parameter characterized sum of the particle-size distribution, according the Kachinskiy method (mm) $\sum <0.01 = <0.001 + (0.005 - 0.001) + (0.01 - 0.005)$

Table 2.5 Variables of humus, pH, carbonates, and particle-size distribution (mm) according to USDA in nine referenced profiles of calcareous regosols in Bulgaria

Profile count 9	Horizon	Upper depth	Lower depth	Humus (%)	pH_{KCl}	Carbonates	Bulk density g.cm[3] calc	Clay < 0.002	Silt 0.002–0.05	Fine sand 0.05–0.2	Total sand 0.05–2.0
Mean	ABk	0	27	1.61	7.6	14.9	1.41	26.7	54.3	18.9	18.9
	BCk	27	53	0.96	7.7	21.6	1.38	25.1	57.5	17.5	17.6
	Ck	53	86	0.40	7.8	23.3	–	21.9	60.1	18.1	18.3
Stdev	ABk	0	3	0.71	0.2	8.7	0.12	9.8	9.2	6.4	6.4
	BCk	3	7	0.50	0.2	6.3	–	9.7	7.3	6.5	6.6
	Ck	7	18	0.06	0.2	6.5	–	8.7	6.2	7.0	7.2

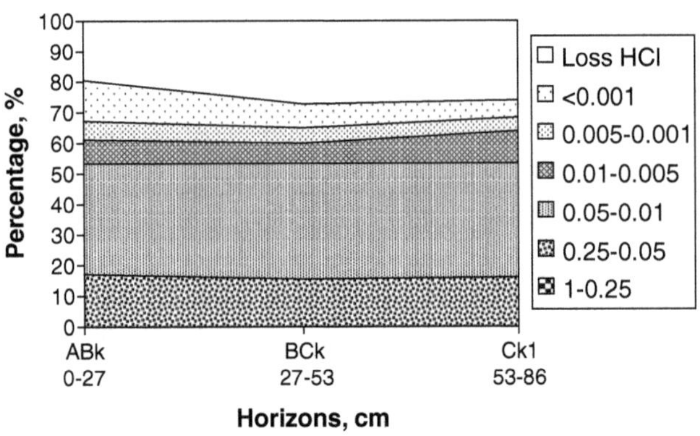

Fig. 2.11 Pattern of the particle-size distribution (mm) in calcareous regosol (according to revised Kachinskiy method)

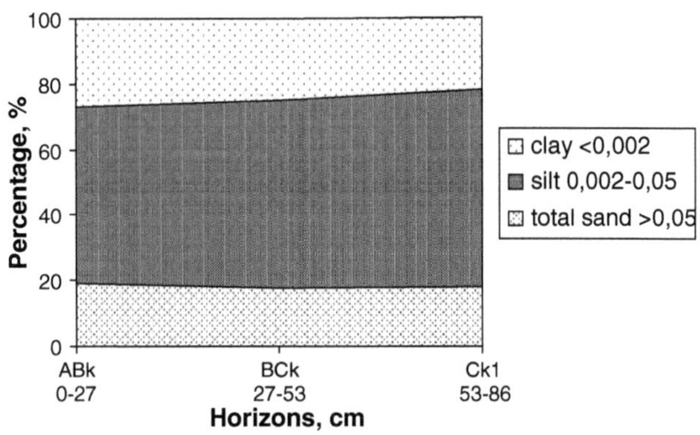

Fig. 2.12 Pattern of the particle-size distribution (mm) in calcareous regosol (USDA)

2.4.3.3 Leached Chernozems

Leached (*subtype*) **chernozems** (*type*), named after the Bulgarian classifications of Koynov et al. (1964), Yolevski and Hadzhiyanakiev (1976), distinguish three groups of leached chernozems (Table 2.6).

Typical distinguishing features of this type of chernozems are humus accumulation, carbonate leaching from the humus horizon, lack of evidence of differentiation of clay and sesquioxides, and high biological activity especially of earthworms to a prominent depth.

The pattern of the horizon sequence is: A(Ap)–(A)–ABk–Bk–BCk–Ck in the slightly leached chernozem. This variety is characterized by carbonates removal to the depth of 65–80 cm, or to the lower part of AB horizon. Carbonates accumulate like pseudomycelium at the metamorphic B horizon.

In leached chernozems the horizon sequence is: A(Ap)–(A)–AB–Bk–BCk–Ck. Carbonates have been removed to the metamorphic B horizon and no pseudomycelium is formed.

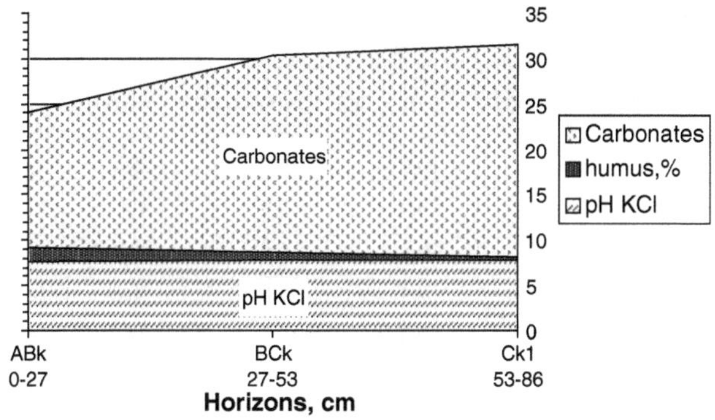

Fig. 2.13 Pattern of pH, humus, and carbonates distribution in calcareous regosol

In strongly leached chernozems the horizon sequence is: A(Ap)–(A)–AB–B–BC–Ck. Carbonates are below 100–120 cm in the parent materials.

The texture is usually loam, silt loam, and silty clay loam. The particle size distribution shows that the silt fraction (particularly coarse silt 0.05–0.01 mm) prevails followed by the clay (<0.001 mm) (Tables 2.7, 2.8, 2.9 and Figs. 2.14, 2.15, 2.16). Texture is homogeneous at the depth of profile (Tables 2.10, 2.11, 2.12 and Figs. 2.17, 2.18, 2.19). The particle-size distribution shows slight clay enrichment below the humus horizon in the strongly leached chernozem variety.

Prevailing water stable aggregates over 1 mm in size are 70 %. The average particle density is 2.55–2.65 g/cm^3 and depends on humus content. Porosity in the humus horizon is high 50–65 % (at field capacity). Field water capacity is 65–75 %. Aeration is more than 30 % of soil volume. Leached chernozems have high water-holding capacity and are characterized by comparatively high water permeability.

In leached chernozems over 15–25 % of pore space is represented by macropores or the pores of aeration (with diameter >300 μm); and over 20 % is mesopores or pores with rapid moisture drainage (>30 μm diameter). Micropore space or water-filled space is about 15–20 %, represented by slow drainage pores (diameter 0.2–30.0 μm) and 15–20 % of micropores is with none remaining available for roots moisture (diameter < 0.2 μm) (Dilkova 1985; Dilkova et al. 1998).

Naturally the spatial relationship of leached chernozems borders with all chernozems varieties, and also with calcareous and eutric regosols.

The thickness of the leached chernozem is 80–120 cm, and the thickness of the humus horizon (A + AB) is 50–80 cm. Content of humus is 2.2–4.5 % at the upper part of the humus horizon and gradually decreases downwards (Figs. 2.20, 2.21, 2.22). In the humus composition humic

Table 2.6 Groups of the leached chernozems in Bulgaria

Bulgarian classification (1964, 1976)	WRB (2006)	Keys to Soil Taxonomy (2010)	Revised FAO Legend (1990)
Slightly leached chernozem	Epicalcic Chernozem	Calcic Haploxerolls	Calcic Chernozem
Leached chernozem	Endocalcic Chernozem	Typic Calciustolls	Haplic Chernozem
Strongly leached chernozem	Bathycalcic Chernozem	Typic Haplustolls	Haplic Chernozem

Table 2.7 Variables of the particle-size distribution in 21 referenced profiles of slightly leached chernozems in Bulgaria

Profile count 21	Horizon	Upper depth	Lower depth	Loss HCl	Particle-size distribution (mm), % (according to method of Kachinskiy revised)							
					>1 mm	1–0.25	0.25–0.05	0.05–0.01	0.01–0.005	0.005–0.001	<0.001	<0.01[a]
Mean	A′p	0	27	2.5	0.0	0.2	13.2	38.4	9.4	8.2	28.1	45.7
	A″	27	49	2.5	0.0	0.3	12.0	39.4	7.5	8.8	29.6	45.8
	ABk	49	73	3.7	0.0	0.2	13.8	38.0	8.5	8.5	27.3	44.3
	Bk	73	99	12.2	0.0	0.2	12.7	38.0	8.7	7.4	20.9	36.9
	BCk	99	118	19.8	0.1	0.3	10.5	38.3	8.7	7.1	15.2	31.1
	Ck	118	145	21.6	0.4	0.5	12.7	35.5	9.2	6.4	14.0	29.7
Stdev	A′p	0	4	1.6	0.0	0.8	2.9	4.5	2.1	2.8	4.5	5.1
	A″	4	7	1.6	0.0	0.9	3.1	4.9	2.3	1.9	3.6	4.8
	ABk	7	13	4.4	0.0	0.8	4.7	5.4	1.7	1.5	6.7	6.9
	Bk	13	16	9.4	0.0	0.9	4.1	4.7	1.9	2.0	8.4	9.5
	BCk	16	18	5.6	0.4	0.8	3.8	5.9	2.1	3.2	6.1	7.7
	Ck	14	14	2.3	0.7	1.3	2.1	3.8	1.7	2.6	4.5	3.8

[a] Physical clay, the parameter characterized sum of the particle-size distribution, according to the Kachinskiy method (mm) $\sum <0.01 = <0.001 + (0.005 - 0.001) + (0.01 - 0.005)$

Table 2.8 Variables of the particle-size distribution in seven referenced profiles of leached chernozems in Bulgaria

Profile count 7	Horizon	Upper depth	Lower depth	Loss HCl	Particle-size distribution (mm), % (according to method of Kachinskiy revised)							
					>1 mm	1–0.25	0.25–0.05	0.05–0.01	0.01–0.005	0.005–0.001	<0.001	<0.01[a]
Mean	A′p	0	27	1.9	0.0	0.0	7.9	42.1	10.0	10.2	28.1	48.2
	A″	27	50	2.1	0.0	0.0	8.8	37.6	9.9	9.9	31.7	51.5
	AB	50	72	2.0	0.0	0.0	8.6	38.2	9.8	9.5	32.1	51.4
	B	72	97	2.4	0.0	0.0	10.0	39.2	9.4	8.8	30.3	48.5
	BCk	97	117	6.4	0.0	0.0	6.9	40.1	10.0	9.7	26.9	46.6
	Ck	117	141	13.4	0.0	0.0	8.2	37.1	8.8	8.3	24.2	41.3
Stdev	A′p	0	2	0.4	0.0	0.0	3.0	3.7	2.5	1.8	3.8	3.5
	A″	2	3	0.7	0.0	0.0	3.8	3.7	1.5	2.3	1.7	1.4
	AB	3	2	0.6	0.0	0.0	3.0	2.7	1.5	0.9	2.4	2.7
	B	2	6	1.3	0.0	0.0	4.5	2.6	2.4	1.8	3.4	5.9
	BCk	6	12	7.9	0.0	0.0	2.0	2.6	1.6	1.3	6.2	7.4
	Ck	6	9	7.1	0.0	0.0	3.7	1.4	1.7	1.3	3.3	2.8

[a] Physical clay, the parameter characterized sum of the particle-size distribution, according to the Kachinskiy method (mm) $\sum <0.01 = <0.001 + (0.005 - 0.001) + (0.01 - 0.005)$

Table 2.9 Variables of the particle-size distribution in 15 referenced profiles of strongly leached chernozem in Bulgaria

Profile count 15	Horizon	Upper depth	Lower depth	Loss HCl	Particle-size distribution (mm), % (according method of Kachinskiy revised)							
					>1 mm	1–0.25	0.25–0.05	0.05–0.01	0.01–0.005	0.005–0.001	<0.001	<0.01[a]
Mean	Ap	0	25	1.7	0.1	3.0	11.4	31.9	9.7	9.3	33.0	52.0
	A″	25	49	2.0	0.0	3.1	10.1	31.3	8.6	8.2	36.7	53.5
	AB	49	73	2.1	0.1	3.1	8.8	29.1	9.6	8.3	39.0	56.9
	B	73	98	2.2	0.2	3.1	8.8	30.0	9.0	9.0	37.8	55.8
	BC	98	122	2.0	1.2	3.3	8.7	29.9	10.0	8.4	37.7	56.1
	Ck	122	148	12.5	1.1	4.9	9.1	29.4	8.7	7.7	27.8	44.2
Stdev	Ap	0	5	0.7	0.2	4.2	4.9	12.1	1.0	1.6	6.8	6.9
	A″	5	3	1.1	0.1	4.5	5.0	11.0	1.8	2.8	7.1	7.3
	AB	3	4	1.4	0.2	4.8	4.6	10.9	1.2	1.5	7.3	7.4
	B	4	5	1.5	0.7	4.9	4.7	11.6	2.0	1.7	7.7	7.1
	BC	5	5	1.1	3.3	5.3	4.7	10.9	1.8	1.4	8.8	8.1
	Ck	5	9	5.2	3.3	9.7	4.6	12.3	2.7	2.2	11.7	11.6

[a] Physical clay, the parameter characterized sum of the particle-size distribution, according to the Kachinskiy method (mm) $\sum <0.01 = <0.001 + (0.005 - 0.001) + (0.01 - 0.005)$

acids (C_h) bound with calcium prevail over fulvic acids. A certain amount of free humic acids as well as those bound with sesquioxides are available in the humus horizon; at the depth 100 % of humic acids are bound with calcium (Shishkov and Petrova 2006).

The process of transformation of organic residues is accompanied by nitrogen enrichment of humic substances shown by the ratio C/N 10-12 which means high rates. The stock of organic matter is from 250 to 450 tons per hectare to a depth of 100 cm (Filcheva et al. 2002).

The qualitative composition of clay minerals, determined by X-ray diffraction analysis and transitional electronic microscopy, has shown a prevalence of illite at rates of 34–45 % in the clay and 12 % in the soil at the surface humus (A) horizon, gradually decreasing at the depth, where in the (C) horizon it is about 23–24 % in the clay and 4–5 % in the soil. This is followed by montmorillonite, which is at rates of 14–32 % in the clay (<0.001 mm) and 4–11 % in the soil at the surface humus (A) horizon, gradually increasing at the depth, where in the (C) horizon it

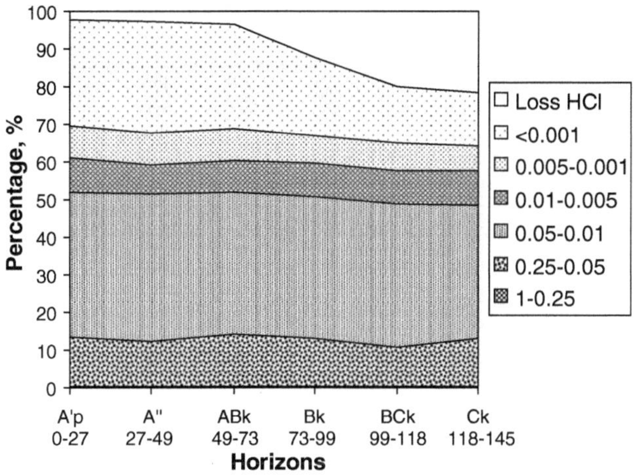

Fig. 2.14 Pattern of the particle-size distribution (mm) in slightly leached chernozem (according to revised Kachinskiy method)

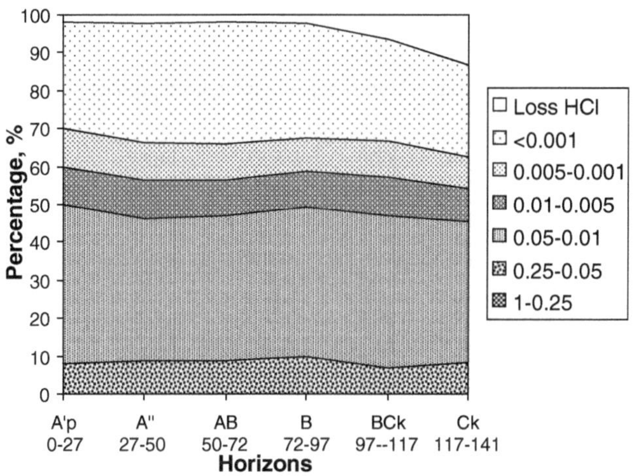

Fig. 2.15 Pattern of the particle-size distribution (mm) in leached chernozem (according to revised Kachinskiy method)

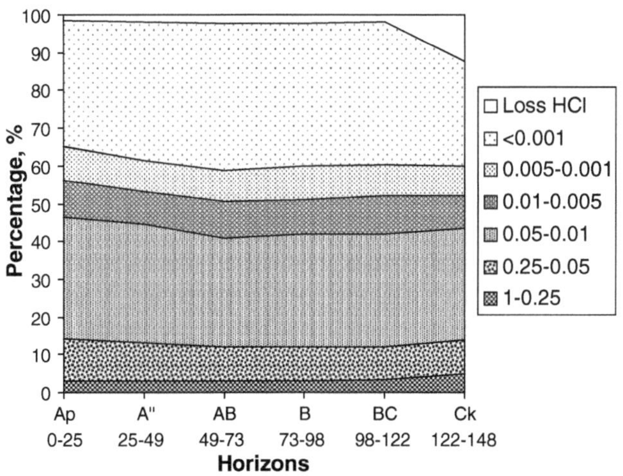

Fig. 2.16 Pattern of the particle-size distribution (mm) in strongly leached chernozem (according to revised Kachinskiy method)

is about 37–56 % in the clay and 2–12 % in the soil. Kaolinite is about 17–23 % in the clay (<0.001 mm) and 3–6 % in the soil. A significant quantity of quartz is at the surface (A) horizon, about 4–11 % in the clay and 1–3 % in the soil, sharply decreasing at the depth (Bechar et al. 1988; Halil et al. 1990).

The profile is not differentiated at SiO_2, Al_2O_3, Fe_2O_3 compounds; there is no evidence of MgO and K_2O migration, only CaO.

The amount of total iron (Fe_2O_3) (Arinushkina 1970) is about 5–6 % with uniform distribution in the profile. In all horizons the silicate form of iron (65–70 % from the total iron Fe_2O_3) greatly dominates.

The non-silicate iron (Mehra and Jackson 1960) is about 30–36 % (from the total iron Fe_2O_3). Almost all non-silicate iron appear in crystallized form (Hadzhiyanakiev 1989).

The amorphous form of iron (Tamm 1922) is only 1.3–3.3 % (from the total iron).

The free amorphous form of iron is only 1–7 % (from the total non-silicate amorphous iron) which is the lowest value found in the Bulgarian soils.

The amorphous form of iron bounded with organic matter (Bascomb 1968) is insignificant and is only 1.5–3.5 % (from the total non-silicate iron). This is due to the fact that humic acids prevail in the humus composition and are characterized by a low ability to make complex formations with iron.

In the clay fraction carbonate crystals occur despite the leaching. Soil reaction in KCl is neutral or slightly acid in the upper part and slightly alkaline in the lower part of the profile. Cation exchange capacity is 40–50 meq/100 g soil in the upper humus horizon and is 30 meq/100 g deeper. Bases saturation is near 90 %, where in cations composition hydrogen appears.

Leached Chernozem-Karasolutcii Variety

Within the subtype of leached chernozem there is a group of soils with pronounced vertic features (Table 2.13). According to the Bulgarian classification these soils are named *Karasolutcy*.

This variety is characterized by high clay content in the profile and high clay content in the parent materials (Tables 2.14, 2.15 and Figs. 2.23, 2.24). The profile is not differentiated in terms of clay and sesquioxides distribution. Morphologically, vertic features are most prominent. The content of humus is 2.5–5 % and gradually decreases downwards (Fig. 2.25). Burrowing features and especially earthworm tunnels are evidence of high biological activity.

2.4.3.4 Degraded (Podzolized) Chernozems

Degraded (podzolized) (*subtype*) **chernozems** (*type*) is named after the Bulgarian classification of Koynov et al. (1964), Yolevski and Hadzhiyanakiev (1976).

Table 2.10 Variables of humus, pH, carbonates, and particle-size distribution (mm) according to USDA in 21 referenced profiles of slightly leached chernozems in Bulgaria

Profile count 21	Horizon	Upper depth	Lower depth	Humus (%)	pH_{KCl}	Carbonates	pH_{H_2O}	Bulk density g.cm^3 calc	Clay < 0.002	Silt 0.002–0.05	Fine sand 0.05–0.2	Total sand 0.05–2.0
Mean	A′p	0	27	2.30	5.6	0.0	6.7	1.31	34.1	51.4	0.6	14.4
	A″	27	49	1.84	5.9	0.0	7.1	1.35	36.4	50.8	0.5	12.8
	ABk	49	73	1.30	6.2	1.4	7.2	1.34	34.5	50.7	0.6	14.8
	Bk	73	99	1.07	6.8	9.4	8.0	1.30	32.1	53.5	0.6	14.4
	BCk	99	118	0.81	7.3	16.5	8.3	–	30.5	56.5	0.7	13.0
	Ck	118	145	0.60	7.4	16.7	–	–	28.9	56.3	0.6	15.0
Stdev	A′p	0	4	0.43	0.6	0.0	0.5	0.03	4.8	7.6	2.0	4.9
	A″	4	7	0.42	0.5	0.0	0.3	0.04	5.9	7.9	1.6	3.9
	ABk	7	13	0.36	0.6	4.1	0.4	0.11	6.6	8.8	1.7	5.3
	Bk	13	16	0.32	0.7	8.6	0.4	0.20	7.1	8.7	1.7	4.6
	BCk	16	18	0.32	0.6	5.8	0.4	–	7.5	9.1	1.8	5.0
	Ck	14	14	–	0.4	6.3	–	–	2.6	3.9	1.3	2.7

Table 2.11 Variables of humus, pH, carbonates, and particle-size distribution (mm) according to USDA in seven referenced profiles of leached chernozem in Bulgaria

Profile count 7	Horizon	Upper depth	Lower depth	Humus (%)	pH_{KCl}	Carbonates	pH_{H_2O}	Bulk density g.cm^3 calc	Clay <0.002	Silt 0.002–0.05	Fine sand 0.05–0.2	Total sand 0.05–2.0
Mean	A′p	0	27	2.7	5.2	0.0	6.2	1.3	34.2	57.8	8.0	8.0
	A″	27	50	1.7	5.3	0.0	7.0	1.4	37.7	53.3	8.9	8.9
	AB	50	72	1.1	5.3	0.0	6.8	1.4	37.9	53.5	8.7	8.7
	B	72	97	0.9	5.3	0.0	6.8	1.5	35.9	53.9	10.2	10.2
	BCk	97	117	0.8	6.0	4.5	7.9	1.6	35.2	57.4	7.4	7.4
	Ck	117	141	–	7.0	16.1	–	–	35.5	55.2	9.3	9.3
Stdev	A′p	0	2	0.6	0.2	0.0	–	0.1	2.8	2.1	3.0	3.0
	A″	2	3	0.4	0.2	0.0	–	0.0	1.1	4.3	3.9	3.9
	AB	3	2	0.1	0.5	0.0	–	0.0	1.9	2.5	3.0	3.0
	B	2	6	0.1	0.5	0.0	–	0.1	3.5	2.0	4.6	4.6
	BCk	6	12	–	0.8	7.5	–	0.0	2.5	2.1	2.4	2.4
	Ck	6	9	–	0.2	5.4	–	–	2.9	2.1	3.2	3.2

Luvic Chernozem is named according to WRB (2006) and *Typic Argialbolls* is according to Keys to Soil Taxonomy (2010).

Luvic Chernozem: Revised Legend of the World Soils (FAO-UNESCO-ISRIC 1990).

It is distinguished from the other chernozem subtypes by the clay differentiation in the profile and compaction pronounce. When dry, the large open cracks are from the surface continuing at the B horizon. To some extent, bleaching in the upper part of the humus horizon, colored dark grayish brown, was previously explained by podzolization process occurence, though this was not proven by facts.

Degraded (podzolized) chernozems are characterized by high humus accumulation to a considerable depth, with strong granular structure, absence of carbonates in the profile, initial development of structural (B) horizon with strong subangular blocky structure, and compaction pronounce.

The reason for strong leaching in this group of soils is open to debate. Carbonates may be removed from a soil in percolating water to a considerable depth, thereby rainwater

Table 2.12 Variables of humus, pH, carbonates, and particle-size distribution (mm) according to USDA in 15 referenced profiles of strongly leached chernozems in Bulgaria

Profile count 15	Horizon	Upper depth	Lower depth	pH_{KCl}	Humus (%)	Carbonates	Bulk density g.cm^3 calc	Clay <0.002	Silt 0.002–0.05	Fine sand 0.05–0.2	Total sand 0.05–2.0
Mean	Ap	0	25	4.9	2.6	0.0	1.28	38.6	46.9	11.5	14.6
	A″	25	49	5..0	2.1	0.0	1.32	41.9	44.9	10.2	13.3
	AB	49	73	5.1	1.3	0.0	1.41	44.3	43.7	8.9	12.0
	B	73	98	5.2	1.1	0.0	1.45	43.6	44.4	9.0	12.2
	BC	98	122	5.5	1.0	0.0	1.46	43.3	45.1	9.0	12.8
	Ck	122	148	7.0	–	11.8	–	38.6	46.9	10.1	15.7
Stdev	Ap	0	5	0.3	0.4	0.0	0.03	6.6	12.4	4.9	8.7
	A″	5	3	0.3	0.5	0.0	0.04	7.4	12.1	5.0	8.1
	AB	3	4	0.4	0.3	0.0	0.05	7.6	11.9	4.5	8.7
	B	4	5	0.3	0.3	0.0	0.04	7.8	13.0	4.7	9.5
	BC	5	5	0.4	0.2	0.0	0.07	8.7	12.1	4.9	9.4
	Ck	5	9	0.3	–	4.0	–	9.8	15.6	4.7	15.0

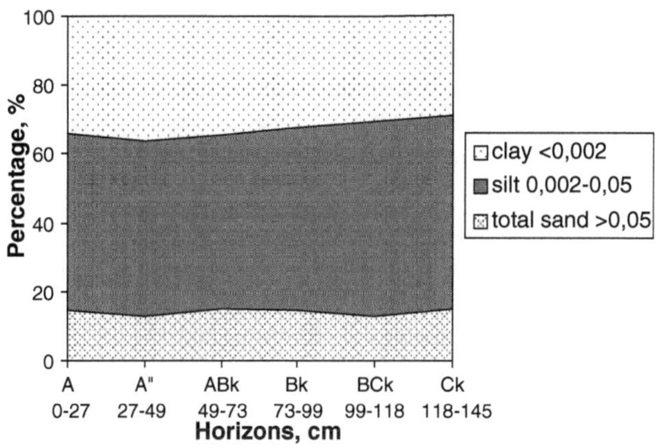

Fig. 2.17 Pattern of the particle-size distribution (mm) in slightly leached chernozem (USDA)

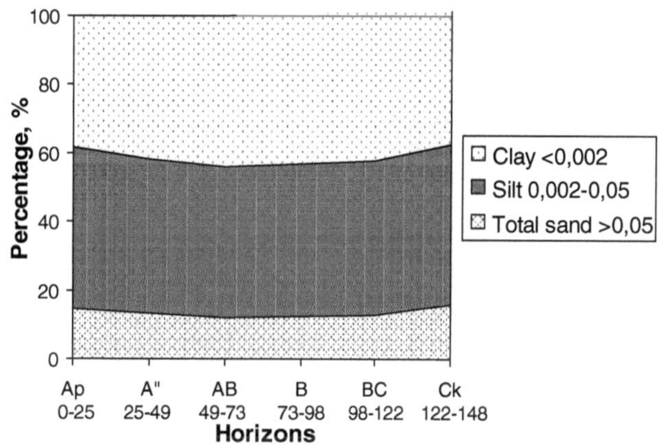

Fig. 2.19 Pattern of the particle-size distribution (mm) in strongly leached chernozem (USDA)

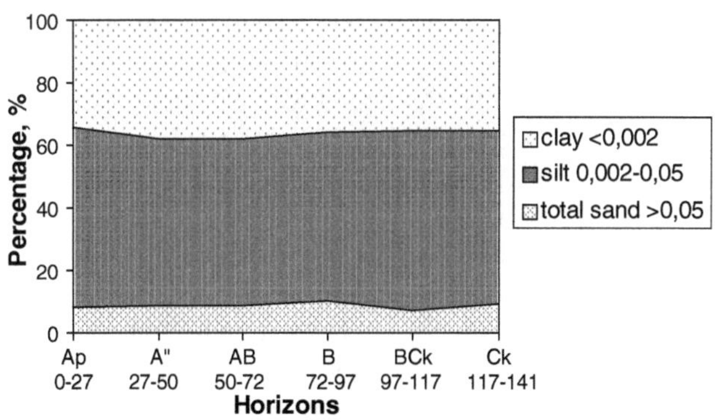

Fig. 2.18 Pattern of the particle-size distribution (mm) in leached chernozem (USDA)

translocates the dissolved material to some depth of subsoil where it is deposited; or depth has always been fixed during soil development.

The explanation of the development of degraded (podzolized) chernozems is not founded on sound scientific arguments. Some hypotheses relate it to the once dense spread of oak, elm, or hornbeam forests in association with the grasses beardgrass (*Andropogon grillus*), fescue (*Festuca pseudovina*). In the past this hypothesis was used to classify the soils as podzolized, claiming bleaching in the humus horizon. Typical podzolization was not proved and the explanation for this was the presence of lessivage and weathering in situ (Angelov and Fotakieva 1982). The

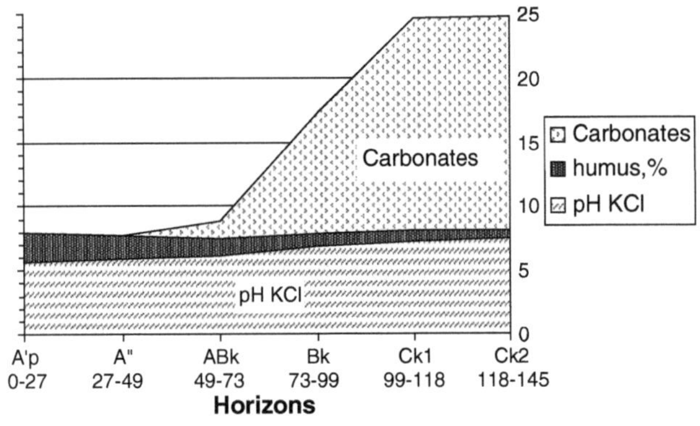

Fig. 2.20 Pattern of pH, humus, and carbonates distribution slightly leached chernozem

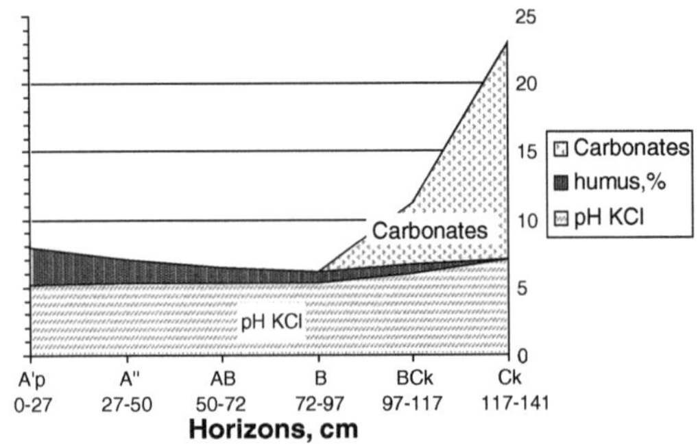

Fig. 2.21 Pattern of pH, humus, and carbonates distribution leached chernozem

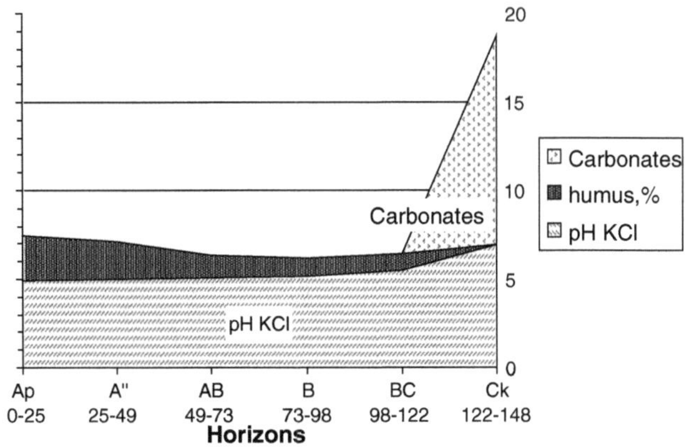

Fig. 2.22 Pattern of pH, humus, and carbonates distribution in strongly leached chernozem

objective fact is that seasonal waterlogging promotes bleaching of the upper layer, mostly washed quartz grains from humus and iron oxide skins. There is absence of destruction in the mineral part and the degradation was applied in terms of change of typical chernozem features. So-called structural (B) horizon pronouncement is obviously linked to structural alteration in the parent materials.

The profile of the degraded (podzolized) chernozems recognizes *mollic* horizon with humus accumulation and *argic* horizon. WRB (2006) has allowed that some soils that respond to the demands of argic horizon presence can be referred to as Luvic Chernozems.

In spatial distribution degraded (podzolized) chernozems are bordered by strongly leached chernozems and gray-brown forest soils.

The texture is silt loam and silty clay loam in the upper part. Particle-size distribution shows that clay fraction (<0.001 mm) prevails followed by the coarse silt fraction (0.05–0.01 mm) (Table 2.16 and Fig. 2.26). Downwards clay distribution in the profile shows enrichment in the (B) horizon (Table 2.17 and Fig. 2.27). The clay differentiation in the profile is expressed by the ratio of clay in the illuvial horizon and that in the overlying horizon which varies from 1.1 to 1.3. The mineral part is stable without destruction. The particle size distribution shows slight clay enrichment below the humus horizon.

Prevailing water stable aggregates (with size > 0.25 mm) comprise 80–87 %. The average particle density is 2.6–2.7 g/cm^3 and depends on humus content. Porosity is high (55 % at field capacity) in the humus horizon and 46 % in the structural horizon. Aeration is more than 30 % of soil volume.

In degraded (podzolized) chernozems 10–15 % of pore space comprises macropores or the pores of aeration (with diameter >300 μm); and 10–25 % is mesopores or pores with rapid moisture drainage (>30 μm diameter). Micropore space or water-filled space is about 20 %, comprising slow drainage pores (diameter 0.2–30.0 μm) and 15–20 % of micropores is with none remaining available for roots moisture (diameter <0.2 μm) (Dilkova 1985) (Dilkova et al. 1998).

The pattern of the horizon sequence is: A(Ap)–AB–B–C(k). The thickness of the humus horizon (A + AB) is 40–60 cm. Humus content varies from 2.0 to 4.0 % at the upper part of the humus horizon where it gradually decreases, and then drops relatively sharply downwards (Fig. 2.28). The stock of organic matter is from 200 to 300 tons per hectare to a depth of 100 cm. In the humus composition humic acids (C_h) bound with calcium prevail over fulvic acids. Relatively more of the free humic acids as well as those bound with sesquioxides are apparent. Carbonates are below 130 cm.

The structural (B) horizon is recognized by clayey texture with display of compaction.

Soil reaction is slightly acid to acid in KCl; the base saturation is high at about 85–95 %; in the cation composition hydrogen H^+ appears Cation exchange capacity is 20–30 meq/100 g soil in humus horizon and 30–40 meq/100 g deeper.

Table 2.13 Groups of leached chernozem-karasolutcii with vertic features

Bulgarian classification (1964, 1976)	WRB (2006)	Keys to Soil Taxonomy (2010)	Revised FAO Legend (1990).
Slightly leached chernozem- karasolutcii	Epicalcic Vertic Chernozem	Vertic Calcixerolls	Verti-Calcic Chernozem
Leached chernozem- karasolutcii	Endocalcic Vertic Chernozem	Vertic Haplustolls	Verti-Haplic Chernozem
Strongly leached chernozem- karasolutcii	Bathycalcic Vertic Chernozem	Vertic Haplustolls	Verti-Haplic Chernozem

Table 2.14 Variables of the particle-size distribution in 10 referenced profiles of leached chernozem-karasolutcii in Bulgaria

Profile count 10	Horizon	Upper depth	Lower depth	Loss HCl	Particle-size distribution (mm), % (according method of Kachinskiy revised)							
					>1 mm	1–0.25	0.25–0.05	0.05–0.01	0.01–0.005	0.005–0.001	<0.001	<0.01[a]
Mean	A′p	0	28	2.6	0.0	0.0	7.1	26.7	8.6	9.2	45.8	63.6
	A″	28	54	2.8	0.0	0.0	6.9	25.0	8.8	8.8	47.7	65.3
	A‴	54	78	2.3	0.0	0.0	7.1	26.6	8.6	9.5	46.1	64.2
	ABk	78	103	5.4	0.0	0.0	6.7	27.0	8.1	8.9	43.7	60.7
	Ck	102	129	17.2	0.0	0.0	6.4	23.7	8.5	8.3	35.7	52.5
Stdev	A′p	0	2	1.1	0.0	0.0	2.3	2.1	2.3	1.0	3.0	1.4
	A″	2	4	1.0	0.0	0.0	3.3	4.1	1.5	0.9	4.0	3.1
	A‴	4	7	1.0	0.0	0.0	3.1	3.5	1.5	1.3	3.9	3.1
	ABk	7	9	4.5	0.0	0.0	2.3	4.7	1.6	1.4	5.8	6.3
	Ck	9	13	4.1	0.0	0.0	3.4	3.8	2.4	1.3	4.9	4.4

[a] Physical clay, the parameter characterized sum of the particle-size distribution, according to the Kachinskiy method (mm) $\sum<0.01=<0.001+(0.005-0.001)+(0.01-0.005)$

Table 2.15 Variables of humus, pH, carbonates, and particle-size distribution (mm) according to USDA in seven referenced profiles of the leached chernozem-karasolutcii variety in Bulgaria

Profile count 10	Horizon	Upper depth	Lower depth	Humus (%)	pH_{KCl}	Carbonates	Bulk density g.cm^3 calc	Clay <0.002	Silt 0.002–0.05	Fine sand 0.05–0.2	Total sand 0.05–2.0
Mean	A′p	0	28	4.1	5.5	0.0	1.1	51.7	40.9	7.3	7.3
	A″	28	54	3.2	5.8	0.0	1.2	53.6	39.3	7.1	7.1
	A‴	54	79	2.4	5.9	0.3	1.3	52.1	40.8	7.2	7.2
	Bk	78	102	1.7	6.4	5.2	1.4	51.1	41.6	7.2	7.2
	Ck	102	129	0.9	7.0	14.4	–	49.1	43.0	7.8	7.8
Stdev	A′p	0	2	0.6	0.7	0.0	0.1	2.9	3.8	2.3	2.3
	A″	2	4	0.8	0.5	0.0	0.1	3.8	4.7	3.3	3.3
	A‴	4	7	0.6	0.4	0.4	0.1	3.8	4.8	3.1	3.1
	Bk	7	9	0.4	0.6	5.1	0.0	3.9	5.1	2.1	2.1
	Ck	9	13	0.0	0.1	6.2	–	3.7	5.8	3.2	3.2

Degraded (podzolized) chernozems have high water-holding capacity and are less water permeable due to compaction. In spring-time an excess of water can be found at the surface.

The qualitative composition of clay minerals, determined by X-ray diffraction analysis and transitional electronic microscopy, has shown a prevalence of montmorillonite at rates of 44 % in the clay and 12 % in the soil at the surface humus (A) horizon, gradually increasing at the depth, where in the (C) horizon it is about 55 % in the clay and 11 % in the soil. This is followed by illite, which is at rates of 36 % in the clay (<0.001 mm) and 15 % in the soil at the surface humus (A) horizon, gradually decreasing at the depth, where in the (C) horizon it is about 23 % in the clay and

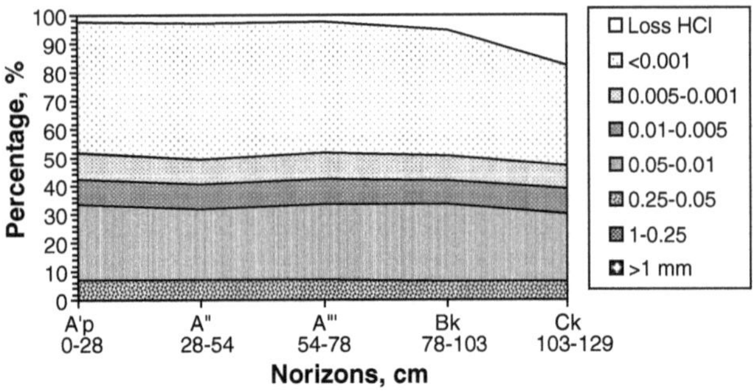

Fig. 2.23 Pattern of the particle-size distribution (mm) in leached chernozem-karasolutcii variety (according to the revised Kachinskiy method)

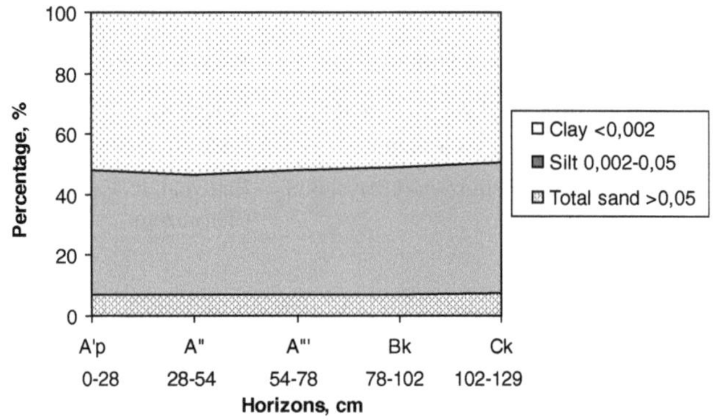

Fig. 2.24 Pattern of the particle-size distribution (mm) in leached chernozem-karasolutcii variety (USDA)

2–12 % in the soil. Kaolinite is about 13–15 % in the clay (<0.001 mm) and 3–6 % in the soil throughout the profile, and quartz is 2–3 % in the clay and 0.6–0.9 % in the soil (Bechar et al. 1988).

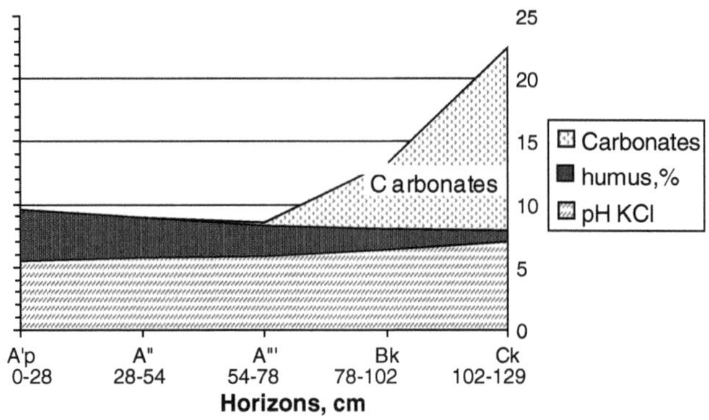

Fig. 2.25 Pattern of pH, humus, and carbonates distribution in the leached chernozem- karasolutcii variety

2.5 Soil Properties and Soil Management

2.5.1 Soil Properties

Chernozems are the finest agricultural soils. Accumulation of humus in the profile is their most significant feature. Despite the relatively low organic matter content of 2–5 %, organic substances have a beneficial effect on the soil properties. Bounded with calcium, the humic substances are characterized by high polymerization, condensation, and a strong link with clay. Forms of humus are composed in equal parts of humin and humic acids (Petrova 1980). Humus has a very low ability to migrate and is resistant to microbial decomposition which is favorable for its accumulation.

Chernozems are characterized by a favorable granular water stable structure, which determines high resistance to compaction, porosity over 40 %, well water capacity and permeability. Its excellent structure means it has a capacity to store large quantities of moisture, good drainage, and good aeration.

High cation exchange capacity is about 26–35 meq/100 g, especially in the humus horizon enriched with organic colloids. In the composition of absorbed cations Ca^{2+} (80 %), Mg^{2+} (10–22 %), exchangeable K^+ and Na^+ (5 %) each prevail. All varieties have base saturation over 85 % and only degraded (podzolized) chernozems have exchangeable hydrogen in the upper horizon (Ganev and Arsova 1980). High cation exchange capacity determines high base buffering capacity.

Its most important chemical property is that the amount of carbonate constantly increases with the depth (Figs. 2.10, 2.20, 2.21, 2.22, 2.25, 2.28). The pH_{KCl} value varies from 5.5 to 7.5 in the upper part and increases with depth up to 8.0. Generally, soil reaction is neutral or slightly alkaline in the upper part of calcareous and typical chernozems and is slightly acid in the strongly leached and podzolized chernozems.

The qualitative composition of clay minerals, determined by X-ray diffraction analysis and transitional electronic microscopy, has shown a prevalence of hydrous micas in the calcareous chernozems located at the strip near the Danube River. Kaolinite prevailed over hydrous micas in calcareous chernozems from southern Dobrudzha and montmorillonite is typical for southern Dobrudzha and central northern Bulgaria (Boneva et al. 1989).

Illite prevailed at rates of 34–56 % in the clay and about 10 % in the soil at the surface humus (A) horizon, gradually decreasing at the depth, where in the (C) horizon it is about 24–40 % in the clay and 3–4 % in the soil. This is followed by montmorillonite, which is at rates of 17–44 % in the clay (<0.001 mm) and 3–12 % in the soil at the surface humus (A) horizon, gradually increasing at the depth, where in the

Table 2.16 Variables of the particle-size distribution in 15 referenced profiles of degraded (podzolized) chernozems in Bulgaria

Profile count 15	Horizon	Upper depth	Lower depth	Loss HCl	Particle-size distribution (mm), % (according to method of Kachinskiy revised)							
					>1 mm	1–0.25	0.25–0.05	0.05–0.01	0.01–0.005	0.005–0.001	<0.001	<0.01[a]
Mean	A	0	27	1.7	0.1	3.3	11.6	33.6	9.5	9.2	31.0	49.7
	AB	27	51	1.7	0.0	1.8	10.5	32.1	8.9	8.7	36.3	54.1
	B1	51	75	1.8	0.0	1.9	10.0	30.6	9.1	8.1	38.5	55.7
	B2	75	102	1.7	0.0	1.7	9.1	32.1	9.7	8.4	37.3	55.4
	BCk	102	124	8.5	0.0	1.5	8.4	30.9	9.6	7.7	33.5	51.1
	Ck	124	151	14.3	0.0	0.3	7.6	30.1	9.7	7.2	30.6	47.6
Stdev	A	0	3	0.6	0.4	6.2	5.8	9.7	1.9	1.6	6.2	6.8
	AB	3	5	0.5	0.0	4.4	2.7	8.6	2.4	1.5	7.3	6.2
	B1	5	7	0.7	0.0	4.2	3.8	9.1	2.0	2.0	8.8	7.3
	B2	7	10	0.7	0.0	3.8	2.9	10.3	1.3	1.4	9.6	8.4
	BCk	10	14	8.8	0.0	3.3	3.2	7.8	2.0	1.1	11.8	10.8
	Ck	12	15	7.0	0.0	0.7	3.1	6.5	0.8	1.0	11.0	10.9

[a] Physical clay, the parameter characterized sum of the particle-size distribution, according to the Kachinskiy method (mm) $\sum<0.01 = <0.001 + (0.005-0.001) + (0.01-0.005)$

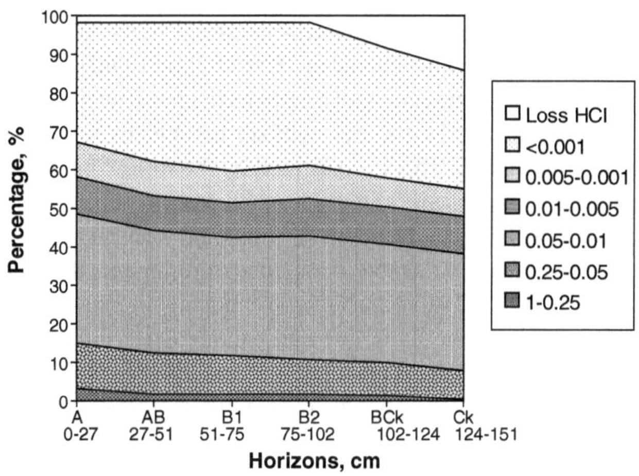

Fig. 2.26 Pattern of the particle-size distribution (mm) in degraded (podzolized) chernozems (according to the revised Kachinskiy method)

(C) horizon it is about 28–56 % in the clay and 2–9 % in the soil. Kaolinite is about 12–20 % in the clay (<0.001 mm) and 2–4 % in the soil. A significant quantity of quartz is at the surface (A) horizon, about 4–11 % in the clay and 1–3 % in the soil, sharply decreasing at the depth (Bechar et al. 1988).

Influence of $CaCO_3$ on extractable forms along depth of soil profiles

In calcereous soils the content of the extractable forms of Fe, Al, and Mn, i.e., the rates of release of their compounds from the primary mineral lattices, are close to the content in the neutral and acid soils (Jokova 1997). This is due to the amphoteric character of these elements. The carbonates of Na and Mg contribute to the increase in the extractable compounds of Fe. The typical calcareous and slightly leached soils, compared with the non-calcareous soils, have a similar content of the dithionite and oxalate extractable compounds of Fe and Al, while the content of the pyrophosphate extractable compounds of Fe_p, Al_p, and Mn_p is higher, and the content of the oxalate extractable Mn compounds is lower. The accumulations of these compounds are observed at the boundaries of the calcic horizons. The precipitation of the migrated extractable compounds is probably a result of the reactions of the complexes of humic acids (or eventually inorganic hydroxides) with $CaCO_3$ or OH^-.

Soils are considered as open systems. Gas CO_2 can escape from soils into the atmosphere and precipitate carbonates in alkaline soils, mainly of alkaline and alkaline earth elements. The carbonates contribute to the release of amphoteric elements from mineral lattices and the increase of the content of their pyrophosphate extractable compounds of Fe_p, Al_p, and Mn_p. Leaching of carbonates and respiratory processes decrease pH and the content of these compounds, probably due to the lower stability of the hydroxy complexes of Fe, Al, and Mn with humic acids in the conditions of higher soil acidity. In the strongly leached soils at more advanced soil development stages the influence of $CaCO_3$ is less expressed because of the interference of the gradual acidification combined with other predominating processes. This is observed in the gray-brown forest (periodically waterlogged), cinnamonic forest (developed under forest), and especially in the leached smolnitsa soils.

Table 2.17 Variables of humus, pH, carbonates, and particle-size distribution (mm) according to USDA in 15 referenced profiles of degraded (podzolized) chernozems in Bulgaria

Profile count 15	Horizon	Upper depth	Lower depth	pH_{KCl}	Humus (%)	Carbonates	Bulk density g.cm^3 calc	Clay <0.002	Silt 0.002–0.05	Fine sand 0.05–0.2	Total sand 0.05–2.0
Mean	A	0	27	4.9	2.1	0.0	1.32	36.5	48.4	11.8	15.2
	AB	27	51	4.9	1.5	0.0	1.38	41.6	46.0	10.6	12.5
	B1	51	75	5.2	0.9	0.0	1.44	43.5	44.4	10.1	12.0
	B2	75	102	5.1	0.8	0.1	1.47	42.4	46.7	9.2	10.9
	BCk	102	124	6.1	0.9	7.5	1.42	41.9	47.6	9.0	10.6
	Ck	124	151	7.1	–	10.5	–	42.0	49.0	8.8	9.1
Stdev	A	0	3	0.5	0.4	0.0	0.04	6.7	11.2	5.8	10.8
	AB	3	5	0.4	0.4	0.0	0.05	7.2	9.5	2.7	5.7
	B1	5	7	0.7	0.2	0.0	0.06	8.5	10.6	3.9	6.8
	B2	7	10	0.5	0.2	0.2	0.07	9.3	11.6	2.9	5.7
	BCk	10	14	1.0	0.2	7.7	–	9.1	10.0	3.4	4.9
	Ck	12	15	0.2	–	6.8	–	7.7	7.9	3.3	3.3

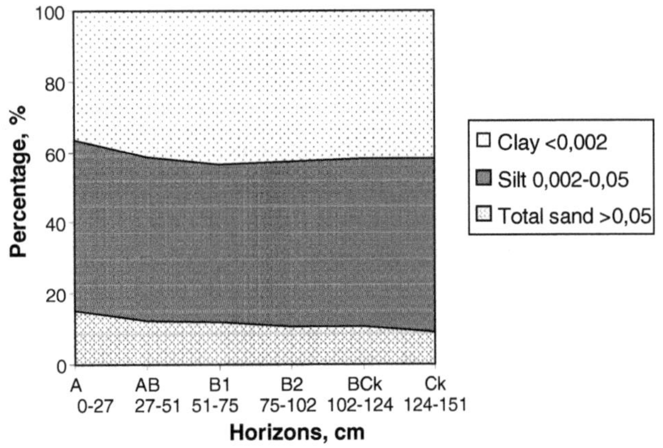

Fig. 2.27 Pattern of the particle-size distribution (mm) in degraded (podzolized) chernozems (USDA)

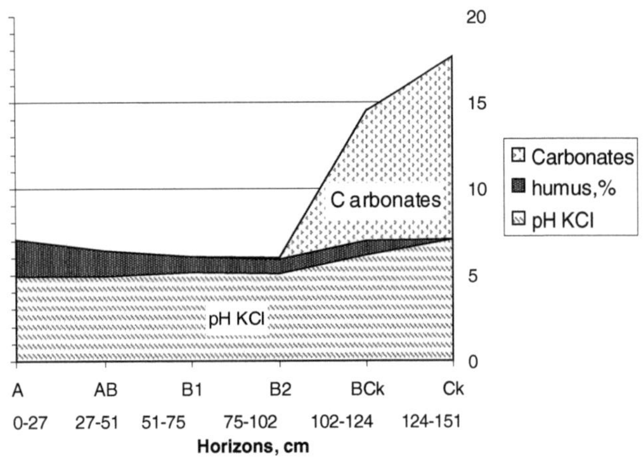

Fig. 2.28 Pattern of pH, humus, and carbonates distribution in degraded (podzolized) chernozems

$CaCO_3$ neutralizes the acid activity and protects minerals from the destruction and subsequent release of any elements with strong metallic character. $CaCO_3$ contributes to the release of the elements with amphoteric character from the primary minerals and to the hydroxy complexes formation with humic acids. This is confirmed by the following: in the typical calcareous and slightly leached soils, compared with the strongly leached and acid soils, the contents of the pyrophosphate extractable compounds are higher, the contents of Mn oxalate extractable compounds are lower, but the contents of the dithionite and oxalate extractable compounds of Fe and Al are similar. The influence of $CaCO_3$ on Mn extractable compounds distributions along depth is related to the redox status of the soils. The calcic horizons result in the accumulation of the extractable compounds along the depth of most soils.

These conclusions are somewhat contrary to the current theories about the role of carbonates in protecting minerals.

2.5.2 Soil Management

Today the main mass of chernozems are arable lands. These soils have a natural high fertility with high nutrient status, good structure, and high water- holding capacity, all of which are favorable for most agricultural practices and crop yields.

Cultivated soils differ from those in pristine lands in that nutrients uptake by crops prevails and is not compensated for by residues that enter back into the soil. As a result, humus content is lower than that in pristine land by more than 30 %. Storage and supply of nutrients, especially organic nitrogen, is a benefit for sustainable land use (Petkova 2011; Petkova and Mitovska 2002).

The change of land use results in deterioration of structure with a reduction of water stable aggregates in the plough layer.

Since summer drought is a natural hazard during the growing season, moisture conservation is crucial. Chernozems are the preferred soils for irrigated agriculture. The moisture regime of chernozems is in close relation with texture and differentiation in the profile. Adequate measures have to be taken to control wind and water erosion.

Winter wheat, barley, rye, and rape are produced in a wheat-fallow system. Maize, sunflower millet, sorghum, soybeans, beans, broad beans, peas, lentils, chick-peas, vetch sugar beet, tobacco, and alfalfa are the major crops grown. Vegetables including tomato, paprika, cucumber, pumpkin, watermelon, melon, and cabbage can be grown under irrigation. Fruit-trees such as apple, cherry, apricot, and pear are successfully cultivated, as well as high-quality vineyards.

References

Angelov E, Fotakieva E (1982) Chernozems in Bulgaria and their classification. The third national conference on soil science 21–23 of November 1981, part 1, Sofia: 30–33 (In Bulgarian)

Arinushkina E (1970) Manual on chemical soil analysis. Moscow State University Press, Moscow (In Russian)

Bascomb CL (1968) Distribution of pyrophosphate extractable iron and organic carbon in soils of various groups. J Soil Sci 19(1):251–268

Bechar A, Boneva K, Hubenov (1988) Mineralogical composition of Bulgarian chernozems. Soil Sci Agrochem XXIII(2):3–10 (In Bulgarian)

Boneva K, Houbenov G, Hanna A (1989) Clayey minerals contained in calcareous chernozems from Northern Bulgaria. Soil Sci Agrochem XXIV(1):44–50 (In Bulgarian)

Bondev I (1991) Vegetation in Bulgaria, map at scale 1:600 000 with explanatory text. St. Kliment Oxridski University Press, Sofia (In Bulgarian)

Boykov Y (1936) Loess in the Northern Bulgaria and soils developed on it. J Bulgarian Geol Soc VII(1) (In Bulgarian)

Dilkova R (1985) Structure and aeration in main soil varieties in Bulgaria. Theses, Institute of Soil Science Nikola Poushkarov, Sofia (In Bulgarian)

Dilkova R, Kerchev Zh, Kercheva M (1998) Evaluating and grouping of soils according to their susceptibility to anthropogenic degradation. Advances in geoecology, vol 31. CATENA VERLAG, Reiskirchen, pp 125–131

Dokuchaev VV (1883) Russian chernozem. Report of free economic society. St. Petersburg (In Russian)

Enculescu P (1924) Zonele de vegetatie lenmoasa din Romania, in raport cu conditiunile oro-hidrografice, climaterice de sol si de subsol. Mem.Inst.Geol.Rom., vol 1, Editura Cartea Romaneasca, Bucuresti

FAO-UNESCO-ISRIC (1990) Revised legend of the soil map of the world. FAO World soil resources report no 60, Rome

Filcheva E, Rousseva S, Kulikov A, Nedyalkov S, Chrnogorova T (2002) Organic carbon stocks in soils of Bulgaria. In: Kimble JM, Lal R, Follett RF (eds) Agricultural practices and policies for carbon sequestration in soils, Lewis Publishers, CRC Press LLC, chapter 45, pp 471–476

Hadzhiyanakiev A (1989) Iron in Bulgarian soils. Zemizdat Press, Sofia (In Bulgarian)

Halil R, Boneva K, Houbenov G (1990) Investigations on the composition of clayic minerals in some Bulgarian soils. Soil Sci Agrochem XXV(4):13–22 (In Bulgarian)

Ganev S, Arsova A (1980) Methods of determining the strongly acid and the slightly acid cation exchange in soil. Soil Sci Agrochem XV(3):22–33 (In Bulgarian)

Jokova M (1997) Influence of $CaCO_3$ on extractable compound distribution along depth of soil profiles. In: Lal R, Kimble J, Follett R (eds) Soil properties and their management for carbon sequestration. USDA, Natural Resources Concervation Services, National Soil Survey Center, Lincoln, NE, pp 19–25

Kachinskiy NA (1965) Soil physics, part I and II. High School Press, Moscow (In Russian)

Keys to Soil Taxonomy (2010) USDA United States Department of Agriculture, NRCS Natural Recourses Conservation Service, 11th edn. Soil Survey Staff, Washington, DC, p 338

Kononova MM (1963) Soil organic matter. Soviet Academy of Sciences Press, Moscow (In Russian)

Kostichov PA (1886) Soils of chernozemic region in Russia, their origin, composition and features (In Russian)

Koynov V (1964) Peculiarities of the main Bulgarian soil types and their basic soil-genetic processes. In: Gerasimov IP et al. (eds) Soils of south–eastern Europe. Proceedings of the international symposium of soil science, June 1963, Sofia. Bulgarian Academy of Sciences Press, Sofia: 15–62 (In Russian)

Koynov V et al. (1964) Classification and systematic of soils in Bulgaria. In: Gerasimov IP et al. (ed) Soils of south–eastern wurope. Proceedings of the international symposium of soil science, June 1963, Sofia. Bulgarian Academy of Sciences Press, Sofia: 231–246 (In Russian)

Koynov V, Trashliev Ch, Yolevski M et al (1972) Soil resources of Bulgaria and their use. First national soil science congress. Bulgarian Academy of Sciences Press, Sofia, 23–25 Sept 1969 (In Bulgarian)

Mehra OP, Jackson ML (1960) Iron oxide removal from soils and clays by dithionite—citrate—system buffered with sodium bicarbonate. Clays Clay Miner 7:317–327

Minkov M (1968) Loess in the Northern Bulgaria, Sofia (In Bulgarian)

Penkov M et al (1992) Soil classification and diagnostic in Bulgaria in relation to the land reform fragmentation. Zemizat Press, Sofia (In Bulgarian)

Petkova Z (2011) Effective use of nitrogen from manure by plants. In: Scientific proceedings "100 years Soil Science in Bulgaria", 2 part, pp 626–630 (In Bulgarian)

Petkova Z, Mitovska R (2002) Nitrogen and carbon mineralizing ability of leached chernozem under different norms and C: N ratio of farmyard manure. Soil Science, Agrochemistry and Ecology XXXVII(1–3):110–112 (In Bulgarian)

Petrova L (1980) State of organic nitrogen in some soils in Bulgaria. Theses, Institute of Soil Science N Poushkarov, Sofia, p 138 (In Bulgarian)

Petrova L, Shishkov T (2001) Comparative characteristics of humus of calcareous chernozems from southern Romania and northern Bulgaria. Soil Sci Agrochem Ecol XXXVI (4–6):51–53 (In Bulgarian)

Rousseva SS (1997) Data transformations between soil texture schemes. Eur J Soil Sci (48):749–758

Sibirtsev NM (1898) Brief review of the general soil types in Russia, Warshava

Shishkov T (1998) An idea concerning genesis and classification of soils from northern part of Danube Valley in Bulgaria. In: 16th World Congress of soil science, 20–26 August 1998, Montpelier, France, 1998, file in CD-Rom

Shishkov T, Petrova L (2006) Humus systems of some leached chernozems from Dobrich region of north eastern Bulgaria. Ecology and Industry, Section 8, No 1–2, Balcanereco 06, Sofia, pp 202–204. 21–23 June 2006 (In Bulgarian)

Tamm O (1922) Eine methode zur bestimung der anorganischen komponente ds gelcomplexes im boden. Statens Skogsforsoksanstalt, Medd 19:387–404 (by Arinoushkina E (1972) Manuel of chemical analysis of soils, Moscow

Yolevski M, Hadzhiyanakiev A (1976) Agro-grouping of soils in Bulgaria, Extended systematic list. Code of groups and soil varieties. Institute of Soil Science Nikola Poushkarov, Sofia (In Bulgarian)

Yolevski M, Ninov N, Krastanov S et al (1983) Brief diagnostic and criterion of the main taxonomic soil divisions in Bulgaria. Precise Classification Scheme, Institute of Soil Science Nikola Poushkarov, Sofia (In Bulgarian)

World Reference Base for Soil Resources (2006) A framework for international classification, correlation and communication. FAO-ISRIC, Rome

3 Gray-Brown Forest Soils

Derived from the Latin word *luo,* meaning to wash.

3.1 Concept of the Type

Main concepts

1. Gray-brown forest soils are mineral soils having clay enrichment in compacted Bt horizon as a result of their evolution;
2. Characterized by high base saturation (>80 %);
3. Gradual decrease of organic matter is to a considerable depth;
4. Annual organic decay of vegetation during the biological cycle promotes the supply of considerable amounts of nitrogen, bases, and calcium back into the soil.

The main processes that have taken place in gray-brown forest soils are textural differentiation of profile, humification, lessivage, and weathering in situ.

3.2 Ecology of Soil Formation

3.2.1 Distribution

Gray-brown forest soils are recognizable on the territory of northern Bulgaria from east to west (Fig. 3.1). They are located in the southern part of the hilly Danube Plain, at the northern foot slopes of the Fore Balkan, partly are on the North-Western and Ludogorie plateaus, at the Kamchia River basin, and north of the city of Varna. In spatial distribution the soils are bordered by chernozems (leached and/or degraded).

Dark gray-brown forest soils occupy about 200,000 ha and are located on the North-Western and Ludogorie plateau or approximately 1.79 % of the entire terrain of the country. The other region where they are situated spreads from east to west at the northern foot slopes of the Balkan Mountains. The region is strongly dissected and affected by erosion.

The areas of gray-brown forest soils location are in the central uplands of the hilly Danube Plain, on the Ludogorie plateau, and in the region of Kamchia River. This variety occupies approximately 14.5 % of the entire territory; of which 1,000,000 ha are occupied by deep soils while the rest are shallow, located on the northern Fore Balkan slopes.

Climatic conditions

Gray-brown forest soils have formed in the temperate continental climate with periodic East European and Atlantic influence. Usually there is a dry season during the summer. Temperature regime and total precipitation are variable in the eastern and western regions of the country. The maximum falls are during the summer.

In the northeastern province in the region of Ludogorie the mean annual temperature is 9.7–11.0 °C. In January the mean annual temperature is −1.1 (−2.4) °C. In July the mean annual temperature is 20.2–22.0 °C. The mean amount of precipitation varies from 564 to 712 mm.

In the southern province in the region of the Danube Plain the mean annual temperature is 11.4–11.6 °C. In January the mean annual temperature is −1.6 (−1.8) °C. In July the mean annual temperature is 22.6–22.9 °C. The mean amount of precipitation varies from 680 to 684 mm.

In the central province in the region of the Fore Balkan the mean annual temperature is 9.4–11.5 °C. In January the mean annual temperature is −1.4 (−3.2) °C. In July the mean annual temperature is 19.7–22.9 °C. The mean amount of precipitation varies from 668 to 939 mm.

In the North-Western province in the region of Danube Plain the mean annual temperature is 10.0–11.1 °C. In January the mean annual temperature is −1.9 (−2.1) °C. In July the mean annual temperature is 20.9–22.2 °C. The mean amount of precipitation varies from 698 to 809 mm.

In the geographic distribution, gray-brown forest soils are found in areas characterized by sufficient water supply during the growing season.

Topography

Gray-brown forest soils are mainly found on the undulated and gently sloping areas of the southern hilly Danube Plain, where the average rates of elevation are 200–600 m. The

T. Shishkov and N. Kolev, *The Soils of Bulgaria*, World Soils Book Series, DOI: 10.1007/978-94-007-7784-2_3,

Fig. 3.1 Landscape of gray-brown forest soils

northern foot slopes of the Balkan Mountains are of dissected relief.

At the Ludogorie plateau relief is hilly dissected and bedrock is often found on the surface. Usually terrains are well drained.

Parent materials
Gray-brown forest soils have formed on the unconsolidated deposits of loess-like materials or on bedrock. Usually, deposits are calcareous being of different origin and age (Pliocene and Quaternary). Extremely acidic parent materials are rare.

Some provincial differences in the soils are determined by the character of these materials.

In the North-Western plateau the dark gray-brown forest soils and eroded varieties have formed on the marine sediments (Fore-Carpathian basin) in North Bulgaria on the Neogene/Miocene (Middle Pontian) Age clays and sands; on the eluvium of the Neogene/Miocene (Middle-Upper Sarmatian) Age limestones and banded clays; and on the Lower–Middle Sarmatian Age clays, sands and limestones, and conglomerates.

In the central part of the hilly Danube Plain the gray-brown forest soils have formed on the marine sediments (Fore-Carpathian basin) in North Bulgaria on the eluvium of the Neogene/Miocene (Middle-Upper Sarmatian) Age limestones and banded clays; on the Lower-Middle Sarmatian Age clays, sands and limestones, and conglomerates; on the marine sedimentary rocks in North Bulgaria of Paleogene/Middle Eocene Age marls, mica sandstones, clays, marls and siltstones; on the eluvium of the Upper Cretaceous (Upper Maastrichtian) Age (Mesozoic Era) limestones and terrigenous-carbonate rocks; on the Lower Cretaceous Age marls, and sandy marls with sandstone interbeds; and on the Lower Cretaceous (Barremian-Aptian, Valanginian-Hauterivian and Berriasian-Hauterivian) Age thick packets of sandstones, marls, sandy marls, limestones, and mixed rocks.

At the Ludogorie plateau the gray-brown forest soils have formed on the Quaternary Age eolian sediments (loess, sandy loess, clayey loess); on the eluvium of the Lower Cretaceous Age (Mesozoic Era) clayey and silicified limestones; on the Lower Cretaceous (Barremian-Aptian, Valanginian-Hauterivian and Berriasian-Hauterivian) Age thick packets of sandstones, marls, sandy marls, limestones, and mixed rocks; and on the Aptian Age marls, and clayey marls with sandstone interbeds.

In the eastern part of the country in the region of the Kamchia River the gray-brown forest soils have formed on the marine sedimentary rocks (of Euxino-Kaspian basin) in North Bulgaria on the Neogene/Miocene (Konkian) Age sands with clay interbeds, sandstones, and limestones; on the Paleogene/Middle-Upper Eocene Age marls locally with interbeds of detrital sandstones.

It should be mentioned that soil texture is unlike that of the parent materials.

Vegetation
Gray-brown forest soils are mostly distributed in the southern forest-steppe zone. This area commonly has a

cover of mixed vegetation, and may have had various sequences of vegetation over a long period of time, but these are predominantly forested soils. The most common plant community is that of deciduous forest, but there may be mixed forests or coniferous trees and shrubs. Species vary from place to place but mainly consist of oak (*Quercus cerris, Quercus petraea, Quercus conferta*), European hornbeam (*Carpinus betulus*), linden (*Tilia argentea*), maple (*Acer campestre*), oriental hornbeam (*Carpinus diunensis*), flowering ash (*Fraxinus ornus*), and beech (*Fagus orientalis*) (Bondev 1991).

Oak forests cover the uplands of the North-Western, Ludogorie, and southern Dobrudzha plateaus. The large territory of Ludogorie plateau is covered with dense oak forest. However, the territory of the North-Western plateau has sparse oak forest cover and single trees on the agricultural lands. Gray-brown forest soils can be found under grasslands.

Time
Most gray-brown forest soils may be relatively young (a 1,000 years or so), their textural differentiation formed in a different time during Holocene. But some soils may be quite old (hundreds of thousands of years) holding features of the late Pleistocene Era when conditions were relatively cold.

Human influence
There was a reduction in the woodlands occupied by gray-brown forest soils when most of the forested areas were turned into arable. Waste forests that had once covered the territory of the gray-brown forest soils disappeared. Nowadays the preserved areas of oak forest on the plateaus help to restore the landscape that existed two centuries ago.

The conversion of the woodlands into arable land has had the negative result of decreasing the amount of organic matter. Structure has deteriorated, becoming loose so that fine soil particles become detached. The subsoil has become more prone to compaction. Pore space has decreased, restricting water movement through the soil. Low infiltration rates mean that soil erosion has accelerated. Application of high rates of fertilizers together with liming has to be applied.

3.2.2 Genesis

The genesis of soils with differentiated profile is still open to discussion (Duchaufour 1951; Kubiena 1953; Rode 1964; Trashliev et al. 1964; Zonn 1973; Gyurov et al. 1978). One hypothesis is about the previous evolution stages, as a result of which there is polygenetic soil development with partly existed clay-rich horizon. The second hypothesis posits a cycle of gradual accumulation of fine material that took place when a continental climate with a marked dry season existed. Annually, during the wet season, the dispersed material was translocated with solutions downwards; during the dry season the fine particles were firmly attached in the pores and on the surfaces of peds.

The third hypothesis is about the prolonged migration downwards. Initially, soluble salts and carbonates were leached, followed by the gradual translocation of clay moved by gravity. The precise mechanism of clay deposition is not understood.

The clay that was removed from the overlying horizon cannot be explained by the thickness of the underlying part. It may be that the amount of cutanic material is less than to be expected from the removed total amount.

It seems probable that the textural differentiation was enhanced by the weathering in situ.

Seasonal waterlogging promotes bleaching of the upper part, mostly washed quartz grains from iron oxides skins. The leaching of carbonates is debatable whether the depth of accumulation is fixed during soil formation.

3.3 Soil Diagnostic and Classification

3.3.1 Morphology

The profile of the gray-brown forest soil consists of humus eluvial and illuvial clay-rich Bt horizon (Fig. 3.2).

O—a thin (<5 cm) loose leafy litter is at the surface composed of partly decomposed leaves and grass remnants and naturally occurs in woodlands; below is clear transition into

A(Ap)—a thin mineral humus accumulative horizon that naturally occurs in the pristine lands or plough; homogeneous, colored dark grayish-brown or dark brown (moist) and dark gray (dry), with crumb to granular structure (spheroids or polyhedrons), medium size (2–5 mm) diameter, strong pedality grade, firm consistence (moist), and slightly hard (weakly resistant to pressure) dry. Large aggregates comprise many small ones. There are many earthworm casts and biological features, many like burrows (unspecified) and open large burrows. There is an abundance of fine roots (size 0.5–2 mm); clear grades into

AE(p)—a mineral moderately thick humus-eluvial horizon. Bleached brownish gray, mostly due to the washed quartz grains from iron oxides skins. Pedality type is granular (spheroids or polyhedrons), with medium size (2–5 mm), strong pedality grade, consistence is firm (moist) and slightly hard (dry), porosity is high; few rounded iron-manganese concretions; there is sharp change beneath into

Bt(g)—a thick illuvial horizon, usually dark reddish brown colored due to the presence of iron oxides, showing an increase of clay with coarse size (20–50 mm) angular blocky structure (blocks having mixed rounded and sharp angled faces of aggregates), where compaction is

Fig. 3.2 Gray-brown forest soil, Bulgarian classification Yolevski and Hadzhiyanakiev (1976), *Cutanic Luvisols,* WRB (2006), *Typic Hapludalfs,* Keys to Soil Taxonomy (2010), *Haplic Luvisol,* FAO-UNESCO-ISRIC (1990)

pronounced, with strong pedality grade, and consistence is firm (moist) and hard (dry). Porosity is medium consisting mainly of fine pores and cracks. The surfaces of peds, pore spaces, and cracks are coated with clay cutans. There is no tonguing penetration of upper humus material in B horizon. This may be due to subsequent root decay. Many rusty colored spots and black iron-manganese concretions are in the soil mass; they grade into

B(g)—underneath is mineral illuvial-metamorthic horizon brighter colored, with subangular blocky structure with coarse size (20–50 mm), strong pedality grade, and very firm in consistence when moist, and hard when dry. There are less skins on ped surfaces and cracks. Many rusty and grayish green spots as well as soft black manganese concretions are in the soil mass; they grade into

BC(k)—a thin, light brown, transitional to the upper part of parent materials horizon, with lower content of clay and absence of cutans. Structure is blocky to massive with medium size (10–20 mm), moderate pedality grade, and consistence is firm (moist) and hard (dry). Porosity is medium consisting mainly of large pores and cracks; clear transition into

C(k)—parent materials altered by the solutions, with reddish yellow or brownish yellow color. There is clearly less clay than in the overlaying horizon, structure is massive (structureless), and secondary carbonates can occur.

In the humus-eluvial (AE) horizon some processes are related to decomposition, mineralization of organic remains, and accumulation of humus. Intensive mineralization of organic matter results in relatively small humus content of mull type. High biological activity occurs mostly at this layer.

Leaching of soluble salts and carbonate is a phase of further lessivage process development. Usually during spring waterlogging occurs at the surface on flat terrains, so anaerobic conditions support the pseudogley process.

Lessivage in the gray-brown forest soils is a factor of illuvial B(t) horizon formation with oriented skins of clays on ped surfaces. These require dispersal of the clay in AE horizon, translocation of clay to the B horizon, and physical precipitation of clay in Bt horizon.

The oxidation and reduction of iron oxides is evident throughout the profile in the form of rusty or gray greenish spots.

Environmental conditions contribute to the prolonged weathering in situ over the whole year.

3.3.2 Classification

Gray forest soils have been an object of research since 1879 when for the first time V. I. Chaslavsky showed the soil on the soil map of European Russia at scale 1:2,500,000.

Back in 1886, the gray forest soil type was included in the Russian classification and referred to as "transitional," meaning a soil whose formation has been influenced by broad-leaved forests.

In Bulgaria after 1947, the soil type of gray forest soils was defined by the Bulgarian-Soviet expedition. Bulgarian soil classification of 1948 (Antipov-Karataev and Gerasimov) defined the type and three soil subtypes, namely dark gray, gray and light gray forest soils. Soil classification of Koynov et al. (1964) shifted the subtype of light gray soils to the type of pseudopodzolic soils. In 1963, at the Soil Science Symposium there was a proposal to change gray forest soil to gray-brown forest soil (Koynov et al. 1963). Soil classification of Yolevski et al. (1983) renamed the type of gray forest soil as gray-brown. There was classification in Penkov et al. (1992) who at the theoretic level named subtypes of gray forest soil (mollic, albic, and haplic), but this has not been applied in practice.

Table 3.1 Variables of the particle size distribution (according to the revised Kachinskiy method) in 18 referenced profiles of the dark gray-brown forest soils in Bulgaria

Profile count 18	Horizon	Upper depth	Lower depth	Loss HCl	Particle size distribution (mm), % (according method of Kachinskiy revised)							
					>1 mm	1–0.25	0.25–0.05	0.05–0.01	0.01–0.005	0.005–0.001	<0.001	<0.01[a]
Mean	Ap	0	26	1.4	0.6	6.3	13.3	26.9	10.5	9.8	31.7	52.0
	AB	26	54	1.4	0.7	5.3	11.2	21.2	9.0	9.0	42.8	60.8
	Bt	54	90	1.5	0.1	4.1	10.5	18.6	8.3	7.7	49.4	65.4
	BC	90	120	2.9	0.0	5.3	11.6	18.5	8.1	8.0	45.5	61.7
	Ck	120	152	9.5	1.4	3.7	9.5	19.2	7.6	8.6	41.8	58.1
Stdev	Ap	0	3	0.6	1.4	4.9	5.4	5.0	1.9	1.8	6.5	6.4
	AB	7	6	0.5	2.0	4.2	4.4	4.9	2.6	1.9	12.4	9.2
	Bt	9	13	0.6	0.3	3.9	4.8	3.4	1.9	2.0	8.5	7.5
	BC	15	22	5.6	0.7	6.1	7.0	4.0	1.7	2.1	9.3	10.8
	Ck	20	23	5.9	2.9	3.2	2.4	3.3	1.0	2.0	4.5	3.9

[a] Physical clay, the parameter characterized sum of the particle-size distribution, according to Kachinskiy method (mm) $\sum < 0.01 = < 0.001 + (0.005 - 0.001) + (0.01 - 0.005)$

The soil is distinguished by the thickness of humus horizon and rate of differentiation in clay-reach Bt horizon. Prevalence of humus accumulation or textural differentiation reflects the soil morphology and is followed by the defining of soil subtypes.

The subtype of dark gray-brown forest soil usually exhibits *Mollic* horizon and is associated with Phaeozems (WRB 2006) or Mollisols (Keys to Soil Taxonomy 2010). In Soil Taxonomy *Mollic* epipedon may occur in other orders without Mollisol classification.

In the subtype of gray-brown forest soils *Argic* horizon took precedence, so these soils are referred to as Luvisols (WRB 2006) or Alfisols (Keys to Soil Taxonomy 2010).

Subtype varieties differ in the thickness of humus horizon (non-eroded, slightly eroded or accumulated), in the degree of erosion (slightly, moderately or severely eroded), and in the degree of gleyic properties occurrence in the profile.

3.3.3 Gray-Brown Forest Soils Subtypes in Bulgaria

3.3.3.1 Dark Gray-Brown Forest Soils

Dark (*subtype*) **gray-brown forest** (*type*) soil is named after the Bulgarian classification of Koynov et al. (1964), Yolevski and Hadzhiyanakiev (1976).

Luvic Grayic Phaeozem is named according to WRB (2006) and *Typic Argiudolls* according to Keys to Soil Taxonomy (2010).

Luvic Phaeozem: Revised Legend of the World Soils (FAO-UNESCO-ISRIC 1990).

Classification after WRB (2006) required *Mollic* diagnostic horizon in the reference group of Phaeozems.

Keys to Soil Taxonomy (2010) required a *Mollic* epipedon for Mollisols order classification. In Soil Taxonomy there are demands for subsurface horizons, allowing for *cambic* expressing mottling, gleying and structure development; *argillic* horizon with high base saturation (>35 %); *calcic or petrocalcic* horizon; and *albic, natric or none*.

The typical distinguishing features are humus accumulation, deep carbonate leaching (below 120–250 cm), low differentiation of clay, low compaction pronounce, and high biological activity, especially of earthworms.

The pattern of the horizon sequence is O–A(Ap)–AB–B(t)–BC(k)–Ck.

The average thickness of humus horizon (A + AE) is 20–45 cm; there is a gradual transition (AB) horizon to the Bt textural horizon which has average thickness over >50 cm.

Texture depends on the parent materials. Particle size distribution shows that the clay fraction (<0.001 mm) prevails followed by the silt fraction (particularly coarse silt 0.05–0.01 mm) (Table 3.1 and Fig. 3.3). The distribution of the clay fraction showed evidence of clay differentiation in the soil (Table 3.2 and Fig. 3.4).

The average particle density is 2.54–2.64 g/cm^3 in the upper part depending on humus content and is 2.6–2.7 g/cm^3 at the depth. Bulk density is about 1.3 g/cm^3 at the upper part and increased to 1.45–1.55 g/cm^3 at the B horizon with evidence of compaction.

Water stable aggregates with size >1 mm are about 55 % at the surface (0–15 cm) and decreased to about 45–50 % at depth (20–30 cm) in the pristine land. Prevailing water stable aggregates with size over 0.25 mm are 70–80 %.

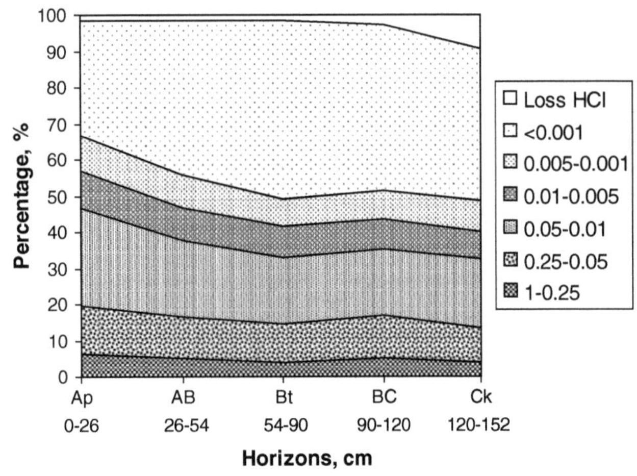

Fig. 3.3 Pattern of the particle size distribution (mm) in dark gray-brown forest soil (according to the revised Kachinskiy method)

In dark gray forest soils 10–15 % of pore space is represented by macropores or the pores of aeration (with diameter >300 μm), and 15–20 % is mesopores or pores with rapid moisture drainage (>30 μm diameter). Micropore space or water-filled is about 20 %, represented by the slow drainage pores (diameter 0.2–30.0 μm) and 15–25 % of micropores is with none remaining available for roots moisture (diameter <0.2 μm) (Dilkova 1985; Dilkova et al. 1998).

The natural spatial relationship of the dark gray-brown forest soils is with leached and podzolized chernozems varieties, the gray-brown forest soils, and with Eutric Regosols.

Content of humus in (A) horizon is 2.0–4.5 %, and in AE horizon humus content is below 2 % (Fig. 3.5). Humic acids prevail over fulvic acids and are bounded with calcium that makes them stable. The stock of organic matter in a 100 cm layer is 200–300 ton per ha, however, it is about 200–250 ton per ha in the arable lands (Filcheve et al. 2002). The upper 20 cm of soil stores more than 40–50 % of the total amount of organic matter to a depth of 100 cm. The pattern of organic matter distribution at the depth is gradual, similar to that of chernozems. The process of transformation of organic residues is accompanied by nitrogen enrichment of humic substances shown by the ratio C/N 11–12 which means high rates. Soil reaction is slightly acid to neutral (pH_{KCl} 4.5–5.5) and is neutral to slightly alkali (pH_{KCL} 6.0–6.5) in the parent materials. Cation exchange capacity is over 75–85 % (Ganev and Arsova 1980) and is 20–30 meq/100 g at the upper part and is 30–40 meq/100 g at the middle part of illuvial B(t) horizon. Exchangeable cations dominate by Ca^{2+}, Mg^{2+}, and exchangeable K^{+}.

In all horizons hydromicas prevail and there is much less montmorillonite, chlorite, kaolinite, quartz, and iron hydroxides occurrence. Clay minerals commonly occur in the soil as mixed layers of two or more different minerals. The qualitative composition of clay minerals, determined by X-ray diffraction analysis and transitional electronic microscopy, has shown prevalence of illite at rates of 43 % in the clay (<0.001 mm) and 12 % in the soil at the surface (A) horizon, gradually decreasing at the depth, where in the (C) horizon it is about 30 % in the clay and 8 % in the soil. Montmorillonite content is at rates of 25 % in the clay and 8 % in the soil at the surface (A) horizon, gradually increasing at the depth, where in the (C) horizon it is about 40 % in the clay and 12 % in the soil. Vermiculite (or hydrated micas) shows slight differentiation in the profile distribution at rates of 11–13 % in the clay and 3 % in the soil in both surface (A) horizon and (C) horizons, and a slight increase to 15 % in the clay and 6 % in the soil at the illuvial (Bt) horizon. Kaolinite is about 15–17 % in the clay (<0.001 mm) and 4–6 % in the soil (Boneva 1998).

Table 3.2 Variables of humus, pH, carbonates, and particle size distribution (mm) according to USDA in 18 referenced profiles of the dark gray-brown forest soils in Bulgaria

Profile count 18	Horizon	Upper depth	Lower depth	Humus (%)	pH_{KCl}	Carbonates	Bulk density g.cm^3 calc	Clay <0.002	Silt 0.002–0.05	Fine sand 0.05–0.2	Total sand 0.05–2.0
Mean	Ap	0	26	2.0	4.3	0.0	1.3	37.5	42.9	13.5	20.0
	AB	26	54	1.2	4.4	0.0	1.4	48.2	35.4	11.3	16.9
	Bt	54	90	0.8	4.4	0.0	1.4	54.1	31.3	10.6	14.7
	BC	90	120	0.0	5.0	0.3	1.4	51.1	31.8	11.8	17.1
	Ck	120	152	0.0	6.4	7.1	1.5	51.6	35.0	10.3	14.7
Stdev	Ap	0	3	0.4	0.6	0.0	0.0	6.7	6.5	5.4	8.2
	AB	7	6	0.3	0.6	0.0	0.0	11.8	7.9	4.4	7.0
	Bt	9	13	0.2	0.4	0.0	0.0	8.1	4.1	4.8	6.8
	BC	15	22	0.1	1.0	0.9	0.0	9.9	4.2	6.9	12.2
	Ck	20	23	0.0	1.0	5.4	0.0	3.5	3.6	2.3	4.5

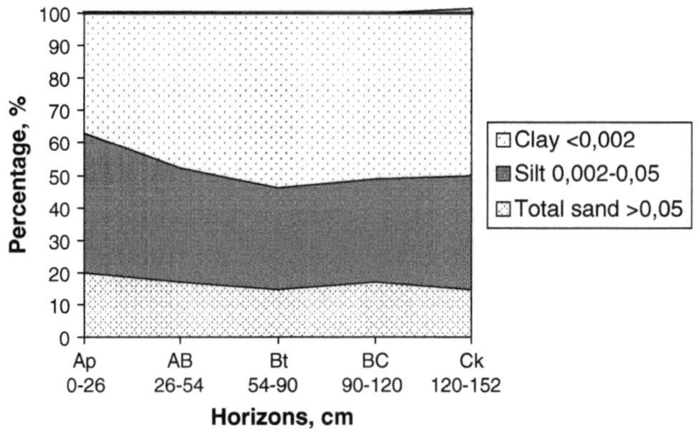

Fig. 3.4 Pattern of the particle size distribution (mm) in the dark gray-brown forest soil (USDA)

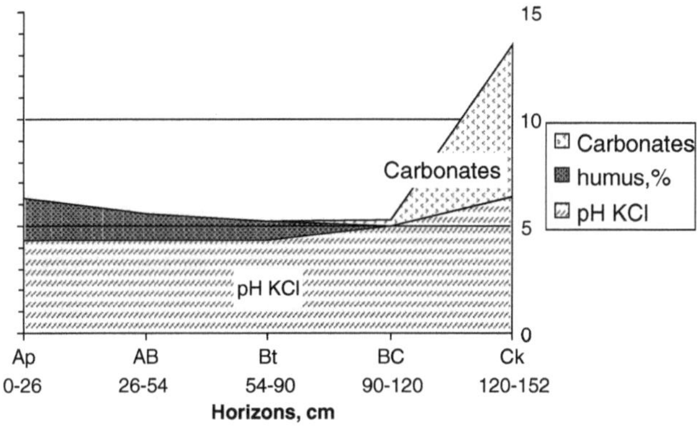

Fig. 3.5 Pattern of pH, humus, and carbonates distribution in the dark gray-brown forest soil

3.3.3.2 Gray-Brown Forest Soils

Gray-brown forest (*subtype, type*) soil is named after the Bulgarian classification of Koynov et al. (1964), Yolevski and Hadzhiyanakiev (1976).

Cutanic Luvisols is named according to WRB (2006) and *Typic Hapludalfs* according to Keys to Soil Taxonomy (2010).

Haplic Luvisol or Gleyic Luvisol: Revised Legend of the World Soils (FAO-UNESCO-ISRIC 1990).

WRB (2006) has allowed that soils that respond to the demands of argic horizon can be referred to as Luvisols.

Keys to Soil Taxonomy (2010) required the *ochric* epipedon and the *argillic* horizon for Alfisols order classification. In Soil Taxonomy there are demands for subsurface horizons, allowing *argillic* horizon with high base saturation (>35 %) expressing mottling, gleying, and coarse structure development; *or kandic* horizon with low (<35 %) base saturation in Ultisols; *calcic or petrocalcic* horizon; and *albic, natric or none*.

Ochric epipedon is most commonly a thin A horizon showing some incorporation of organic matter.

Typical distinguishing features are differentiation of clay in the profile and evidence of compaction in the middle part. Morphologically, these features are represented as clay-reach compacted Bt horizon with strong coarse subangular blocky structure.

Particle size distribution shows that the clay fraction (<0.001 mm) prevails followed by the silt fraction (particularly coarse silt 0.05–0.01 mm) (Table 3.3 and Fig. 3.6). The distribution of clay fraction showed the evidence of clay differentiation in the soil (Table 3.4 and Fig. 3.7).

The average particle density is 2.45–2.55 g/cm^3 at the upper part and is 2.65–2.70 g/cm^3 at the depth. Bulk density is near 1.3–1.4 g/cm^3 at the upper part and is about 1.5–1.6 g/cm^3 at the middle part of the profile with evidence of compaction. When dried the textural Bt horizon is very hard in consistence.

Water stable aggregates of size >1 mm are about 50–60 % at the surface (0–15 cm) and sharply decreased to about 30–45 % at the depth of 20–30 cm in pristine land. Prevailing water stable aggregates of size over 0.25 mm are over 75 %.

In gray-brown forest soils 5–15 % of pore space is represented by macropores or the pores of aeration (of diameter >300 μm); and 15–20 % is mesopores or pores with rapid moisture drainage (>30 μm diameter). Micropore space or water-filled is about 15–20 %, comprising slow drainage pores (diameter 0.2–30.0 μm) and 10–20 % of micropores is with none remaining available for roots moisture (diameter <0.2 μm) (Dilkova 1985; Dilkova et al. 1998).

Total porosity is high at the surface (50–55 % at field capacity) and decreases with the depth. Soils are characterized by low water permeability.

In spatial distribution gray-brown forest soils are bordered by the leached and degraded chernozems, dark and light gray-brown forest soils, as well as by Eutric Cambisols and Regosols.

The pattern of horizon sequence is O–AE–Bt(g)–B(g)–C(k).

The average thickness of humus-eluvial horizon (A + AE) is 25–35 cm, and the average thickness of Bt textural horizon is 100–165 cm. Humus content at the upper part of the humus horizon varies from 1.5 to 3.5 %, with higher values in forest land. The pattern of organic carbon and nitrogen distribution in the profile showed a sharp decrease below 25 cm, which is typical for the soils formed under forests (Fig. 3.8). The stock of organic matter in a 100 cm layer is 180–200 ton per ha in pristine land and is much less in arable land. The upper 20 cm of soil store 60–70 % of the total amount of organic matter. Forms of

Table 3.3 Variables of the particle size distribution (according to revised Kachinskiy method) in 20 referenced profiles of gray-brown forest soils in Bulgaria

Profile count 20	Horizon	Upper depth	Lower depth	Loss HCl	Particle size distribution (mm), % (according to method of Kachinskiy revised)							
					>1 mm	1–0.25	0.25–0.05	0.05–0.01	0.01–0.005	0.005–0.001	<0.001	<0.01[a]
Mean	AEp	0	28	2.0	1.0	7.9	16.4	25.4	8.8	9.1	29.3	47.2
	Bt1	28	55	2.1	0.8	6.4	14.1	20.9	7.2	7.0	41.7	55.8
	Bt2	55	81	2.4	1.4	5.9	12.9	22.2	7.3	7.4	40.7	55.4
	Bg	81	114	4.7	1.5	5.9	14.1	22.0	7.5	7.6	37.1	52.1
	Ck	114	142	10.0	1.0	10.4	11.5	21.4	7.5	7.0	31.2	45.7
Stdev	AEp	0	5	1.1	3.8	10.3	9.6	8.6	2.7	3.0	8.5	10.9
	Bt1	8	11	1.3	2.4	9.4	9.6	7.5	3.4	2.5	11.4	12.0
	Bt2	11	14	1.5	4.5	8.3	9.4	8.8	3.1	2.1	10.8	11.8
	Bg	14	20	5.3	2.9	9.3	11.2	10.1	3.0	2.9	11.4	13.3
	Ck	20	23	7.6	2.9	15.7	4.6	9.1	3.8	2.4	10.6	13.4

[a] Physical clay, the parameter characterized sum of the particle-size distribution, according to Kachinskiy method (mm) $\sum < 0.01 = < 0.001 + (0.005 - 0.001) + (0.01 - 0.005)$

Table 3.4 Variables of humus, pH, carbonates, and particle size distribution (mm) according to USDA in 20 referenced profiles of gray-brown forest soils in Bulgaria

Profile count 20	Horizon	Upper depth	Lower depth	Humus (%)	pH KCl	Carbonates	Bulk density g.cm^3 calc	Clay <0.002	Silt 0.002–0.05	Fine sand 0.05–0.2	Total sand 0.05–2.0
Mean	AEp	0	28	1.9	4.6	0.0	1.3	34.6	38.8	17.1	26.7
	Bt1	28	55	0.9	4.5	0.1	1.4	45.8	31.8	14.7	22.5
	Bt2	55	81	0.8	4.6	0.2	1.4	45.6	33.5	13.6	20.9
	Bg	81	114	0.0	5.3	0.7	1.5	43.5	34.4	15.0	22.4
	Ck	114	142	0.0	6.1	7.0	1.5	40.0	35.3	12.8	24.7
Stdev	AEp	0	5	0.5	0.7	2.4	0.1	9.1	11.7	9.6	17.1
	Bt1	8	11	0.5	0.6	2.6	0.1	11.9	10.7	9.7	16.4
	Bt2	11	14	0.2	0.7	2.7	0.0	11.0	11.9	9.5	15.8
	Bg	14	20	0.0	1.1	4.3	0.1	11.7	12.4	11.3	16.7
	Ck	20	23	0.0	1.3	5.7	0.0	11.0	13.1	4.8	20.2

humus are composed in equal parts of both humin and humic (C_h) or fulvic (C_f) acids (Petrova 1980). A significant proportion of humic acids are free or bound with sesquioxides which characterizes them as mobile. The process of transformation of organic residues is accompanied by nitrogen enrichment of humic substances shown by the ratio C/N 11–12 meaning high rates.

Soil reaction is acid to slightly acid at the surface pH_{KCl} 4.5–5.0, changed to strongly acid in the upper Bt horizon and is neutral at the parent materials pH_{KCl} 5.5–7.5. Exchangeable cations Ca^{2+}, Mg^{2+} and K^+ prevail in the cation exchange capacity. Some exchangeable H^+ is available. Base saturation is over 75–85 % (Ganev and Arsova 1980) and the cation exchange capacity is about 15–25 meq/100 g in the upper part and is significantly higher at the middle part of a soil.

Clay minerals commonly occur as mixed layers in the soil of two or more different minerals. The qualitative composition of clay minerals, determined by X-ray diffraction analysis and transitional electronic microscopy, has shown a prevalence of illite at rates of 38 % in the clay (<0.001 mm) and 9 % in the soil at the surface (AE) horizon, gradually decreasing at the depth, where in the (C) horizon it is about 31 % in the clay and 10–12 % in the soil. Montmorillonite content is at rates of 19 % in the clay and 6 % in the soil at the surface (AE) horizon, gradually increasing with depth, where in the (C) horizon it is about 30 % in the clay and 10 % in the soil. Vermiculite (or

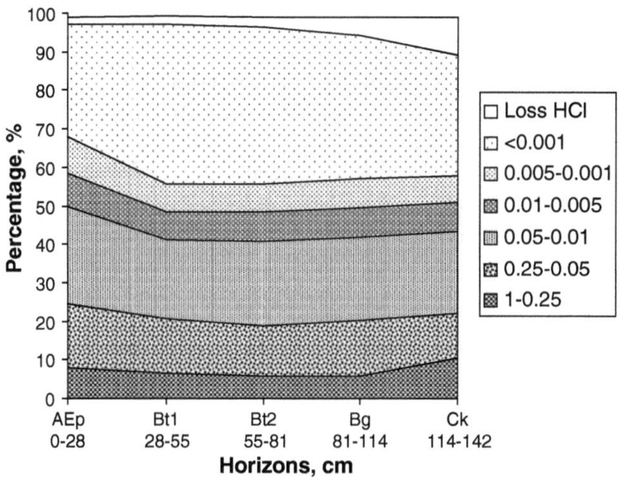

Fig. 3.6 Pattern of the particle size distribution (mm) in gray-brown forest soil (according to revised Kachinskiy method)

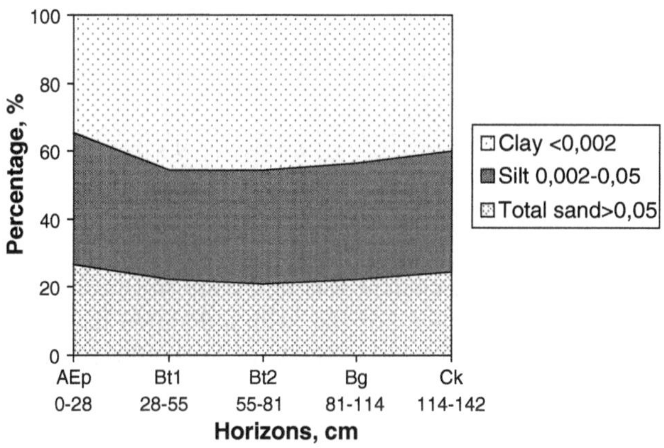

Fig. 3.7 Pattern of the particle size distribution (mm) in forest soil (USDA)

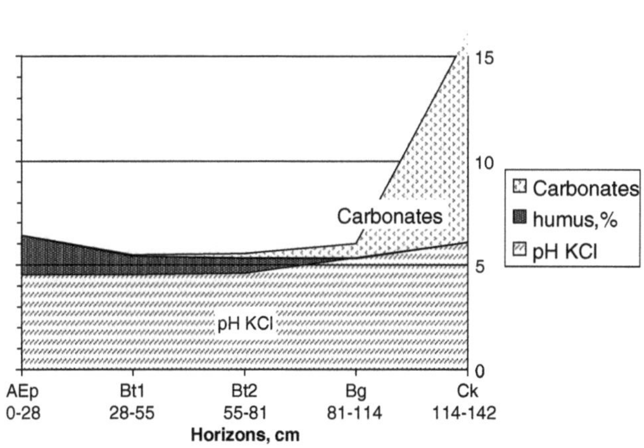

Fig. 3.8 Pattern of pH, humus, and carbonates distribution in gray-brown forest soil

hydrated micas) is at rates of 11–12 % in the clay and 2–4 % in the soil. Kaolinite is about 27 % in the clay (<0.001 mm) and 7–10 % in the soil at the surface (AE) horizon, gradually decreasing at the (C) horizon 20 % in the clay and 6 % in the soil (Boneva 1998).

The amount of total iron (Fe_2O_3) (Arinushkina 1970) distribution in the profile is differentiated in different soil horizons. The amount of total iron in the humus eluvial horizon is about 4.5–5.5 % and is 6–8 % in the illuvial Bt horizon.

Dominant in all horizons is the silicate form of iron 52–72 % (from the total iron Fe_2O_3). The silicate iron lowest values are in the humus eluvial horizon and the highest values are in the illuvial Bt horizon.

The non-silicate iron (Mehra and Jackson 1960) is about 38–48 % (from the total iron Fe_2O_3) in the humus eluvial horizon gradually decreasing downwards. Compared to the chernozems this indicates more advanced weathering. Almost all non-silicate iron appears in slightly crystallized form which is evidence of wetter conditions.

The amorphous form of iron (Tamm 1922) is commonly higher in the humus eluvial horizon and gradually decreases with depth. The amount in the humus eluvial horizon varies from 9 to 21 % (of the total iron); in the Bt horizon it varies from 3 to 18 % (of the total iron). This fact is explained by advanced clay formation at the Bt horizon (Hadzhiyanakiev 1989).

The free amorphous form of iron is differentiated; in the eluvial horizon it is about 2–9 % (of the total non-silicate amorphous iron) and in the illuvial Bt horizon it is about 0.7–8.0 %. This fact is explained by periodic water saturation (Hadzhiyanakiev 1989).

The amorphous form of iron bounded with organic matter (Bascomb 1968) varies significantly. The highest values are in the upper humus eluvial horizon, where it is 1.8–12.8 % (of the total non-silicate amorphous iron), gradually decreasing downwards.

In spatial distribution, gray-brown forest soils are bordered by the leached and degraded chernozems, dark and light gray-brown forest soils, as well as by Eutric Cambisols and Regosols. Shallow varieties of Eutric Cambisols are widely distributed throughout (Tables 3.5, 3.6 and Figs. 3.9, 3.10, 3.11).

3.4 Soil Properties and Soil Management

3.4.1 Soil Properties

Translocation of the fine material without further destruction and permanent weathering in situ are pronounced in the profile of gray-brown forest soils. In the illuvial Bt horizon

Table 3.5 Variables of the particle size distribution (according to revised Kachinskiy method) in 13 referenced profiles of Eutric Cambisols in Bulgaria

Profile count 13	Horizon	Upper depth	Lower depth	Loss HCl	Particle size distribution (mm), % (according method of Kachinskiy revised)							
					>1 mm	1–0.25	0.25–0.05	0.05–0.01	0.01–0.005	0.005–0.001	<0.001	<0.01[a]
Mean	Ap	0	27	1.9	3.2	7.0	21.9	21.7	7.2	8.7	29.0	44.9
	B	28	53	1.7	0.8	7.9	21.3	19.4	7.5	8.2	33.3	49.1
	Ck	57	78	12.4	2.3	8.7	22.2	14.5	6.7	7.8	26.9	41.4
Stdev	Ap	0	3	0.9	9.4	4.9	10.7	5.4	2.7	3.7	11.0	13.5
	B	4	8	0.9	2.7	7.2	9.9	5.8	2.4	2.6	9.8	12.1
	Ck	4	9	11.2	3.7	8.3	17.6	4.4	3.7	3.6	10.2	14.1

[a] Physical clay, the parameter characterized sum of the particle-size distribution, according to Kachinskiy method (mm) $\sum < 0.01 = < 0.001 + (0.005 - 0.001) + (0.01 - 0.005)$

Table 3.6 Variables of humus, pH, carbonates, and particle size distribution (mm) according to USDA in 13 referenced profiles of Eutric Cambisols in Bulgaria

Profile count 13	Horizon	Upper depth	Lower depth	Humus (%)	pH KCl	Carbonates	Bulk density g.cm^3 calc	Clay <0.002	Silt 0.002–0.05	Fine sand 0.05–0.2	Total sand 0.05–2.0
Mean	Ap	0	27	2.0	5.1	0.2	1.33	34.9	34.4	22.3	31.3
	B	28	53	1.0	5.2	0.2	1.40	38.5	31.8	21.5	29.8
	Ck	57	78	0.7	6.0	10.4	1.48	38.0	30.2	23.4	33.2
Stdev	Ap	0	3	0.8	1.0	0.4	0.07	11.2	7.3	10.5	15.6
	B	4	8	0.5	1.0	0.6	0.06	10.3	7.6	9.8	14.9
	Ck	4	9	0.3	1.6	11.0	–	14.9	10.6	16.9	22.9

there is some amount of accumulated sesquioxides (oxides, hydrous oxides, hydroxides), and the clay is hydromicas and less montmorillonite.

Gray-brown forest soils are fertile agricultural soils. Accumulation of humus is of mull type in the upper 20 cm. Despite the low organic matter content (2.0–3.5 %) gray-brown forest soils have favorable soil properties for agricultural practices. Biological activity is high.

The highest values of sorbtion capacity are in the middle part of the soil. In the composition of adsorbed cations Ca^{2+}, Mg^{2+} prevail, exchangeable K^+ and H^+. Base saturation is over 70 %. Soils have high base buffering capacity.

Carbonates are deeply leached. Generally, soil reaction is slightly acid at the upper part which is followed by a steady increase with depths up to a maximum of about pH_{KCl} 6.5.

Water permeability is low. Commonly, waterlogging occurs on the soil surface on flat terrains.

The assessment of waterloging in the gray-brown forest soils based on two decades of soil moisture regime data has shown a relation with impermeable layer occurrence in the profile. In the sandy loamy gray-brown forest soil, developed in hard sandstone, there was excessive moisture throughout the profile, mainly in winter and much less in early spring. However in the loamy gray-brown forest soil, excessive moisture content was in the top soil layer, overlying the impermeable horizon, with a duration that lasted from early autumn to spring (Dilkova et al. 1986).

3.4.2 Soil Management

The natural potential of these soils for agricultural purposes varies from moderate to good. The environment conditions have a favorable effect on the temperature and moisture regimes in these soils. Because of the temperature inversion occurrence, the start of the spring tillage period comes earlier than it does on the lowlands of the hilly Danube Plain.

These soils have a natural high water-holding capacity that makes them suitable for agriculture. The problem is aeration of the subsoil. Seasonable soils have poor technologic plough characteristics which means that they are only suitable for ploughing for short periods, resulting in higher expenses.

Dark gray-brown forest soils have a well-pronounced humus horizon and are in the category of productive soils. In the maintenance of this soil group crop rotation is important. Winter wheat, barley, maize, rye, oats, sunflower, sorghum, millet, beet, tobacco, alfalfa, and

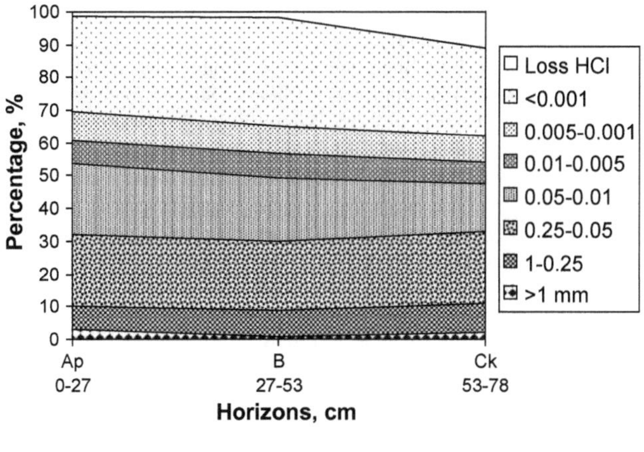

Fig. 3.9 Pattern of the particle size distribution (mm) in Eutric Cambisol (according to revised Kachinskiy method)

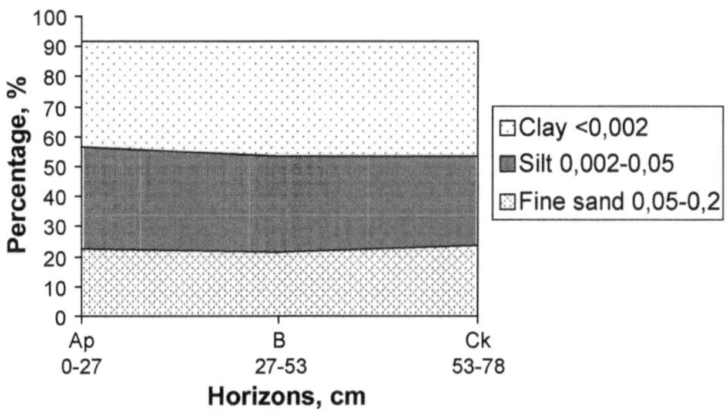

Fig. 3.10 Pattern of the particle size distribution (mm) in Eutric Cambisol (USDA)

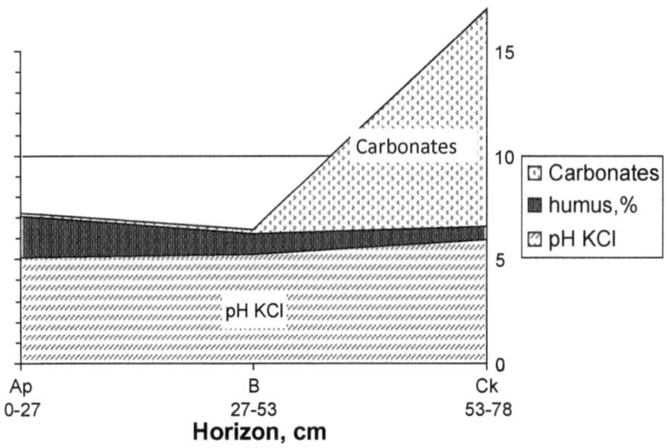

Fig. 3.11 Pattern of pH, humus, and carbonates distribution in Eutric Cambisol

leguminous crops can be grown and viticulture can be established. Fruits such as apples, pears, plums, blackcurrents, and raspberries can also be successfully grown.

Gray-brown forest soils are common on lands with dissected topography and in the presence of deciduous forests. These soils have the potential to be very productive if conserved, but can also degrade rapidly if eroded. Adequate measures should be taken to control erosion. The removal of woodland has resulted in poor nutrient balance.

Because they occur under moist conditions, these soils are frequently used for mixed farming, dairy farming, or horticulture. The moisture retained during the summer is sufficient for the growing of winter crops. Winter ray, spring oats, leguminous crops, alfalfa, and rye-grass can be grown and viticulture can be established. Horticulture is common. Frequently, it is maintained by the liming and fertilizer application.

References

Arinushkina E (1970) Manual on chemical soil analysis. Moscow State University Press, Moscow (In Russian)

Bascomb CL (1968) Distribution of pyrophosphate extractable iron and organic carbon in soils of various groups. J Soil Sci 19(1):251–268

Boneva K (1998) Clay minerals in gray forest soils from north Bulgaria and an attempt to group the soils according to the mineralogical composition. Soil Sci Agrochem Ecol J 33(3):23–25 (In Bulgarian)

Bondev I (1991) Vegetation in Bulgaria, map at scale 1:600,000 with explanatory text. St. Kliment Oxridski University Press, Sofia (In Bulgarian)

Dilkova R (1985) Structure and aeration in main soil varieties in Bulgaria. Theses, Institute of Soil Science Nikola Poushkarov, Sofia (In Bulgarian)

Dilkova R, Kerchev G, Nedyalkov S (1986) Waterlogging degree for two gray forest soils with different textures. Soil Sci Agrochem Plant Prot J 21(6):27–32 (In Bulgarian)

Dilkova R, Kerchev Zh, Kercheva M (1998) Evaluating and grouping of soils according to their susceptibility to anthropogenic degradation. Adv GeoEcol 31:125–131 (Catena Verlag, Reiskirchen)

Duchaufour Ph (1951) Lessivage et Podzolisation. Revue forestiere francaise, no 10

FAO-UNESCO-ISRIC (1990) Revised Legend of the soil map of the world. FAO world soil resources report no 60, Rome

Filcheva E, Rousseva S, Kulikov A, Nedyalkov S, Chrnogorova T (2002) Organic carbon stocks in soils of Bulgaria. In: Kimble JM, Lal R, Follett RF (eds) Agricultural practices and policies for carbon sequestration in soils, chapter 45. Lewis Publishers, CRC Press LLC, New York, p 471–476

Ganev S, Arsova A (1980) Methods of determining the strongly acid and the slightly acid cation exchange in soil. Soil Sci Agrochem J XV(3):22−33 (In Bulgarian)

Gyurov G et al (1978) Some problems of genesis and classification of gray forest soils, brown forest and pseudopodzolic soils in Bulgaria. In: Second scientific conference on soil science, Sofia (In Bulgarian)

Hadzhiyanakiev A (1989) Iron in Bulgarian soils. Zemizdat Press, Sofia (In Bulgarian)

Keys to Soil Taxonomy (2010) USDA United States Department of Agriculture, NRCS Natural Recourses Conservation Service, 11th edn. Soil Survey Staff, Washington, DC, p 338

Koynov et al (1963) On the question of soil classification and systematic in the country. 50th anniversary Bulgarian soil science, Sofia, p 87–97 (In Bulgarian)
Koynov V (1964) Peculiarities of the main Bulgarian soil types and their basic soil-genetic processes. In: Gerasimov IP et al. (eds) Soils of south–eastern Europe. Proceedings of the international symposium of soil science, June 1963, Sofia. Bulgarian Academy of Sciences Press, Sofia, p 15–62 (In Russian)
Koynov V et al (1964) Classification and systematic of soils in Bulgaria. In: Gerasimov IP et al. (eds) Soils of south–eastern Europe. Proceedings of the international symposium of soil science, June 1963, Sofia. Bulgarian Academy of Sciences Press, Sofia, p 231–246 (In Russian)
Kubiena WL (1953) The soils of Europe. Thomas Murby & Co, London, p 314
Mehra OP, Jackson ML (1960) Iron oxide removal from soils and clays by dithionite–citrate-system buffered with sodium bicarbonate. Clays Clay Miner 7:317–327
Penkov M et al (1992) Soil classification and diagnostic in Bulgaria in relation to the land reform fragmentation. Zemizat Press, Sofia (In Bulgarian)
Petrova L (1980) State of organic nitrogen in some soils in Bulgaria. Thesis, Institute of Soil Science N Poushkarov, Sofia, p 138 (In Bulgarian)
Rode AA (1964) On question of podzolization and lessivage. J Soil Sci 7:660–671 (In Russian)
Tamm O (1922) Eine methode zur bestimung der anorganischen komponente ds gelcomplexes im boden. Statens Skogsforsoksanstalt, Medd. 19:387–404 (by Arinoushkina E (1972) Manuel of chemical analysis of soils, Moscow)
Trashliev H, Krastanov S, Hadjiyanakiev A (1964) Genesis and classification of the gray forest soils in Bulgaria. Soils of south-east Europe. Bulgarian Academy of Sciences Press, Sofia, p 143–158
World Reference Base for Soil Resources (2006) A framework for international classification, correlation and communication. FAO-ISRIC, Rome
Yolevski M, Hadzhiyanakiev A (1976) Agro-grouping of soils in Bulgaria. Extended systematic list. Code of groups and soil varieties. Institute of Soil Science Nikola Poushkarov, Sofia (In Bulgarian)
Yolevski M, Ninov N, Krastanov S et al (1983) Brief diagnostic and criterion of the main taxonomic soil divisions in Bulgaria. Precise classification scheme, Institute of Soil Science Nikola Poushkarov, Sofia (In Bulgarian)
Zonn SV (1973) Environmental settings of the processes of lessivage, pseudo-podzolization and podzolization during the Quaternary period in the western and northwestern regions of the USSR. J Soil Sci 116:211–217

4 Pseudopodzolic Soils

Derived from the Latin word *planus,* meaning flat.

Pseudopodzolic forest soils are differentiated soils with sharp texture change. The profile has an illuvial horizon as a result of a process of enrichment of subsoil with fine soil material.

4.1 Concept of the Type

Main concepts

1. Pseudopodzolic soils are acid mineral soils that have a sharp increase in clay content at the illuvial (Bt) horizon, as a result of soil evolution;
2. Have an eluvial horizon due to movement out of fine soil material;
3. Chemical weathering in situ significantly contributes to the texture differentiation.

The main processes that have taken place are leaching and lessivage, oxidation and reduction, pseudogley and weathering in situ, complex formation, and migration.

4.2 Ecology of Soil Formation

4.2.1 Distribution

A broad belt of the Fore Balkan is occupied by pseudopodzolic soils. Three soil varieties with similar morphology and soil properties are considered as pseudopodzolic forest soils in Bulgaria. They occupy nearly 9.75 % of the country's territory.

Light gray (pseudopodzolic) forest soils are spread from east to west in northern Bulgaria. They are located in a central belt comprising the great part of the Fore Balkan and in the region of Kamchia River. Of these, 130,000 ha are deep soils and the rest are shallow. They are distributed over the south line of the sub-zone of gray-brown forest soils (with elevation 450–500 m) and between the zones of brown forest soils (with elevation over 700–800 m).

Cinnamonic podzolic (pseudopodzolic) forest soils are located at the sub-Balkan and the southern slopes of the Fore Balkan spreading from east to west. The deep soils are about 200,000 ha and the rest are shallow. They are distributed in the Sredna Gora Mountain, at the low elevated areas of the Rhodope Mountains, Sakar Mountain, and on the uplands of Strandzha Mountain as well as in the Thracian Lowland and in the region of the Struma and Mesta River valleys. In the southeastern part of the country, a single region of zheltozem (podzolic) pseudopodzolic soils occurs in the Strandzha Mountain.

Climatic conditions

The temperature regime and total precipitation are variable from one region to another where pseudopodzolic soils occur.

Light gray (pseudopodzolic) forest soils are located in the regions with humid temperate continental conditions that have a warm summer. In the central province in the region of the Fore Balkan the mean annual temperature is 9.8–10.3 °C. In January the mean annual temperature is −1.8(2.5) °C. In July the mean annual temperature is 20.1–20.9 °C. The mean amount of precipitation varies from 760 to 949 mm.

Cinnamonic podzolic (pseudopodzolic) soils are located in the regions with a transitional continental climate influenced by the Mediterranean. In the central province in the region of the Thracian Lowland the mean annual temperature is 11.9–12.1 °C. In January the mean annual temperature is −0.5 (0.1) °C. In July the mean annual temperature is 22.9–23.6 °C. The mean amount of precipitation varies from 548 to 609 mm.

In the central province in the region of the sub-Balkan the mean annual temperature is 9.8–10.3 °C. In January the mean annual temperature is −1.8(2.5) °C. In July the mean annual temperature is 20.1–20.9 °C. The mean amount of precipitation varies from 760 to 949 mm.

In the western province the mean annual temperature is 8.8–10.7 °C. In January the mean annual temperature is −1.9(−2.9) °C. In July the mean annual temperature is

T. Shishkov and N. Kolev, *The Soils of Bulgaria*, World Soils Book Series, DOI: 10.1007/978-94-007-7784-2_4,

17.0–20.8 °C. The mean amount of precipitation varies from 585 to 787 mm.

In the eastern province the mean annual temperature is 11.3–12.4 °C. In January the mean annual temperature is 0.1–1.7 °C. In July the mean annual temperature is 21.6–22.6 °C. The mean amount of precipitation varies from 571 to 593 mm.

In the southeastern province the mean annual temperature is about 12.8 °C. In January the mean annual temperature is 1.7 °C. In July the mean annual temperature is 23.4 °C. The mean amount of precipitation varies from 587 mm.

Zheltozem podzolic (pseudopodzolic) soils are located in the region of transitional continental climate influenced by the Black Sea.

Topography

Typically, pseudopodzolic soils in Bulgaria are situated at the older stable landforms (Fig. 4.1). These soils may occur on flat or gently sloping terrains, old terraces and valleys, or on gently steeply sloping sites. Usually, low slopes may be found in sites with very poor drainage.

The foot slopes of the Balkan Mountains, the low elevated Sredna Gora Mountain, the Rhodope Mountains, and Sakar Mountain are characterized by moderate mountainous steep topography. These regions are of dissected relief and are all affected by erosion. The topography in the western part of the country in the region of Lyulin, Lozen, and Konyavska Mountains is mountain-valley dissected and affected by erosion. The topography of the Strandzha Mountain is elevated hilly and slightly dissected.

The relief of the Thracian Lowland and valleys of the Struma and Mesta Rivers are smooth to moderately undulating.

The topography of Strandzha Mountain ranges from gently sloping to rolling.

Parent materials

Parent materials are consolidated rock and sediments from different ages. Some provincial differences in soils are determined by the character of these materials. Mineralogy is variable especially in the mountainous regions. On high plateaus and mountains base saturation in rock materials is usually <50 %. Parent materials are various, mostly unconsolidated deposits of non-calcareous acidic rocks, non-calcerous and calcareous sandstone, granite, and reddish brown clays. Some sites are with sandy gravel materials a few meters thick and below that calcareous rock material.

On the northern foot slopes of the Fore Balkan Mountains the light gray (pseudopodzolic) forest soils have formed on the Lower Cretaceous Age (Aptian and Valanginian-Hauterivian) (Mesozoic Era) sandstones and marls; on the Barremian Age limestones, clayey limestones, and marls; on the Jurassic-Cretaceous Age (Mesozoic Era) sandy flysch, shales, marls, flysch interbedding of sandstones, conglomerates, marls, and shales.

In the western region on the inter-mountain plains the cinnamonic podzolic (pseudopodzolic) forest soils have developed on the Quaternary Age alluvial-drift and talus-drift sediments (boulders, pebbles, sands); and on the

Fig. 4.1 Woodland with cinnamonic pseudopodzolic soils

Neogene Age continental sediments in the superimposed depressions (conglomerates, breccia-conglomerates, sandstones, siltstones, shales, sands, clays).

In the central Thracian Lowland and in the southern foot slopes of the Sredna Gora and Central Balkan Mountains the cinnamonic podzolic (pseudopodzolic) forest soils have developed on the Quaternary Age alluvial-drift and talus-drift sediments (boulders, pebbles, sands); on the Neogene Age continental sediments in the superimposed depressions (conglomerates, sands, clays); on the eluvium of the Upper Cretaceous Age (Campanian) (Mesozoic Era) volcanic andesites, tuffs and tephroidal rocks interbedded with sedimentary rocks sandstones, siltstones, marls, and limestones; and on the Carboniferous Age (Paleozoic Era) granites and granodiorites.

In the eastern Thracian Lowland and the Balkan Mountains the cinnamonic podzolic (pseudopodzolic) forest soils have developed on the Paleogene/Middle Eocene Age flysch alternation of sandstones, siltstones, clays, sandy limestones with flint, clayey marls, and packets of breccias-conglomerates.

In the southeast region the cinnamonic podzolic (pseudopodzolic) forest soils have developed on the eluvium of the Upper Cretaceous Age (Mesozoic Era) intrusive magmatic rocks (gabbros, syenomonzonites, granodiorites, quartzmonzonites).

In the southeast region of Strandzha Mountain cinnamonic and zeltozem podzolic (pseudopodzolic) soils have developed on the Paleogene/Upper Eocene Age breccia, conglomerates, sandstones, limestones, marls; and on the Upper Cretaceous Age (Mesozoic Era) volcanic gabbros, syenomonzonites, granodiorites, quartzmonzonites, basalts, andesitobasalts, rare packets of sediments sandstones, siltstones, and shales.

Vegetation

Pseudopodzolic soils are typically forested soils. The most common plant community is deciduous forest, but it may be mixed forest or coniferous trees and shrubs, and it may have had various sequences of vegetation over a long period of time. Species vary from place to place but the most common are deciduous beech and oak forests comprising oak (*Quercus sesiliflora*, *Quercus conferta*, *Quercus cerris*, *Quercus pubesccens*, *Quercus pedunculata*, *Quercus stranjensis*, *Quercus orientalis*), hornbeam (*Carpinus orientalis, Carpinus betulus* L.), beech (*Fagus orientalis*), aspen (*Populus tremula* L.), almond (*Amygdalus nana*), and fig trees (*Ficus carica*) and a shrub cover of hawthorn (*Crataegus monogyna*), privet (*Ligustrum vulgare*), and rose-bush (*Rosa canina*) (Bondev 1991). Usually, mosses are at the soil surface.

In the region of Strandzha Mountain, rhododendron (*Rhododendron ponticum),* which is a relict plant from previously existent conditions is widespread.

Time

The time required for the formation of textural differentiation depends on the parent materials. The precise time required for clay-rich horizon formation is not clear.

Most soils are quite old (hundreds of thousands of years) and could retain features formed during the late Pleistocene Era when relatively cold conditions existed. Some pseudopodzolic soils may be relatively young a thousand years or so, with textural differentiation forming at a different time during Holocene.

Human influence

The change of woodlands into arable areas has led to difficulties that have to be solved by maintaining the practice of these soils. Enhanced mineralization of organic matter usually has negative impacts on the soil, and results in homogenization of the plough horizon and weak (noncoherent) structure.

For agricultural production, pseudopodzolic soils need special maintenance in the form of liming and the application of high rates of fertilizers.

4.2.2 Genesis

The genesis of soils with sharp textural change is not unique in Bulgaria (Rode 1937, 1964; Fridland 1958; Kaurichev 1972; Zonn 1973). How the soils named pseudopodzolic came to be formed is still open to discussion (Dyudal 1972). The evolution of the soil took place during a different time, and environment conditions are complicated by a great range in parent materials. For this reason a unique scheme for the processes of interrelation seems to be impossible.

In Russia, Glinka (1924) proposed the idea that podzolization could sometimes occur with clay movement without its chemical destruction. For the European conditions, Duchaufour (1951) and Kubiena (1953) formulated a conception of the lessivage process as a complex of physico-chemical phenomena, where dispersed clay particles have been translocated in the form of complexes with organic compounds and where iron is leaching. The presumption was that the occurrence of lessivage is mostly under deciduous forests where a lower amount of aggressive organic acidic components existed. Gerasimov (1959) considered that pseudopodzolic process were related to the presence of surface gleyic process (pseudogley) and lessivage.

Back in 1968, during the first national soil science congress, soils were grouped so as to indicate the paleohydromorphic stage of evolution and characterized by relict features.

In the pseudopodzolic group of soils organic, organic-mineral and mineral compounds differ in their mobility. Seasonal waterlogging is typical in the upper part, where it promotes bleaching of the upper layer followed by translocation of the fine particles downwards. There is lack of destruction at the mineral part. The sequence of oxidation and reduction in the wet and dry conditions in the upper part results in the mobilization of free iron and manganese compounds, usually in the form of organic mineral compounds and their segregation as diffuse spots or tiny concretions.

The freezing and thawing of water saturating the upper layers forms a spongy to platy structure. The light "ash" color is due to the presence of SiO_2, mostly washed quartz grains from iron oxide skins.

The bulk chemical composition of the colloids shows a much higher content of sesquioxides (oxides, hydrous oxides, hydroxides) in the middle part of the soil. The ratio of SiO_2/R_2O_3 of the clay fraction shows differentiation which is evidence of movement of iron and aluminum down the profile.

Slow permeability is due to the impermeable horizon. Its formation is open to discussion but it could be a result of soil evolution. The details of this process are not clear. Its development could be a result of parent material discontinuity and relict in previously developed soils.

Obviously, periods of wetting and drying seem necessary for clay translocation. Some non-crystalline aluminum and iron organic complexes also leach and depose as oxides in the middle part of the soil. As soil solutions percolate in the zone where pH changes oxides precipitate, these complexes cover a wide variety of the structural components as skins. This leads to the formation of oriented skins on ped faces, pore spaces, cracks, etc., in the illuvial part of the profile and to an increase in the content of clay.

Phyllosilicate clays may be transported but not necessarily.

Chemical weathering in situ is responsible for the clay formation; however it is possible that the formation of 1 % clay took thousands of years.

Pseudogley

Pseudogley has developed due to seasonal waterlogging together with alteration of reduction and oxidation conditions.

Seasonal waterlogging at the surface layer is typical for texturally differentiated soils with subsurface (Bt) horizon. Texturally differentiated soils can be quite old.

In better drained areas, the genesis of these soils is not hydromorphic. The textural class is important. The upper water-bearing layer is a pseudogley with leakage water and the eluvial process of mineral destruction and substances transportation to the deeper horizon. Iron and manganese are mobile during the wet period. Some clay can also be distributed. The alternation of wetness and dryness is connected with the reduction and oxidation of iron and manganese, which results in their mobility or segregation as ferromagnesian nodules. Relict features are quite often observed in the pseudogley layer.

Pseudogley is characterized by the color pattern of rusty mottles, rusty and/or black concretions, and evidence of iron segregation in the wet conditions. There are many transitional forms of pseudogley color pattern, but typical is homogeneous blue-grayish, reddish, and rusty mottles color with abundance of tiny black or rusty concretions.

So-called degradation is manifested in the eluvial layer. Being analogous to fully drained soils, these soils are not typically hydromorphic.

Soils with pseudogley pattern occurrence during the wet period are characterized by anaerobic conditions and excess of moisture; during the dry period the leakage water disappears and drought affects plants.

Temporary hydromorphic conditions with pseudogley can be as follows:

a. Surface pseudogley, meaning that the top soil is loaded with temporary hydromorphic conditions. Soil Taxonomy defined the suborder *Epiaquepts, Epiaquerts, Epiaquolls,* analogous to the Bulgarian classification of soil with clayey texture throughout the profile.
b. Lessivage pseudogley, meaning that the surface (AE + E) horizon is loaded with the temporary leakage water and there is textural differentiation inside. Soil Taxonomy defined the suborder *Aqualfs,* analogous to the Bulgarian classification of all pseudopodzolic (podzolic) soils.

The surface waterlogging can be modeled based on the set of data that describes regimes through the water-bearing horizon.

4.3 Soil Diagnostic and Classification

4.3.1 Morphology

The profile of pseudopodzolic soil consists of upper eluvial layer and clay-rich illuvial horizon (Fig. 4.2).

O—a thin loose litter, at the top of which are visible leaf tissues, and at the bottom of which are moderate to strongly decomposed remnants, naturally occurring in woodlands; below is

AE(p)—a humus-eluvial horizon, dark grayish brown or brown (moist) to pale gray (dry) colored, about 10 cm thick. Structure of aggregates is crumb with medium size (2–5 mm), moderate pedality grade, consistence is friable (moist) and slightly hard (weakly resistant to pressure) dry; some fine roots (0.5–2 mm) occur; biological features are a few burrows (unspecified); clear change beneath into

E(p),(g)—a thin to thick eluvial horizon; light grayish, pale brown to light yellowish brown (moist) and gray to pinkish gray (dry) colored, where coarse fragments are

Fig. 4.2 Cinnamonic podzolic (pseudopodzolic) forest soil, Bulgarian classification Yolevski and Hadzhiyanakiev (1976), *Luvic Endogleyic Planosol (Albic, Dystric),* WRB (2006), *Vertic Albaqualfs,* Keys to Soil Taxonomy (2010), *Verti-Dystric Planosol,* FAO-UNESCO-ISRIC (1990)

coated with bleached silt material. Pedality type is partly platylike (arranged around a horizontal plane) and subangular blocky (blocks having mixed rounded and plane faces of aggregates), with very fine size (<5 mm), weak pedality grade, consistence soft (very weakly coherent and fragile) dry and friable (moist), or single grain (no structure, individual grains). Large aggregates are strung up on the roots. There are many black manganese concentrations in the soil mass as well as rusty spots. Abundance and size of roots is very few and fine (0.5–2 mm); sharp change into

Btg(s)—a thick illuvial-gleyic horizon usually redder, browner or yellower colored due to the presence of iron oxides. Structure is angular blocky (blocks having mixed, rounded and sharp angled faces of aggregates), with medium (10–20 mm) to coarse size (20–50 mm), strong pedality grade, consistence is very firm (soil material crushes under strong pressure) moist and very hard (very resistant to pressure) dry, showing an increase of clay by compaction pronounce. Porosity is medium consisting mainly of fine pores and large cracks. The surfaces of peds and pores are coated with clay cutans, which have a linear concentration. There is no tonguing penetration of the upper material in Bt horizon; it may occur by root penetration and subsequent decay. There is an abundance of tiny rounded black iron-manganese concentrations in the soil matrix and many rusty spots; mottles can be gray and greenish colored; some cracks may occur, grade into

BCg—a transitional to the upper part of the parent materials horizon, which has lower clay content and few or no cutans. Structure is subangular blocky(blocks having mixed rounded and plane faces of aggregates), with fine size (5–10 mm), moderate to strong pedality grade, consistence is firm (moist) and hard (dry); few or no black iron-manganese concentrations; clear change to

C(k)—parent materials, brownish yellow or yellowish brown often with gray greenish spots colored, structure is massive (structureless).

There is no tonguing silica penetration at the Bt horizon.

Low molecular weight organic substances are formed in the humus-eluvial (AE) horizon. Evidence of biological activity occurs mostly at this layer.

Translocation of the colloids from (AE + E) horizon results in some concentration of sand and silt size particles of quartz and other resistant minerals. The fact is that the amount of clay is significantly low at the eluvial E horizon compared to that in the parent materials.

Aluminum and iron oxide coatings on sand grains in the E horizon are dissolved by the organic acids.

The lessivage process is related with dispersal of clay in eluvial E horizons, translocation of the clay to the illuvial Bt horizon, and physical precipitation of the clay at the middle part. These lead to the formation of the oriented cutance.

Oxidation and reduction of iron oxides is apparent throughout the profile in the form of rusty or gray greenish mottles.

The conditions that existed in the country contributed to the prolonged weathering in situ.

4.3.2 Classification

Typical distinguishing features are abrupt textural change, compaction pronounce, and strong ash color in the upper layer.

The upper layer is associated with the *Albic* horizon but not all soils have developed it. No tonguing of albic material is admitted in the clay-rich *argic* horizon.

Classification after WRB (2006) required *albic* and *argic* diagnostic horizons in the reference group of Planosols.

Table 4.1 Varieties of soils represented the pseudopodzolic type in Bulgaria

Bulgarian classification (1964 and 1976)	WRB (2006)	Keys to Soil Taxonomy (2010)	Revised Legend of the world soils (FAO-UNESCO-ISRIC 1990)
Light gray forest soils (pseudopodzolic)	Luvic Endogleyic Planosol (Albic, Dystric)	Typic Albaqualfs	Verti-Eutric Planosol Verti-Dystric Planosol
Cinnamonic podzolic soils (pseudopodzolic)	Luvic Endogleyic Planosol (Albic, Dystric, Chromic)	Chromic Vertic Albaqualfs	Verti-Eutric Planosol Verti-Dystric Planosol
Zheltozem podzolic (pseudopodzolic) soils	Stagnic Cutanic Acrisol (Albic)	Typic Albaquults	Verti-Dystric Planosol

Keys to Soil Taxonomy (2010) required the *ochric* epipedon and the *argillic* horizon for Alfisols order classification. In Soil Taxonomy there are demands for subsurface horizons allowing a*rgillic* horizon with high base saturation (>35 %) expressing mottling, gleying and coarse structure development; *or kandic* horizon with low (<35 %) base saturation in Ultisols; *calcic or petrocalcic* horizon; and*albic, natric or none*.

Ochric epipedon is most commonly a thin A horizon showing some incorporation of organic matter.

Back in 1948 the Bulgarian soil classification (Antipov-Carataev and Gerasimov) defined the podzolic soil type in the same way as Russia. The soil classification of 1964 (Koynov et al.) shifted the subtypes of light gray soils and podzolic cinnamonic soils to the single type named pseudopodzolic soils. In 1976 (Yolevski and Hadzhiyanakiev) soils were renamed as pseudopodzolic forest soils including the same subtypes. Gyurov et al. (1978) proposed the single soil type of pseudopodzolic surface gleyic soils with three subtypes. In the soil classifications of 1983 (Yolevski et al.) the former existed subtypes were rejected and renamed as gleyic-like, somewhat gleyic and gleyic. In 1989 Hadzhiyanakiev proposed the highest level of classification of the three single types. There was classification in 1992 (Penkov et al.) which at the theoretic level named subtypes (Glossic, Dystric, Eutric) have not been applied in practice.

There is a significant difference between soil processes in various soils. Consequently, within the type pseudopodzolic soils, the second level (subtype) includes soils with different pedogenesis (Table 4.1).

Pseudopodzolic soils encompass the follows arguments: similarity in profile construction, in morphology, need of sound chemical analyses to support soil defining, same natural low fertility, and maintain practices of arable lands.

Subtype varieties differ in the thickness of humus horizon (non-eroded, accumulated or slightly eroded), in the degree of erosion (slightly, moderately, or not eroded), and in the degree of gleyic properties occurrence in the profile.

4.3.3 Distribution of Extractable Compounds of Fe, Al, and Mn in Pseudopodzolic Soils

In pseudopodzolic soils the total content of iron (Fe_t) and aluminum (Al_t), expressed as a % of soil mass, is highest along the profiles at the richest in clay fraction Bt horizons (Table 4.2).

In the pseudopodzolic soils the ratios between total content of iron and clay content (Fe_t/clay) decrease at the Bt horizons, since some of the released iron compounds have migrated out of the profiles. In these soils the transport of the free Fe forms is independent of the clay fraction movement. The clay formation process is less active than the weathering process (Jokova and Boyadjiev 1993).

Manganese is a biogenic element and the maximum of the total content Mn_t coincides with the surface horizons in most Bulgarian soils.

Since the free (non-silicate) compounds of these elements are results of the weathering and soil forming processes, their distributions along the profile (depth functions) illustrate the intensity and duration of these processes (Zonn 1982). The more advanced the processes, the higher the content of the free compounds.

Different forms of the free compounds are extractable in different reagents.

Oxalate extractable forms of Fe and Al (determined by the method of Tamm 1922), are amorphous (poorly ordered), mainly as a result of the modern weathering process.

The dithionite extractable form of Fe (determined by the method of Mehra-Jakson 1960) is the total content of free iron compounds, including both amorphous and crystalline.

Pyrophosphate extractable forms of Fe, Al, and Mn (determined by the method of Bascomb 1968) are organic matter bound compounds.

The content of amorphous forms of Fe and Al is favored by higher soil moisture and organic matter content. The negative influence of these forms on soil aggregate water stability was described by their relationships with main Bulgarian soils, including pseudopodzolic ones (Dilkova et al. 1999).

Table 4.2 Content of extractable forms of Fe, Al, and Mn (as % of the total content), author Jokova M

Horizon	Depth (cm)	[a]Fe_t (percentage of soil)	[a]Al_t (% of soil)	Mn_t (% mg/kg soil)	Fe_d (%)	Fe_o (%)	Fe_p (%)	Fe_c (%)	Fe_t/clay	Al_o (%)	Al_p (%)	Mn_d (%)	Mn_p (%)
Light gray pseudopodzolic forest soil, *Luvic Endogleyic Planosol* (*Albic, Dystric*), dystrict Troyan, village Oreshak													
Ap	0–16	5.09	8.28	2525	21.21	22.21	6.09	0.00	0.33	3.62	8.33	–	6.60
AE	16–29	4.91	10.94	1659	23.01	20.77	5.09	2.24	0.35	2.65	5.30	91.50	4.58
E	29–37	5.11	6.99	1130	23.48	19.96	8.41	3.52	0.23	3.48	20.31	74.69	3.03
EB	37–48	5.31	9.06	386	25.61	9.98	12.05	15.63	0.13	2.09	24.39	65.02	6.10
Btg1	48–62	6.17	13.70	262	32.73	8.42	11.83	24.31	0.14	1.45	14.37	43.89	0.00
Btg2	62–98	6.30	11.89	384	32.53	8.88	10.15	23.65	0.15	1.76	13.70	54.16	0.00
Btg3	98–136	6.24	15.50	845	38.46	12.33	6.08	26.13	0.15	1.35	6.12	74.31	0.00
BC	136–170	5.78	13.77	757	39.10	9.51	2.54	29.59	0.16	1.30	2.54	63.82	–
Cinnamonic pseudopodzolic forest soil, *Luvic Endogleyic Planosol* (*Albic, Dystric, Chromic*), dystrict Burgas, town Primorsko													
Ap	0–20	3.60	10.76	696	32.77	10.56	3.05	22.21	0.16	2.50	1.67	64.37	13.02
AE	20–30	4.28	10.26	606	27.80	7.47	2.10	20.33	0.14	2.53	1.46	56.73	7.52
Bt1	30–41	5.45	13.14	494	23.48	7.88	14.31	15.60	0.13	2.96	14.45	35.13	7.90
Bt2	41–82	5.35	11.44	349	23.92	6.91	12.52	17.01	0.12	2.97	13.28	50.00	16.72
BC1	82–115	4.91	11.53	222	23.21	3.86	2.64	19.35	0.13	1.73	2.60	85.32	11.90
BC2	115–136	3.58	10.52	262	25.97	2.51	0.83	23.46	0.34	1.23	1.23	87.86	–
Cinnamonic pseudopodzolic forest soil, *Luvic Endogleyic Planosol* (*Albic, Dystric, Chromic*), dystrict Plovdiv, village Zlatosel													
A	0–8	1.94	10.43	600	34.53	10.99	3.09	23.54	0.26	0.86	0.96	42.00	13.33
AE	8–22	2.29	10.32	600	37.55	12.22	6.99	25.33	0.35	1.55	3.20	30.00	7.00
Eg	22–33	4.86	14.67	440	24.07	6.17	3.50	17.90	0.61	1.16	2.25	–	2.05
Btg1	33–60	4.46	14.67	520	30.04	7.39	10.76	22.65	0.11	2.79	9.34	11.54	6.72
Btg2	60–96	3.92	11.87	600	29.59	5.35	2.30	24.24	0.09	2.27	1.09	44.00	10.00
BCkg	96–110	3.83	11.91	640	21.93	2.34	3.13	19.59	0.08	1.68	2.77	28.12	–
Ck	110–159	3.80	13.23	680	12.10	2.11	2.37	9.99	0.11	1.28	2.04	24.70	29.41
Yellow pseudopodzolic forest, Cutanic Acrisol (Albic), village Bulgari, dystrict Burgas													
A	0–3	3.86	9.68	3649	40.11	16.58	8.29	23.83	0.29	2.47	3.30	–	–
E	3–20	4.05	8.15	2512	47.90	15.55	9.38	32.35	0.28	3.55	9.06	100.0	–
EB	20–30	4.21	10.23	2355	47.74	13.06	6.71	34.68	0.22	2.73	4.59	93.71	–
Btg1	30–42	4.79	10.13	1576	49.89	11.69	5.21	38.20	0.17	2.56	3.45	86.73	–
Btg2	42–60	6.63	14.33	633	48.11	7.08	3.46	41.03	0.15	2.86	3.17	58.92	–
CR	60–79	6.56	14.00	592	42.68	7.62	2.74	35.06	0.19	2.78	3.21	84.45	–

Confidence interval at 0,05 level of Fe, Al, and Mn is: 0.06 %, 0.06 %, and 61 mg. kg^{-1}
d dithionite extractable, *o* oxalate extractable, *p* pyrophosphate extractable, *c* crystalline
[a] expressed as R_2O_3, *t* total content

The ratios between the extractable forms mainly depend on their chemical and physical characteristics, as well as on the soil moisture and temperature regimes under which the soil profiles were formed. Therefore, these forms are influenced by paleo and modern processes that have occurred during the soil development: weathering, reduction, complex formation with organic matter, migration, and clay formation.

The content of the extractable forms, expressed as a percentage of the total content of elements (relative content) indicates the parent material transformation during the soil processes and is useful for comparison of their stages.

Pseudopodzolic soils are the richest Bulgarian soils in terms of the free compounds of Fe, Al, and Mn (as relative content), where the processes are more advanced.

The metallic character of the studied elements increases in the range Al < Fe < Mn. The activity of the inclusion of Al released compounds in the clay mineral lattices is the highest, followed by that of Fe and Mn. It decreases with the increase in the metallic character of these elements and with the increase in the radius of the ion spheres.

Pseudopodzolic soils are periodically surface waterlogged in different degrees according to the depletion of oxygen in the soils and the release of electrons by plants roots and other organisms as a result of reduction processes (Lindsay 1979).

In pseudopodzolic soils alternate changes in moisture, temperature, and reduction status are observed. This has led to more intensive processes such as destruction of primary minerals, reduction of the easily reducible elements, oxidation and migration of the released compounds, and clay fraction. The extractable compounds are reactants of the reduction processes (Kanivets 1987).

Manganese is a more easily reducible element than Fe. The distributions of the extractable forms of Mn are the most indicative of the occurrence of the gley and migration processes. In the pseudopodzolic soils in the zone between Bt and the upper horizons the reduction of Mn has occurred and the obtained Mn^{2+} compounds are more soluble. They have moved to the lower but mainly to the upper part of the profile. With the soil development the maximum corresponding to the surface horizons and the minimum corresponding to the gley horizons are better expressed. The minimum of Mn_t, and especially of the dithionite and oxalate extractable forms corresponding to the middle part of the soil solum, show the gley horizons (Jokova 1994).

Iron is in the middle place in the range of the relative content of the extractable forms and the depth functions are less indicative of the reduction and clay formation processes than those of Mn and Al. As a result of the intensity and duration of the weathering and soil formation processes, the content of the dithionite extractable free Fe compounds depends on the soil development stages and total iron content (Fe_t) of the parent materials. The amorphous (poorly ordered) iron compounds have gradually crystallized over time. The content of the crystalline form is characterized by higher temperature, lower moisture, and low organic matter content. The ratio between amorphous and crystalline forms is a result of the pedoclimate conditions and development stages. As the processes of the release of Fe and crystallization of the obtained compounds have lasted longer, this ratio is higher (Alexander 1985).

The content of the organic matter bound forms of Fe (pyrophosphate extractable), as well as those of Al and Mn, decreases with depth in most soils. In the profile of pseudopodzolic soils the maximum of their content is apparent at the Bt horizons due to the active migration of the organic matter substances downward. The distributions of the pyrophosphate extractable forms along depth illustrate the migration processes.

The oxalate extractable forms, which have resulted from contemporary weathering and soil forming processes, are accumulated at the surface horizons of most soils, as are the dithionite extractable forms. In pseudopodzolic soils with more advanced processes of migration and crystallization most of the dithionite extractable forms and crystalline Fe are at the Bt horizons (Jokova and Boyadzhiev 1993).

Iron and manganese compounds reflect the soil color, but the contribution of the different iron forms is higher because their amounts and specific surface are much higher than those of Mn. More than 70 % of the free iron compounds are crystalline, related to the red soil color. Hematite gives pink to bright red color (finely divided—the color is more intensive), goethite gives brown and dark reddish brown color (finely divided—the color grades toward yellowish).

The surface horizons of light gray (pseudopodzolic) soils are richer in amorphous forms of Fe and poorer in crystalline forms of Fe than the horizons of the cinnamonic podzolic (peudopodzolic) soils and especially of the zheltozem podzolic (pseudopodzolic) ones.

The increase in pH during the gley processes is due mainly to the reduced amounts of Fe and Mn, but the part of Fe^{2+} is much higher than the parts of Mn^{2+}. After oxidation of the reduced compounds Fe–Mn concretions have been formed and their amounts are higher in soils with more advanced soil development. These concretions are much richer in Co, Pb, Cu, Zn, Ni, P than those found in other soils due to the binding role of Mn (Jokova and Dilkova 1997).

Because of the high activity of inclusion of Al in the clay mineral lattices, the processes of formation, transport, and accumulation of the clay fractions are better described by the depth functions of the respective extractable forms. The contents of the dithionite and oxalate extractable compounds of Al are almost identical. They are accumulated in the richest with clay fraction horizons. The maximum

Table 4.3 Variables of the particle size distribution (mm) in 19 referenced profiles of light gray (pseudopodzolic) forest soils in Bulgaria

Profile count 19	Horizon	Upper depth	Lower depth	Loss HCl	Particle size distribution (mm) (%) (according to method of Kachinskiy revised)							
					>1 mm	1–0.25	0.25–0.05	0.05–0.01	0.01–0.005	0.005–0.001	<0.001	<0.01[a]
Mean	AEp	0	27	1.3	0.0	6.1	20.1	29.3	10.8	12.3	19.8	42.9
	Btg1	27	57	1.3	0.0	3.5	16.8	23.5	9.7	9.8	35.4	53.1
	Btg2	57	89	1.6	0.0	2.9	14.3	20.7	8.4	9.0	43.8	61.2
	Btg3	89	126	1.7	0.0	2.7	17.7	19.5	9.1	9.3	40.5	58.9
	BC	126	162	6.6	0.2	4.2	15.2	20.6	8.4	9.4	35.1	53.2
Stdev	AEp	0	4	0.7	0.0	7.8	7.3	7.6	2.5	2.4	4.2	7.0
	Btg1	4	7	0.9	0.0	3.6	7.0	7.6	2.9	2.8	14.2	11.0
	Btg2	7	13	1.0	0.0	3.2	6.0	4.6	2.3	2.7	8.3	7.5
	Btg3	13	17	1.1	0.1	3.4	7.3	3.9	3.5	2.9	7.6	6.8
	BC	17	27	8.0	1.0	7.1	8.0	5.5	3.0	2.9	8.6	9.9

[a] Physical clay, the parameter characterized sum of the particle-size distribution, according to Kachinskiy method (mm) $\sum < 0.01 = < 0.001 + (0.005 - 0.001) + (0.01 - 0.005)$

corresponds to Bt horizons of the pseudopodzolic soils, as well as the maximum of the pyrophosphate extractable compounds of Al (Jokova and Iolevsky 1988).

The influence of tillage on the distribution of the extractable forms is due mainly to the higher permeability of the arable layers. Fe and Mn depth functions are more affected than Al ones. The arable layers of the most cultivated soils are richer in the dithionite extractable forms of Fe_d than the surface horizons of pristine soils, which is associated with the increase of iron oxalate extractable form Fe_o in most soils, i.e., tillage makes more iron compounds available to plants. In most cultivated soils Mn extractable compounds content increases in the arable layers and decreases in the gley horizons more strongly than in the horizons of virgin soils.

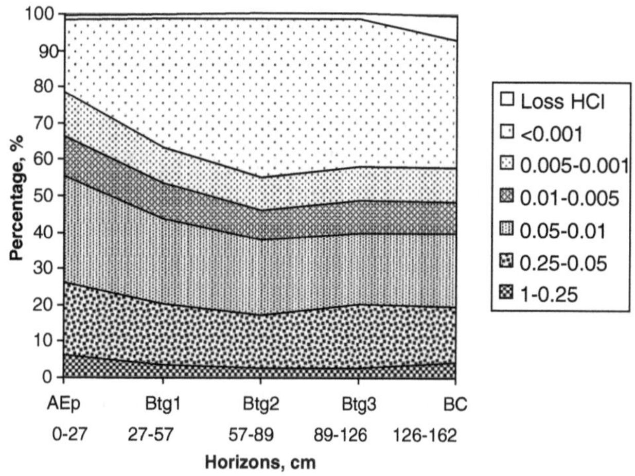

Fig. 4.3 Pattern of the particle size distribution (mm) in light gray (pseudopodzolic) soil (according to revised Kachinskiy method)

4.3.4 Pseudopodzolic Soil Subtypes in Bulgaria

4.3.4.1 Light Gray (Pseudopodzolic) Forest Soils

Light gray (*subtype*) **(pseudopodzolic) forest** (*type*) soil is named after the Bulgarian classification of Koynov et al. (1964), and Yolevski and Hadzhiyanakiev (1976).

Luvic Endogleyic Planosol (Albic, Dystric) is named according to WRB (2006) and *Typic Albaqualfs* according to Keys to Soil Taxonomy (2010).

Verti-Eutric Planosol, *Verti-Dystric Planosol:* Revised Legend of the World Soils (FAO-UNESCO-ISRIC 1990).

Average thickness of the eluvial layer (AE + E) is 15–35 cm, and average thickness of the Bt textural horizon exceeds 1 m.

The particle size distribution shows that the clay fraction (<0.001 mm) prevails followed by the coarse silt fraction (0.05–0.01 mm) and fine sand (0.25–0.05 mm) (Table 4.3 and Fig. 4.3).

The distribution of the clay fraction shows a sharp increase at the Bt horizon (Tables 4.3, 4.4 and Figs. 4.3, 4.4). Most cutans are observed on ped surfaces, pore spaces, and cracks in this horizon. Below 150 cm at the parent materials where weathering in situ takes place, the amount of clay is less.

The qualitative composition of clay minerals, determined by X-ray diffraction analysis and transitional electronic microscopy, has shown montmorillonite at rates of 50 % in the clay (<0.001 mm) at the surface (AE + E) horizons, gradually increasing at the (Btg) horizon to about 54 % in the clay, and 59 % in the clay in the (C) horizon. Hydrated micas content is at rates of 20 % in the clay at the surface (AE + E) horizon, gradually increasing at the depth, where in the (C) horizon it is about 26 % in the clay. Kaolinite is about 19 % in the clay, gradually decreasing at the depth, where at the (C) horizon it is 13 % in the clay. There was a

Table 4.4 Variables of humus, pH, carbonates, and particle size distribution (mm) according to USDA in 19 referenced profiles of light gray (pseudopodzolic) forest soils in Bulgaria

Profile count 19	Horizon	Upper depth	Lower depth	Humus (%)	pH_{KCl}	Car-bona-tes	Bulk density g.cm^3 calc	Clay <0.002	Silt 0.002–0.05	Fine sand 0.05–0.2	Total sand 0.05–2.0
Mean	AEp	0	27	1.9	4.6	–	1.3	26.7	46.7	20.2	26.3
	Btg1	27	57	0.8	4.4	–	1.4	41.0	38.5	16.9	20.4
	Btg2	57	89	1.1	4.6	–	1.4	49.1	34.1	14.5	17.3
	Btg3	89	126	0.6	5.0	1.1	–	46.1	33.9	17.8	20.5
	BC	126	162	0.5	5.9	11.3	–	43.5	36.3	15.7	20.0
Stdev	AEp	0	4	0.4	0.7	–	0.0	4.9	9.3	7.3	11.7
	Btg1	4	7	0.3	0.5	–	0.1	13.5	10.1	7.0	8.4
	Btg2	7	13	2.2	0.6	–	0.2	8.0	6.7	6.0	8.0
	Btg3	13	17	0.2	0.8	–	–	7.5	6.1	7.3	8.1
	BC	17	27		1.3	10.5	–	8.2	8.0	7.8	13.8

significant content of quartz at rates of 11 % in the clay at the surface (AE + E) horizons, with a sharp decrease at the depth, where at the (C) horizon it is 2 % in the clay (Boneva et al. 1986).

Clay minerals distribution in the profile manifested differentiation similar to the clay distribution. Mineral association at the illuvial part (Btg) is over 80 % in the clay consisting of montmorillonite and illite. This mineral composition determined many soil properties.

The average particle density is 2.45–2.50 g/cm^3 at the surface and 2.65–2.72 g/cm^3 at the illuvial horizon. Bulk density is near 1.2 g/cm^3 in the upper part and about 1.5–1.6 g/cm^3 in the middle part of the profile, where there is evidence of compaction. When dried, textural Bt horizon is very hard in consistence.

Water stable aggregates with size >1 mm are about 50–65 % at the surface (0–15 cm) and sharply decrease to 30–45 % at the depth of 20–30 cm in pristine land. Prevailing water stable aggregates with size over 0.25 mm are over 60 %.

In the light gray (pseudopodzolic) soil 15–25 % of pore space is represented by macropores or the pores of aeration (with diameter >300 μm); and 15–25 % is mesopores or pores with rapid moisture drainage (>30 μm diameter). Micropore space or water-filled is about 20–25 %, represented by slow drainage pores (diameter 0.2–30.0 μm) and 10–15 % of micropores is with none remaining available for roots moisture (diameter <0.2 μm), (Dilkova 1985; Dilkova et al. 1998).

Total porosity is high at the surface (near 55 % at field capacity) and decreases with depth. Soils are characterized by low water permeability.

Naturally, the spatial relationship of light gray (pseudopodzolic) forest soils is with gray-brown and brown forest soils and also with dystrict, eutric regosols.

The content of humus is 1.0–2.5 %, where the sharp decrease is below the upper 10 cm (Fig. 4.5). Prevailing fulvic acids are bounded with sesquioxides (Boyadgiev et al. 1994). Low molecular organic acids are strong complexes of aluminum and iron and are mobile.

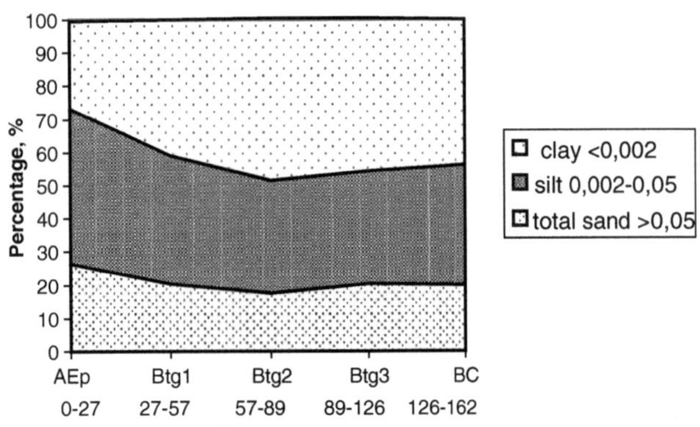

Fig. 4.4 Pattern of the particle size distribution (mm) in light gray (pseudopodzolic) soil (USDA)

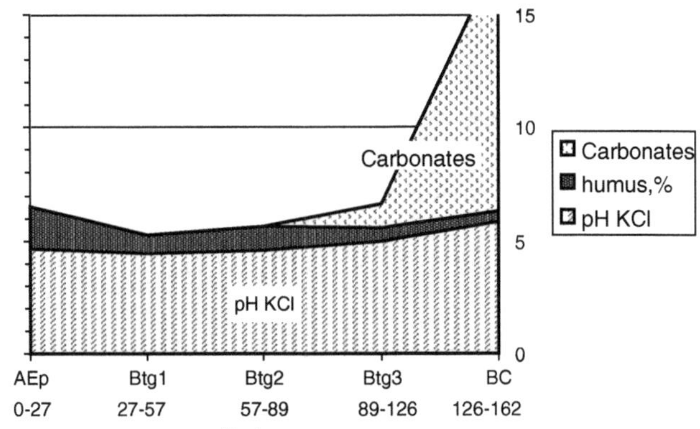

Fig. 4.5 Pattern of pH, humus, and carbonates distribution in light gray (pseudopodzolic) soil

Table 4.5 Variables of the particle size distribution (mm) in 29 referenced profiles of cinnamonic podzolic (pseudopodzolic) forest soils (eutric) in Bulgaria

Profile count 29	Horizon	Upper depth	Lower depth	Loss HCl	Particle-size distribution (mm) (%) (according to method of Kachinskiy revised)							
					>1 mm	1–0.25	0.25–0.05	0.05–0.01	0.01–0.005	0.005–0.001	<0.001	<0.01[a]
Mean	AEp	0	26	1.0	2.4	18.3	22.9	22.1	8.4	8.9	16.0	32.6
	Bt1	26	49	1.3	2.8	14.6	18.0	16.9	6.2	7.6	32.5	46.7
	Bt2	49	75	1.5	1.6	11.9	15.4	14.3	5.1	6.8	43.4	55.3
	Bt3	75	102	1.9	2.9	13.2	17.3	14.4	5.6	6.5	38.1	50.2
	Bg	102	129	3.8	4.5	13.6	17.0	14.1	5.4	7.1	34.4	46.8
	Ck	129	158	7.6	5.3	15.6	17.7	14.1	5.9	6.0	27.9	39.8
Stdev	AEp	0	4	0.7	5.6	11.1	8.3	8.3	3.5	3.1	5.7	8.8
	Bt1	4	8	0.9	4.7	10.2	4.6	7.3	3.3	3.2	15.9	14.1
	Bt2	8	12	0.7	2.8	7.8	2.8	5.5	2.2	2.7	6.8	7.2
	Bt3	12	18	2.0	5.2	8.9	5.5	5.6	2.2	2.3	7.5	9.0
	Bg	19	23	4.6	8.2	9.0	5.5	5.7	2.3	3.1	9.3	11.6
	Ck	24	32	8.3	9.6	12.0	5.6	6.5	2.9	2.5	13.1	14.9

[a] Physical clay, the parameter characterized sum of the particle-size distribution, according to Kachinskiy method (mm) $\sum < 0.01 = <0.001 + (0.005 - 0.001) + (0.01 - 0.005)$

The stock of organic matter in a 100 cm layer is 100–150 ton per ha in forest land and is much less in arable land. The upper 10–15 cm of soil store more than 80–90 % of the total amount of organic matter.

In the cation exchange capacity exchangeable H^+ (50–60 %) and Al^{3+} prevails at the upper eluvial layer, and decreases with depth.

The cation exchange capacity is <50 %, and is 5–15 meq/100 g in the upper layer, and is 75–90 % or about 30 meq/100 g at the depth.

The primary calcium carbonates have been leached out from the soil profile. The soil reaction is acid at the surface pH_{KCL} 4.0, strong acid pH_{KCL} 3.3–3.7 in the illuvial horizon, followed by an increase at the parent materials.

4.3.4.2 Cinnamonic Podzolic (Pseudopodzolic) Forest Soils

Cinnamonic (*subtype*) **podzolic (pseudopodzolic) forest** (type) soil is named after the Bulgarian classification of Koynov et al. (1964), and Yolevski and Hadzhiyanakiev (1976).

Luvic Endogleyic Planosol (Albic, Dystric, Chromic) is named according to WRB (2006) and *Vertic Albaqualfs* according to Keys to Soil Taxonomy (2010).

Verti-Eutric Planosol, Verti-Dystric Planosol: Revised Legend of the World Soils (FAO-UNESCO-ISRIC 1990).

Soils are formed by the active process of lessivage and pseudogley in the upper layer and weathering in situ, oxidation and reduction in the middle part.

Average thickness of the eluvial layer (AE + E) is 15–35 cm, and average thickness of the illuvial Bt horizon exceeds 1 m.

The particle size distribution shows that the clay fraction (<0.001 mm) prevails followed by the fine sand fraction (1.25–0.05 mm) and coarse silt fraction (0.05–0.01 mm) (Tables 4.5, 4.6 and Figs. 4.6, 4.7).

Clay content sharply increases to a maximum at the illuvial Bt horizon (Tables 4.7, 4.8 and Figs. 4.8, 4.9). Most cutans are observed on ped surfaces, pore spaces, and cracks in this horizon. Below the depth of 150 cm clay decrease is significant at the parent materials, where the weathering in situ takes place.

The average particle density is 2.50–2.6 g/cm^3 at the surface and is near 2.65 g/cm^3 at the illuvial horizon. Bulk density is near 1.2–1.3 g/cm^3 in the upper part and is about 1.5–1.7 g/cm^3 in the middle part of profile, where there is evidence of compaction. The dried textural Bt horizon is very hard in consistence.

Water stable aggregates with size >1 mm are about 40–60 % at the surface (0–15 cm) and increased at the depth in pristine land. Prevailing water stable aggregates with size over 0.25 mm are 20–40 % at the surface.

In cinnamonic (pseudopodzolic) soil 6.5–20.5 % of pore space is represented by macropores or pores of aeration (with diameter >300 μm); and 5–25 % is mesopores or pores with rapid moisture drainage (>30 μm diameter). Micropore space or water-filled is about 10–20 %, represented by slow drainage pores (diameter 0.2–30.0 μm); and micropores with none remaining available for roots moisture (diameter < 0.2 μm) is 5–10 % at the surface and 15–30 at the illuvial horizon (Dilkova 1985; Dilkova et al. 1998).

Total porosity is near 40–50 % (at field capacity). Soils are low water permeable.

Table 4.6 Variables of the particle size distribution (mm) in 21 referenced profiles of cinnamonic podzolic (pseudopodzolic) forest soils (dystric) in Bulgaria

Profile count 21	Horizon	Upper depth	Lower depth	Loss HCl	Particle size distribution (mm) (%) (according to method of Kachinskiy revised)							
					>1 mm	1–0.25	0.25–0.05	0.05–0.01	0.01–0.005	0.005–0.001	<0.001	<0.01[a]
Mean	AEp	0	27	1.0	1.6	18.9	23.0	21.0	9.6	8.3	16.4	34.4
	Bt1	27	52	1.3	1.1	10.6	15.9	16.0	6.8	7.2	41.2	55.1
	Bt2	52	79	1.6	1.5	11.3	16.0	14.9	6.9	6.7	41.2	54.3
	Bg	79	122	1.7	2.7	13.0	17.3	15.2	6.3	6.9	36.9	50.1
	CR	122	153	5.6	3.3	14.4	20.2	15.0	6.2	6.8	28.6	41.5
Stdev	AEp	0	3	0.6	4.2	13.8	7.2	7.4	4.9	3.8	7.8	13.1
	Bt1	4	8	0.7	2.6	7.4	4.6	8.3	3.3	3.9	11.9	9.4
	Bt2	8	11	1.0	2.9	8.9	5.8	5.0	3.9	4.6	12.1	10.8
	Bg	11	22	1.0	4.7	10.4	7.1	5.3	2.7	3.1	10.4	13.5
	CR	22	28	6.1	7.2	15.2	9.6	6.3	4.1	4.0	11.1	15.6

[a] Physical clay, the parameter characterized sum of the particle-size distribution, according to Kachinskiy method (mm) $\sum < 0.01 = < 0.001 + (0.005 - 0.001) + (0.01 - 0.005)$

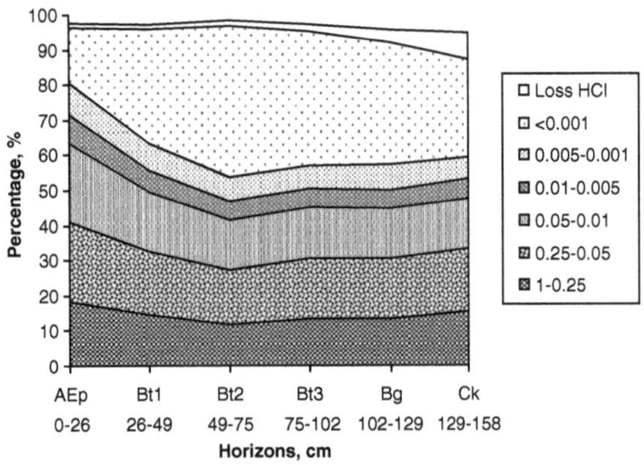

Fig. 4.6 Pattern of the particle size distribution (mm) in cinnamonic podzolic (pseudopodzolic) forest soils (eutric), (according to revised Kachinskiy method)

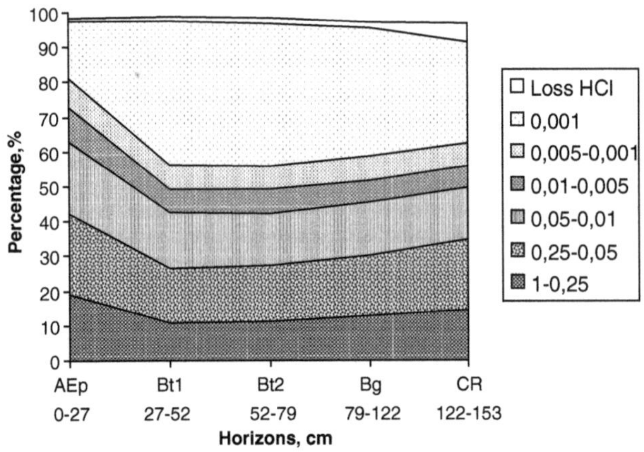

Fig. 4.7 Pattern of the particle size distribution (mm) in cinnamonic podzolic (pseudopodzolic) forest soils, dystric (according to revised Kachinskiy method)

In spatial distribution cinnamonic podzolic (pseudopodzolic) forest soils are bordered by the leached cinnamonic forest soils, brown forest soils and Dystric, Eutric Regosols.

The content of humus is less than 2 %, where sharp decrease is below 10 cm (Figs. 4.10, 4.11). Stock of organic matter in a 100 cm layer is below 100 t per ha in forest land and is less in arable land. The upper 10–15 cm of soil stores more than 80–90 % of the total amount of organic matter (Artinova 1977). Prevailing fulvic acids are bounded with sesquioxides. Low molecular organic acids are strong complexes of aluminum and iron and are mobile.

The process of the transformation of organic residues is accompanied by nitrogen enrichment of humic substances shown by the ratio C/N and varies from 10 to 16 which means medium to low rates. Soil reaction is acid pH_{KCl} 3.8 at the surface, strongly acid pH_{KCl} 3.0–3.5 in the illuvial horizon, followed by an increase at the parent materials.

In the cation exchange capacity the exchangeable Al^{3+} and H^{+} prevails in the eluvial layer and decreases at the depth. The cation exchange capacity is 30–60 % or about 5.0–20 meq/100 g at the eluvial layer and is 65–95 or about 20–40 meq/100 g in the illuvial (Bt) horizon (Ganev and Arsova 1980; Ganev et al. 1990).

The qualitative composition of clay minerals, determined by X-ray diffraction analysis and transitional electronic microscopy, has shown montmorillonite at rates of 23 % in the clay (<0.001 mm) and 5 % in the soil at the surface (AE + E) horizons, gradually increasing at the (Bt) horizon to about 53 % in the clay (<0.001 mm) and 26 % in the soil, and is 56 % in the clay and 22 % in the soil in the (C) horizon. Vermiculite shows significant content at rates of 33 % in the clay and 7 % in the soil at the surface (AE + E)

Table 4.7 Variables of humus, pH, carbonates, and particle size distribution (mm) according to USDA in 29 referenced profiles of cinnamonic podzolic (pseudopodzolic) forest soils (eutric) in Bulgaria

Profile count 29	Horizon	Upper depth	Lower depth	humus %	pH_{KCl}	Carbonates	Bulk density g.cm^3 calc	Clay < 0.002	Silt 0.002–0.05	Fine sand 0.05–0.2	Total sand 0.05–2.0
Mean	AEp	0	26	1.45	4.8	0.0	1.47	21.2	35.7	23.3	43.1
	Bt1	26	49	0.86	5.0	0.0	1.41	37.5	27.8	18.4	34.6
	Bt2	49	75	0.71	5.0	0.0	1.42	48.0	23.5	15.7	28.4
	Bt3	75	102	0.57	5.4	0.0	1.49	43.1	24.3	17.8	32.6
	Bg	102	129	0.47	6.0	0.0	–	40.9	25.1	17.8	34.0
	Ck	129	158	0.27	6.2	11.6	–	35.9	26.4	18.8	37.7
Stdev	AEp	0	4	0.60	0.6	–	0.11	65	11.7	8.2	15.5
	Bt1	4	8	0.37	0.7	–	0.08	15.4	11.3	4.7	14.7
	Bt2	8	12	0.24	0.7	–	0.06	6.8	7.8	2.9	97
	Bt3	12	18	0.14	0.6	–	0.09	7.6	7.9	5.5	12.0
	Bg	19	23	016	0.8	–	–	9.2	8.1	5.5	13.7
	Ck	24	32		1.0	–	–	14.0	8.9	5.5	16.5

Table 4.8 Variables of humus, pH, carbonates, and particle-size distribution (mm) according to USDA in 21 referenced profiles of cinnamonic podzolic (pseudopodzolic) forest soils (dystric) in Bulgaria

Profile count 21	Horizon	Upper depth	Lower depth	Humus (%)	pH_{KCl}	Carbonates	Bulk density g.cm^3 calc	Clay <0.002	Silt 0.002–0.05	Fine sand 0.05–0.2	Total sand 0.05–2.0
Mean	AEp	0	27	1.4	4.1	0.0	1.5	21.3	35.5	23.4	43.3
	Bt1	27	52	0.6	3.6	0.0	1.4	45.8	27.0	16.1	27.2
	Bt2	52	79	0.5	3.5	0.0	1.5	45.7	25.9	16.2	28.4
	Bg	79	122	0.5	4.1	0.0	1.5	40.5	25.1	18.1	34.4
	CR	122	153	0.3	5.8	0.0	–	35.7	27.0	20.9	37.3
Stdev	AEp	0	3	0.8	0.8	0.0	0.1	8.9	12.6	7.5	18.7
	Bt1	4	8	0.2	0.3	0.0	0.1	11.1	12.7	4.6	10.4
	Bt2	8	11	0.2	0.2	0.0	0.1	11.3	10.6	5.8	12.5
	Bg	11	22	0.1	0.5	0.0	0.1	10.5	8.6	7.4	15.4
	CR	22	28	–	1.4	0.0	–	11.5	11.4	9.4	18.5

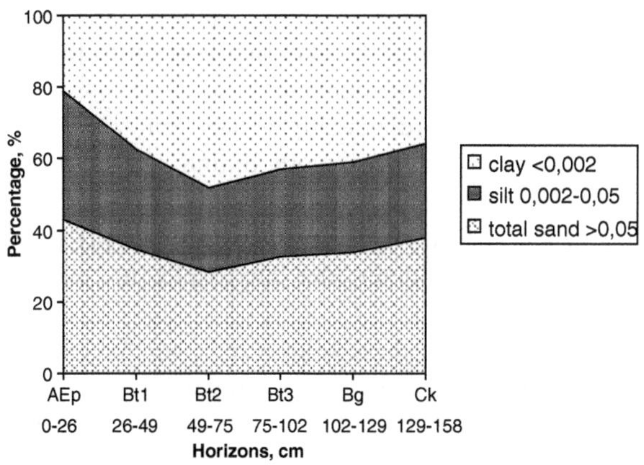

Fig. 4.8 Pattern of the particle size distribution (mm) in cinnamonic podzolic (pseudopodzolic) forest soils, eutric (USDA)

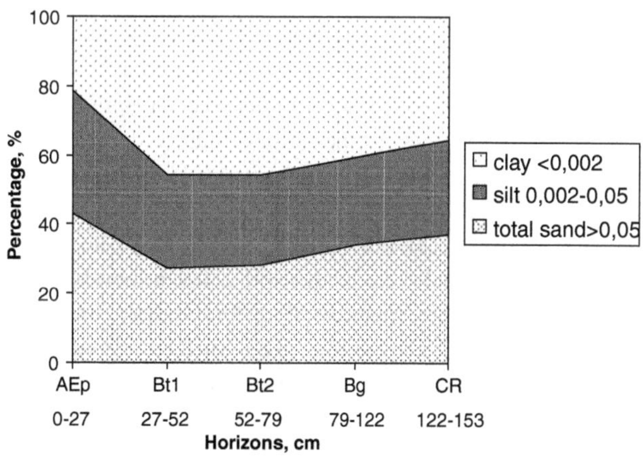

Fig. 4.9 Pattern of the particle-size distribution (mm) in cinnamonic podzolic (pseudopodzolic) forest soils, Dystric (USDA)

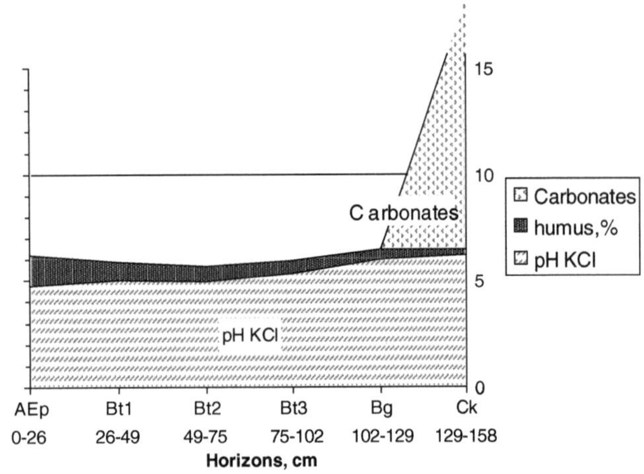

Fig. 4.10 Pattern of pH, humus, and carbonates distribution in cinnamonic podzolic (pseudopodzolic) forest soils, eutric

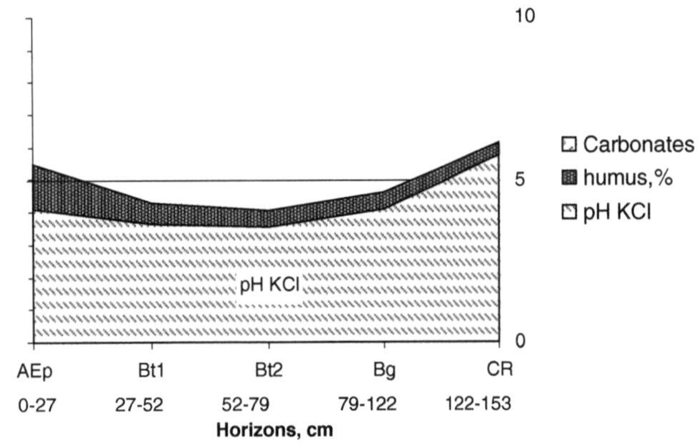

Fig. 4.11 Pattern of pH, humus, and carbonates distribution in cinnamonic podzolic (pseudopodzolic) forest soils, Dystric

horizons, gradually decreasing at the depth, where at the (C) horizon it is 29 % in the clay and 12 % in the soil. Hydrated micas content is at rates of 15 % in the clay and 3 % in the soil at the surface (AE + E) horizon, gradually decreasing at the depth, where in the (C) horizon it is about 7 % in the clay and 3 % in the soil. Kaolinite is about 23 % in the clay (<0.001 mm) and 5 % in the soil, gradually decreasing at the depth, where at the (C) horizon it is 6 % in the clay and 2 % in the soil (Halil et al. 1990). Other resistant minerals like fine quartz occur only at the surface (AE) horizon—6 % in the clay and 1 % in the soil. Clay minerals distribution in the profile manifested differentiation similar to that of the clay distribution. The surface eluvial part (AE + E) is dominated by the presence of mixed layers clay minerals, of two or more different minerals like montmorillonite-kaolinite-vermiculite. However the mineral association of pure montmorillonite-vermiculite at the illuvial part (Btg) is a significant 73–77 % in the clay and 37–38 % in the soil. This mineral composition determines many soil properties and resembles that which occurs in the group of Vertisols.

4.4 Soil Properties and Soil Management

4.4.1 Soil Properties of Pseudopodzolic Soils

Pseudopodzolic forest soils often have high water-holding capacity and low infiltration rates. Commonly, waterlogging occurs at the surface (Dilkova et al. 2002). Soils are often seasonally in aquic conditions.

Pseudopodzolic forest soils are characterized by sharp texture change at the subsoil. The physical precipitation of clays is in the middle part. Chemical weathering in situ takes place during the whole year in wet conditions, and is naturally less prominent in freely drained soils.

The amount of water stable aggregates >1 mm is low, characterizing the soils as vulnerable to structure destruction.

The chemical characteristic of the eluvial layer (AE + E) is acid with low presence of organic matter (<2 %) being of moder type. Organic matter is generally in the upper 10–15 cm of the soil. The loss of bases is due to leaching. In the adsorbed cations the compositions H^+ and Al^{3+} prevail. Soil reaction is acid with value pH_{KCl} 3.9–4.5.

4.4.2 Soil Management

Natural pseudopodzolic soils have low fertility because of their physical and chemical properties. The soils are mostly used for forestry or rough grazing. Most of them are waterlogged.

Erosion strongly affects the surface horizons resulting in loss of structure. In the past, these soils were called by their local folk names. For example, at sites where the surface horizon is preserved they were named "spongy soil," and at sites where erosion brings the clay-rich horizon to the surface they were named "sodden soil."

Cultivated soils are commonly located on smooth terrains, and generally require more liming and fertilizers containing nitrogen, phosphorus, potassium, and possibly other elements. Their low nutrient status and low pH values make these soils difficult or impossible to use for agriculture.

The nutrient pool of these soils is too poor for growing most plants. Soils are suitable for growing potatoes and other acid-loving crops with deep penetrating roots such as winter ray, oats (spring), alfalfa, and a grass-legume mix. Due to bad soil nutrition and late dates of summer harvesting, spring wheat is not suitable.

Agricultural practices like ploughing may cause degradation leading to loss of nutrients. The arable soil surface horizon has been homogenized by regular cultivation.

These soils have a natural high water-holding capacity, and the main problem in agricultural practice is air filled porosity of the subsoil.

References

Alexander EB (1985) Estimating relative ages from iron oxide total iron ratios of soils in the Western Po valley, Italy. Geoderma 35:257–259

Antipov-Karataev IN, Gerassimov IP (1948) The soils of Bulgaria. Zemsnab Press, Sofia (In Bulgarian)

Antipov-Karataev IN, Galeva V, Gerassimov IP et al (1960) The soils in Bulgaria. Zemizdat Press, Sofia (In Bulgarian)

Arinushkina E (1970) Manual on chemical soil analysis. Moscow State University Press, Moscow (In Russian)

Artinova N (1977) Organic matter in pseudopodzolic soils in Bulgaria. Agricultural University, Plovdiv, p 178 (In Bulgarian)

Bascomb CL (1968) Distribution of pyrophosphate extractable iron and organic carbon in soils of various groups. J Soil Sci 19(1):251–268

Boneva K, Houbenov G, Behar A (1986) Chemical and mineralogical composition of soils of the Central Balkan Mountains and the Eastern Rhodopes. Soil Sci Agrochem Plant Prot J XXI(6):3–11 (In Bulgarian)

Bondev I (1991) Vegetation in Bulgaria, map at scale 1:600 000 with explanatory text. St. Kliment Oxridski University Press, Sofia (In Bulgarian)

Boyadgiev T, Filcheva E, Petrova L (1994) Organic matter composition of the main soil groups in Bulgaria. In: Senesi N, Miano TM (eds) Humic substances in the global environment and implications on human health. Elsevier Science B V, Amsterdam, pp 535–540

Dilkova R (1985) Structure and aeration in main soil varieties in Bulgaria. Theses, Institute of Soil Science Nikola Poushkarov, Sofia (In Bulgarian)

Dilkova R, Kerchev Zh, Kercheva M (1998) Evaluating and grouping of soils according to their susceptibility to anthropogenic degradation, Advances in GeoEcology, 31. CATENA VERLAG, Reiskirchen, pp 125–131

Dilkova R, Kerchev G, Jokova M et al (1999) Physical characteristics and agronomic potential of red soils in Bulgaria. In: Extended abstracts soils with Mediterranean type of climate. 6th international meeting, Barcelona, Spain, 4–9 July 1999, pp 28–30

Dilkova R, Filcheva E, Kercheva M et al (2002) Changes of organic matter and aggregate stability of arable waterlogged soils. In: Kimble J et al. (eds) Agricultural practices and policies for carbon sequestration in soil. Lewis Publishers, Crc Press LLC, USA, Chapter 30, pp 315–319

Duchaufour Ph (1951) Lessivage et podzolisation. Revue Forestiere Francaise, Paris, pp 647–652

Dyudal R (1972) Ninety years since the introduction of term « podzolic soils ». In: First national soil science congress, Bulgarian Academy of Sciences Press, Sofia, 23–25 Sept 1969, pp 87–101

FAO-UNESCO-ISRIC (1990) Revised legend of the soil map of the world. FAO World Soil Resources Report No 60, Rome

Fridland VM (1958) Podzolization and illimerization. Soil Sci 1:27–38 (In Russian)

Ganev S, Arsova A (1980) Methods of determining the strongly acid and the slightly acid cation exchange in soil. Soil Sci Agrochem J XV(3):22–33 (In Bulgarian)

Ganev S et al (1990) Physic-chemical properties of soils in Bulgaria and other world's regions. Department of chemistry, Institute of Soil Science Nikola Poushkarov, Sofia (In Bulgarian)

Gerasimov IP (1959) Gleyic pseudopodzols in Central Europe and the formation of bi-component sediment cover. Izvestia of Academy of Sciences of USSR, series Geography, 3:20–30

Glinka KD (1924) Die degradation und der podzolige prozess. Int Mitt Bodenk 14:40–49

Gyurov G et al (1978) Some problems of genesis and classification of gray forest soils, brown forest and pseudopodzolic soils in Bulgaria. Second Scientific Conference on Soil Science, Sofia (In Bulgarian)

Hadzhiyanakiev A (1989) Iron in Bulgarian soils. Zemizdat Press, Sofia (In Bulgarian)

Halil R, Boneva K, Houbenov G (1990) Investigations on the composition of clayic minerals in some Bulgarian soils. Soil Sci Agrochem J XXV(4):13–22 (In Bulgarian)

Jackson MS (1958) Soil chemistry analysis. Prentic-Hall Inc., Engelowood Cliffs

Jokova M, Iolevsky M (1988) Forms of aluminium compounds in some soils of Bulgaria. Soil Sci Agrochem J 23(1):69–74 (In Bulgarian)

Jokova M, Boyadzhiev T (1993) Forms of compounds of iron, aluminium and manganese in some Bulgarian soils. J For Sci 30(2): 62–72 (In Bulgarian)

Jokova M (1994) Manganese of some Bulgarian soils—dependence of C Humic: C Fulvic acids on the content of manganese. In: Lal R, Kimble J, Lavine E (eds) Soil processes and greenhouse effect. USDA, Soil Conservation Service, National Soil Survey Centre, Lincoln, NE, 3, pp 24–30

Jokova M, Dilkova R (1997) Influence of tillage on iron, aluminium, and manganese distribution along depth of soils. In: Proceedings 14th ISTRO conference, vol 2A, July 27 - August 1, 1997, Pulawy, Poland, pp 319–322

Jokova M, Boyadzhiev T (1998) Extractable compounds of Fe, Al, and Mn in Bulgarian textural differentiated soils. Bulgarian J Agric Sci 4(1):17–24

Jokova M (1999) Extractable compounds of Fe, Al, and Mn as participants and indicators of soil processes. J Soil Sci Agrochem Ecol J 34(4/5):95–102 (In Bulgarian)

Kachinskiy NA (1965) Soil physics, part I and II. High school Press, Moscow (In Russian)

Kanivets V (1987) Soil Sci J N(11) (In Russian)

Kaurichev I (1972) General characteristics of the semihydromorphic soils with eluvial horizon. In: First national soil science Congress, Bulgarian Academy of Sciences Press, Sofia, 23–25 Sept 1969

Keys to Soil Taxonomy (2010) USDA United States Department of Agriculture, NRCS Natural Recourses Conservation Service, 11th edn. Soil Survey Staff, Washington, DC, p 338

Kononova MM (1963) Soil organic matter. Its nature, its role in soil formation and in soil fertility. Soviet Academy of Sciences Press, Moscow

Koynov V (1964) Peculiarities of the main Bulgarian soil types and their basic soil-genetic processes. In: Gerasimov IP et al. (ed) Soils of south–eastern Europe. Proceedings of the international symposium of soil science, June 1963, Sofia. Bulgarian Academy of Sciences Press, Sofia: 15–62 (In Russian)

Koynov V et al (1964) Classification and systematic of soils in Bulgaria. In: Gerasimov IP et al. (ed) Soils of south–eastern Europe. Proceedings of the international symposium of soil science, June 1963, Sofia. Bulgarian Academy of Sciences Press, Sofia: 231–246 (In Russian)

Kubiena WL (1953) The soils of Europe, illustrated diagnosis and systematics. Thomas Murby & Co, London, p 314

Lindsay WL (1979) Chemical equilibrium in soils. Wiley, New York, pp 128–160

Mehra OP, Jackson ML (1960) Iron oxide removal from soils and clays by dithionite—citrate—system buffered with sodium bicarbonate. Clays Clay Mineral 7:317–327
Penkov M et al (1992) Soil classification and diagnostic in Bulgaria in relation to the land reform fragmentation. Zemizat Press, Sofia
Rode AA (1937) Podzol process development. Soviet Academy of Sciences Press, Moscow
Rode AA (1964) On question of podzolization and lessivage. J Soil Sci 7:23–33 (In Russian)
Rousseva SS (1997) Data transformations between soil texture schemes. Eur J Soil Sci 48:749–758
Shishkov T (2011) Implementation of World Reference Base and Soil Taxonomy within the framework of Bulgarian soil classification. In: International conference "100 years Bulgarian Soil Science", Sofia, 16–20 May 2011
Tamm O (1922) Eine methode zur bestimung der anorganischen komponente ds gelcomplexes im boden. Statens Skogsforsoksanstalt Medd 19:387–404 (by Arinoushkina E (1972) Manuel of chemical analysis of soils, Moscow)
World Reference Base for Soil Resources (2006) A framework for international classification, correlation and communication. FAO-ISRIC, Rome
Yolevski M, Hadzhiyanakiev A (1976) Agro-grouping of soils in Bulgaria, extended systematic list. Code of groups and soil varieties. Institute of Soil Science Nikola Poushkarov, Sofia
Yolevski M, Ninov N, Krastanov S et al (1983) Brief diagnostic and criterion of the main taxonomic soil divisions in Bulgaria. Precise Classification Scheme, Institute of Soil Science Nikola Poushkarov, Sofia
Zonn SV (1973) Environmental settings of the processes of lessivage, pseudo-podzolization and podzolization during the Quaternary period in the western and northwestern regions of the USSR. J Soil Sci 116:211–217
Zonn SV (1982) Iron on soils. Nayka Press, Moscow

Zheltozem Soils

5

Derived from the Russian words *zhoeltiy,* meaning yellow and *zem,* meaning soil.

Zheltozem soils are deeply weathered; some are texturally differentiated soils, formed in a wetter conditions dry period alternation. Hydrated and oxidized iron, and goethite have a yellow or yellowish-brown color and various combinations of iron oxides result in yellow–brown or reddish color.

5.1 Concept of the Type

Main concepts

1. Zheltozem soils are old mineral soils with a clay enrichment mainly of kaolinite-micas, as a result of hydrolysis of the primary silicates during evolution;
2. Characterized by low base saturation (<50 %);
3. Deeply weathered rock, it has a well-preserved structure and may be soft.

The main processes that have taken place in the development of Zheltozem soils are chemical weathering in situ and permanent leaching of bases.

5.2 Ecology of Soil Formation

5.2.1 Distribution

Zheltozem soils are located in the southeastern part of the country within the Strandzha Mountain area in the zone of the middle stream of Veleka River (Figs. 5.1, 5.2). The length of the Veleka is 147 km. Usually, these soils occur in some places in the central part of the Strandzha Mountain area and not in compact territory. The territory of Strandzha Mountain is a Natural Park, and includes the southern Black Sea province.

The Strandzha Natural Park was established in 1995 on a territory of 1,61,362 ha in the water basins of the Veleka, Mladezhka, and Rezovska rivers in the areas around the Malko Turnovo and Tsarevo municipalities. Its goal is the long-term protection of the rare and extensive biodiversity and cultural heritage of Strandzha Mountain.

Climatic conditions

In Bulgaria the development of zheltozem soils was influenced by the Mediterranean type of climate that existed during the Pliocene.

Specificity of the atmospheric circulation, solar radiation, and geographical location renders the region of the Black Sea, a continental–Mediterranean climatic zone. The significant peculiarity of the Black Sea climatic region is that is has comparatively warm and wet winters and hot dry summers. The influence of the sea is evident 15–20 km inland. Seasonal maximum precipitation is during the winter–autumn and the dry period is during summer. The mean annual temperature is 11.3–13.1 °C. The mean annual temperature in January is 1.7(3.2) °C. The mean annual temperature in July is 20.6–23.7 °C. The frost-free period with temperatures over 0 °C lasts for 272 days. The mean annual precipitation varies from 560 to 792 mm. The amount of precipitation increases with altitude. Spring is rather cold due to the influence of frequent northeast air translocations. Autumn is warm and dry. During summer there are dry periods that last for 16–30 days, increasing the threat of drought.

Unlike the peripheral and coastal parts, the central inner part of the Strandzha Mountain has high atmospheric humidity. Soils are formed in local conditions of constant high soil and air moisture. Mist stretches in the valley for much of the year resulting in high atmospheric humidity. Sea breeze brings the mist along the river valley and onto the mountain slopes. It was previously thought that high humidity was due to the amount of precipitation of nearly 1,000 mm. However, the amount of rainfall is the same as that on the south Black Sea coast and higher values increase with altitude. Actually, the high humidity in the central inner part of the Strandzha Mountain is due to the presence of constant mist. As the slopes climb humidity decreases and zheltozem soils diminish.

T. Shishkov and N. Kolev, *The Soils of Bulgaria*, World Soils Book Series, DOI: 10.1007/978-94-007-7784-2_5,

Fig. 5.1 The Strandzha mountain

Fig. 5.2 The misty valley of river Veleka

Topography
Zheltozem soils are usually situated on old stable landforms in the low-mountain region of the Strandzha Mountain. These soils are much more common on the lowlands and hilly terrains with elevation from 100 to 300 m. They occur on low slopes or at the lower ends of slopes, and are usually in well-drained areas, in places protected from the wind.

Parent materials
The formation of zheltozem soils took place on the Lower Triassic (Anisian) Age (Mesozoic Era) low-grade metamorphic rocks, metasandstones, metapelites, and subordinate marbles and metavolcanics; on the Upper Cretaceous (Turonian-Coniacian) Age sediments fine to block breccias with packets of sandstones, siltstones, shales, marls, and clayey limestones; on volcanic rocks such as basalts, trachybasalts, trachyandesite basalts in extrusive, explosive and subvolcanic facies; and on the Turronian-Santonian Age (Mesozoic Era) sandtones breccia conglomerates, flysch and flyschlike sediments.

Vegetation
Strandzha Mountain is a Natural Park where natural forest areas are preserved and controlled. The Natural Park includes large oak and beech forests with unique southern Pontic ecosystems.

Zheltozem soils are generally formed under deciduous forest (Fig. 5.3). The region is a part of the sub-Mediterranean forest zone with deciduous forests comprising beech (*Fagus orientalis*), oak (*Quercus sesiliflora*, *Quercus conferta, Quercus polycarpa*, *Quercus cerris*, *Quercus frainetto*, *Quercus pubesccens*, *Quercus pedunculata*, *Quercus pedunculata*, *Quercus pedunculiflora*, *Quercus hartwissiana*, *Quercus stranjensis*, *Quercus orientalis*), hornbeam (*Carpinus betulus* L.), aspen (*Populus tremula* L.), almond (*Amygdalus nana*), fig tree (*Ficus carica*), and other species (Bondev 1991).

The central internal, more humid part of the mountain is inhabited/occupied by *Rhododendron ponticum, Epimedium pubigerum, Laurocerasus officinalis, Acer pseudoplatanus, Vaccinium arctostaphylos, Ilex colchica, Hypericum androsaemum, Daphne pontica.*

The relict evergreen rhododendron (*Rhododendron ponticum*) is unique vegetation preserved since the Tertian period. In areas of zheltozem occurence the soil surface is covered with mosses and lichens (Fig. 5.4).

Time
The majority of zheltozem soils are the oldest on the territory of Bulgaria and hold relict features of previously existent conditions. The younger age soils occur on sediments of the Pleistocene Age. Soils formed on consolidated rocks such as granite date back to at least the late Tertiary period.

Human influence
Forest areas have always existed in the region of Strandzha Mountain. Since the establishment of the Natural Park in

Fig. 5.3 Typical deciduous forest in the Strandzha mountain

Fig. 5.4 Soil surface under the deciduous forest

1995 the planting of conifers and non-native plants has been banned. Land use is controlled by national legislation.

5.2.2 Genesis

The genesis of the zheltozem soil group is little understood and the interrelation of the processes is not clear. Zheltozem soils are developed in response to atmospheric moisture mainly on the foot slopes in transient landscapes. Soil formation takes place in well-drained and permanently leaching soils with high intensity and duration of weathering.

Zheltozem soils are formed with gradual loss of rock structure as a result of deep chemical weathering in situ (lithomarge). The minerals are weathered in such a way that the thin dispersed material slowly peels off with practically no mechanical destruction (Figs. 5.5, 5.6).

Under conditions of high humidity most of the basic cations and iron have been leached.

In the literature these soils are referred to as ferralsols, meaning more humid subtropical soils.

In other parts of the world these soils occur in the transitional zone from the humid tropics to the dry savanna, where rainfall is 1,000–1,300 mm, and the dry season lasts 3–4 months.

This soil group is characterized by a high degree of weathered soil material and the residual accumulation of quartz, kaolinite, and sesquioxides with a significant depletion of silica and bases. The ratio SiO_2: Al_2O_3 is below 2 and there is a predominance of kaolinite in the clay fraction. Hydrolysis of primary silicates forms kaolinite, mica, goethite, and gibbsite. The products of hydrolysis, mostly kaolinite, have been transported to a considerable depth in the soil.

Initially, the intense hydrolysis of the primary silicates releases silica and bases, followed by the formation of montmorillonite. Usually, some montmorillonite occurs in the lithomargic zone. The mineralogical composition is differentiated as the lower part of the profile is clay enriched with smectites. Further, part of the montmorillonite is transformed into kaolinite in situ. Finally, clayey material is formed consisting of quartz, kaolinite, gibbsite, and goethite, with low capability for weathering minerals. The yellow color is due to the presence of goethite which indicates that the soils were formed in a moist environment.

The quick drying of moisture-saturated soil that takes place during the hot period results in the forms of optically oriented clay observed in thin sections.

Differentiation of the clay is due to lessivage, manifested by the appearance of clay skins, including clay cutans, on the surfaces of the structural aggregates and by thin pores in the Bt horizon. The Fe^{2+} form has a clear relation to the humus profile and the gleyic grayish mottled zones.

Fig. 5.5 Lithomarge or rotten stone

5.3 Soil Diagnostic and Classification

5.3.1 Morphology

Morphological characteristics of zheltozem soils are intense red or crimson and bright yellow to brown color of the profile or of its parts; in other words they are particolored. Also, there is a gradual transition boundary between the horizons and clayey texture of the soil (Figs. 5.7, 5.8).

The profile of the zheltozem podzolic (pseudopodzolic) soil is formed by deep hydrolysis (Figs. 5.9, 5.10). There are three clearly distinguishable parts—the upper humus-eluvial, lower illuvial-etamorphic, and zone of intense weathering—known as lithomarge; below is a rottenstone of weathering rock. Lithomarge is a rock fragment with brightly colored layers which retains its shape and structure but is quite soft and is transformed mineralogically.

Often, clayey yellowish material has a lesser pronounced blocky structure and grades at the massive unified clayey mass. Sometimes the structure of aggregates is not defined and is referred as structureless (no structure).

The majority of described zheltozem soils have a clear concretion iron accumulation.

O—A thin (3–8 cm) loose litter, at the top of which leaf tissues are visible, and at the bottom of which are moderate to strongly decomposed remnants, naturally occurs in woodlands; below

A—a thin (<5 cm) mineral humus horizon, dark brown or reddish brown colored, homogeneous. Structure is fine granular (spheroids or polyhedrons) with size 1–2 mm, medium pedality grade, consistence friable (moist) and soft (dry); grades into

AE—a mineral humus-eluvial horizon, of about 10–20 cm thickness, light reddish brown to bright brown colored. Structure of aggregates is subangular blocky of very fine size (<5 mm), moderate pedality grade, consistence is friable (moist) and slightly hard (weakly resistant to pressure) dry. The biological features are few burrows (unspecified) and the rare occurrence of worm casts. There is gradual change beneath into

Eg—a thin to thick eluvial-gleyic horizon which may be not expressed clearly and which can only be pronounced in the zheltozem podzolic (pseudopodzolic) soil variety. Pedality type is partly subangular blocky with very fine size (<5 mm), weak pedality grade, consistence is soft (very weakly coherent and fragile), dry, and friable (moist). There are many black manganese concretions in the soil mass. There is change to

Btg—an illuvial-metamorphic horizon, usually of mixed yellowish red color due to the presence of iron oxides or particolored, compacted. Structure is subangular blocky (blocks having mixed rounded and plane faces of aggregates), with medium (10–20 mm) size, medium pedality grade, consistence is firm (soil material crushes under

Fig. 5.6 Subangular blocky aggregate with lithomargic and gleyic pattern

Fig. 5.7 Zheltozem soil Bulgarian classification Yolevski and Hadzhiyanakiev (1976), *Haplic Acrisol,* WRB (2006), *Typic haplustults*, Keys to Soil Taxonomy (2010), *Haplic Acrisol,* Revised Legend of FAO-UNESCO-ISRIC (1990)

strong pressure) when moist, and hard (very resistant to pressure) dry or massive (no structure). Porosity is low. The surfaces of peds and pores are coated with stress clay cutans rather than illuviated. Commonly, there are fine rock fragments in the soil mass, many rusty mottles and many clay-iron and iron-manganese concretions that differ in size, shape, and surface; they grade into

The lower part which is enriched with clay (kaolinite), cutans are absent. Structure is massive preserved parent rock. Gradual change to the lower

CR—is motley, light gray, pinkish gray to reddish yellow colored, with many rusty and gray spots. Parent materials, deeply weathered rock with well-preserved initial structure, can be soft and easily diggable, gradual change to

R—continuous hard rock.

The patterns of the horizon sequence in the profile of zheltozem-like soil are O–AE–Bg–CR–R; and in the zheltozem podzolic (pseudopodzolic) soil O–AE(g)–Eg–Btg–CR–R.

The zheltozem podzolic (pseudopodzolic) soil morphology is as follows: darker colored, thin (less than 10 cm) humus-eluvial horizon (AE) lies over a moderately thick 15–25 cm eluvial (E) horizon. Light colored (E) horizon (may be albic, but not always the case) is over the relatively thick (20–200 cm) illuvial-matamorphic (Bt) horizon showing an increase in clay content, which is evidenced by oriented clay on ped faces or around pores and cracks.

An indication of the hydromorphic process is the macro and micro morphologic occurrence of iron-manganese nodules and gleyic mottles, which are most pronounced at the lower part of the illuvial-metamorphic horizon.

In comparison with the soil group of Luvisols where the clay enrichment is in the middle part of the profile, the clay enrichment in the zheltozem-like soils is most pronounced at the lower part of the illuvial horizon.

In the eluvial layer (AE) processes are related to low bioactivity. The reason for this is strongly acid pH and the occurrence of seasonal anaerobic conditions due to poor drainage.

Neoformation of clays appears to play a larger part in these usually more highly weathered soils. The other main process is leaching.

Fig. 5.8 Zheltozem-like soil Bulgarian classification Yolevski and Hadzhiyanakiev (1976), *Endogleyic Cambisol Dystric Clayic,* WRB (2006), *Ultic haplustalfs,* Keys to Soil Taxonomy (2010), *Haplic Acrisol,* Revised Legend of FAO-UNESCO-ISRIC (1990)

Fig. 5.9 Zheltozem podzolic (pseudopodzolic) soil, Bulgarian classification Yolevski and Hadzhiyanakiev (1976), *Endogleyic Cutanic Acrisol (Albic)*, WRB (2006), *Typic Albaquults*, Keys to Soil Taxonomy (2010), *Haplic Acrisol* (Revised Legend FAO-UNESCO-ISRIC (1990)

Oxidation and reduction of iron oxides is apparent throughout the profile in the form of rusty mottles (Figs. 5.11, 5.12, 5.13). Chemical weathering in situ is favorable during the whole year.

5.3.2 Classification

Due to the complexity of the processes that occur in the zheltozem soils, their classification has been problematic. Around the world at different times, this group of soils has been referred to as krasnozems, kaolisols, red-yellow podzolic soils, red and yellow ferrallitic soils, and ferralsols. Ferralsols, the name of which derives from the Latin "ferrum," meaning iron and "aluminum," are humid subtropical soils, deeply weathered with an undifferentiated profile that may be qualified as yellow or red, umbric.

Zheltozem podzolic soils were described in Russia for the first time by Dokuchaev (1899), who referred them to the soil group of Podzols. In Russia at the beginning of the last century Glinka and Zacharov carried out a systematic study of subtropical soils in the Black Sea coastal region (West Georgia) which was continued by Romashkevich (1979). Three zones of pedogenesis were formulated in the subtropical soils: the upper zone with current soil development, the middle transitional yellowish clayey zone, and the lower zone that was not affected by the solutions.

Glinka was the first to call soil-krasnozem meaning a red soil. In Russia, zheltozems were initially considered as a subtype of the krasnozem-zheltozem (red-yellow) soil type. Actually, the whole range of colors from dark red to bright yellow should be taken into consideration.

In Bulgaria, the term zheltozem was adopted from Russian soil classification to denote a specific soil formation. However, the fact is that it refers to different soils. Bulgarian zheltozems do not correspond to the morphological features of the other typical yellow subtropical soils distributed in China, Florida (USA), and Georgia. Aiming at precise correlation, it seems to be correct to apply the term

Fig. 5.10 Zheltozem podzolic (pseudopodzolic) soil, Bulgarian classification Yolevski and Hadzhiyanakiev (1976), *Endogleyic Cutanic Acrisol (Albic)*, WRB (2006), *Typic Albaquults*, Keys to Soil Taxonomy (2010), *Haplic Acrisol* (Revised Legend FAO-UNESCO-ISRIC (1990)

of type krasnozem (red soil) or type of reddish-yellow soil, both of which are used in the Russian soil classification.

In Bulgaria, the first scientific argument for the existence of light soils was made by Yordanov in 1938. Zheltozem was not included in the first systematic list of soils (1948). It was defined for the first time in 1954, and in 1955 Gerasimov defined the type of podzolic zeltozem. Koynov (1964) and Koynov et al. (1964) classified the zheltozem podzolic (pseudopodzolic) soil as a single type. Angelov (1966) named the zheltozem subtypes as typical, podzolic-like, and podzolized. Gurov et al. (1978) shifted podzolized zheltozems to the type of pseudopodzolic soils (Planosol), together with the light gray pseudopodzolic and cinnamonic pseudopodzolic forest soils. The soil classification of 1983 (Yolevsky et al.) split the group of zheltozems into two types: the type of zheltozems with lessive and haplic subtypes, and the podzolic zheltozem-like type. Hadjiyanakiev (1989) confirmed the zheltozem pseudopodzolic type. Penkov et al. (1992) classified the soil as Ultic Luvisol (Gleyic and Haplic).

On the Soil Map of Europe 1:1,000,000, the soil was referred to as Stagnic Luvisol, and in the European Soil Database it is Haplic Acrisol according to the FAO Revised Legend, 1990.

Subtype varieties differ in terms of the thickness of humus horizon (non-eroded or slightly eroded), and in the degree of erosion (slightly, moderately or not eroded).

WRB (2006) allowed that soils that respond to the demands of *argic* horizon with base saturation less than 50 % in the major part between 50 and 100 cm can be referred to as Acrisols.

Keys to Soil Taxonomy (2010) required the *ochric* or *umbric* epipedon and the *argillic* or *kandic* horizon for Ultisols order classification. In Soil Taxonomy there are demands for subsurface horizons allowing an *argillic* horizon with base saturation (<35 %) at a particular depth, whose depth depends on color or lithic contact; or *kandic* horizon with low (<35 %) base saturation in Ultisols; *fragipan or plinthite* horizon; *albic, or none*.

Ochric epipedon is most commonly a thin A horizon showing some incorporation of organic matter.

5.3.3 Distribution of Extractable Compounds of Fe, Al, and Mn in Zheltozem Soils

Since the stages of soil development are advanced in pseudopodzolic soils, the maximum of Fe_t, Al_t and Mn_t are in the zheltozem podzolic (pseudopodzolic) forest soil followed by the light gray pseudopodzolic forest soil and cinnamonic podzolic (pseudopodzolic) soil.

At the illuvial Bt horizon is observed the maximum of the content of the following extractable forms: the total content of Fe and Al, the organic matter bound forms (pyrophosphate extractable) of Fe, Al, and Mn, the dithionite and oxalate extractable forms of Al, as well as of the crystalline forms of Fe, which is linked with the red soil color (Jokova and Boyadzhiev 1993, 1998).

The ratios between total content of iron and clay content Fe_t/clay decrease at the Bt horizon, since some of the released iron compounds have migrated out of the profiles.

The minimum of Mn_t, and especially of the dithionite and oxalate extractable forms of Mn along depth of the profile at the zone between Bt and the upper horizons, indicate the gley horizons (Jokova 1994).

Soil aggregate water stability in zheltozem podzolic (pseudopodzolic) soils is negatively influenced by the amorphous forms of Fe and Al. The content of the dithionite extractable iron compounds and the content of the crystalline forms are high and indicate more advanced stages of weathering, migration, and crystallization.

Fig. 5.11 Subangular clocky aggregate (blocks having mixed rounded and plane faces of aggregates) from the upper Btg horizon

Fig. 5.12 Subangular blocky aggregate (blocks having mixed rounded and plane faces of aggregates) from the lower Btg horizon

Fig. 5.13 Massive structure from BCg horizon

The amorphous form of iron bounded with organic matter showed lack of mobility down the profile and is not influenced by waterlogging at the surface layer.

A comparison between pseudopodzolic soil subtypes shows that the content of the free Fe forms and of the total Mn at the surface horizon is enriched in the zheltozem podzolic (pseudopodzolic) forest soils, which is related to the more advanced weathering and soil forming processes determined by the conditions of the region of these soils (Table 4.28).

Generally, the distribution of Fe forms strongly reflects the thermo-moisture regime of the soils; the distribution of Mn forms reflects the gley process; and the distribution of Al forms reflects migration and lessivage processes.

5.3.4 Zheltozem Soils Subtypes in Bulgaria

5.3.4.1 Zheltozem-Like Soils

Zheltozem-like soil (*subtype*, *type*) is named after the Bulgarian classification of Yolevski, and Hadzhiyanakiev (1976).

Endogleyic Cambisol Dystric Clayic is named according to WRB (2006) and *Typic Distrustepts* according to Keys to Soil Taxonomy (2010).

Dystric Cambisol: Revised Legend of the World Soils (FAO-UNESCO-ISRIC 1990).

Zheltozem-like soils are transient soils with clayey texture, pronounced hydration, and yellowish color. Soils are characterized by the absence of illuvial horizon. There is evidence of advanced metamorphic process.

5.3.4.2 Zheltozem Soils

Zheltozem soil (*subtype*, *type*) is named after the Bulgarian classification of Yolevski, and Hadzhiyanakiev (1976).

Haplic Acrisol is named according to WRB (2006) and *Ultic Haplustalfs or Typic Haplustults* according to Keys to Soil Taxonomy (2010).

Haplic Acrisol Revised Legend of the World Soils (FAO-UNESCO-ISRIC 1990).

Zheltozem soils have inherited from the parent materials clayey texture, and strongly pronounced hydration together with a high content of iron and yellowish color. Soils are slightly differentiated. The structural rearrangement of the minerals and the presence of stress cutans are mostly evidence of the advanced metamorphic process rather than illuviation. Released iron oxides from the weathered rock alter the getite are a factor for the yellowish color in the lower horizon.

5.3.4.3 Zheltozem Podzolic (Pseudopodzolic) Forest Soils

Podzolic (pseudopodzolic) (*subtype*) **zheltozem** (*type*) forest soils are named after the Bulgarian classification of

Table 5.1 Variables of the particle-size distribution (mm) in 6 referenced profiles of zheltozem podzolic (pseudopodzolic) forest soils in Bulgaria

Profile count	Horizon	Upper depth	Lower depth	Loss HCl	Particle-size distribution (mm), % (according method of Kachinsky revised)							
6					>1 mm	1–0.25	0.25–0.05	0.05–0.01	0.01–0.005	0.005–0.001	<0.001	<0.01[a]
Mean	AEg	0	24	0.9	7.9	11.0	21.4	21.1	9.8	8.7	19.1	37.6
	Btg1	24	47	1.1	3.1	8.0	15.5	19.1	9.5	10.8	33.0	53.3
	Btg2	47	70	2.1	3.6	7.5	15.4	19.5	7.7	9.3	35.0	51.9
	CR	70	97	1.4	1.4	15.3	19.7	20.6	9.0	7.9	24.6	41.7
Stdev	AEg	0	4	0.4	6.6	6.0	10.9	7.9	1.5	1.9	6.7	8.3
	Btg1	4	7	0.6	4.1	6.8	5.3	2.6	3.2	3.0	8.7	8.3
	Btg2	7	6	1.5	5.3	7.1	9.0	4.4	3.9	2.8	11.8	12.5
	CR	6	8	1.1	2.2	15.3	8.2	7.4	6.9	1.9	22.2	20.1

[a] Physical clay, the parameter characterized sum of the particle-size distribution, according to Kachinsky method (mm) $\sum < 0.01 = <0.001 + (0.005 - 0.001) + (0.01 - 0.005)$

Koynov et al. (1964), Yolevski, and Hadzhiyanakiev (1976).

Cutanic Acrisol (Albic) is named according to WRB (2006) and *Typic albaquults* according to Keys to Soil Taxonomy (2010).

Haplic Acrisol Revised Legend of the World Soils (FAO-UNESCO-ISRIC 1990).

The texture depends on the parent materials. Particle size distributions show that clay fraction (<0.001 mm) prevails followed by the silt fraction (particularly coarse silt 0.05–0.01 mm) at the illuvial Bt horizon. The fractions of fine sand (0.25–0.05 mm) and coarse silt (0.05–0.0 mm) prevail at the upper eluvial layer (Table 5.1 and Fig. 5.14). Distribution of clay fraction shows a clear increase at the lower part of the Bt horizon (Table 5.2 and Fig. 5.15).

The average particle density is 2.4–2.50 g/cm^3 at the surface and 2.65 g/cm^3 at the illuvial horizon. Bulk density is near 1.1 g/cm^3 at the upper part and is about 1.5–1.6 g/cm^3 at the middle part of the profile with evidence of compaction. When dried, textural Bt horizon is very hard in consistence.

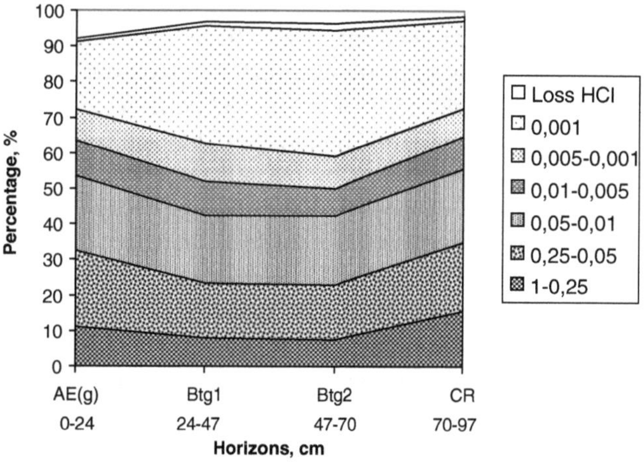

Fig. 5.14 Pattern of the particle size distribution (mm) in zheltozem podzolic (pseudopodzolic) forest soils (according to revised Kachinsky method)

Water stable aggregates with size >1 mm are about 55–65 % at the surface (0–15 cm) and sharply decreased to

Table 5.2 Variables of humus, pH, carbonates, and particle-size distribution (mm) according to USDA in 6 referenced profiles of zheltozem podzolic (pseudopodzolic) forest soils in Bulgaria

Profile count	Horizon	Upper depth	Lower depth	Humus (%)	pH_{KCl}	Carbonates	Bulk density g cm^3 calc	Clay <0.002	Silt 0.002–0.05	Fine sand 0.05–0.2	Total sand 0.05–2.0
6											
Mean	AEg	0	24	1.52	3.9	0.0	1.40	25.1	37.5	22.4	37.3
	Btg1	24	47	0.80	4.0	0.0	1.46	39.7	35.1	15.7	25.2
	Btg2	47	70	0.56	4.1	0.0	1.48	41.7	33.4	15.7	24.9
	CR	70	97	–	4.5	0.0	–	29.6	34.5	19.9	35.9
Stdev	AEg	0	4	0.54	0.2	0.0	0.07	7.1	8.5	10.7	13.2
	Btg1	4	7	0.34	0.4	0.0	0.05	9.1	4.6	5.1	7.9
	Btg2	7	6	0.21	0.5	0.0	0.08	12.9	6.3	8.9	11.6
	CR	6	8	–	1.2	0.0	–	22.0	13.1	8.2	19.3

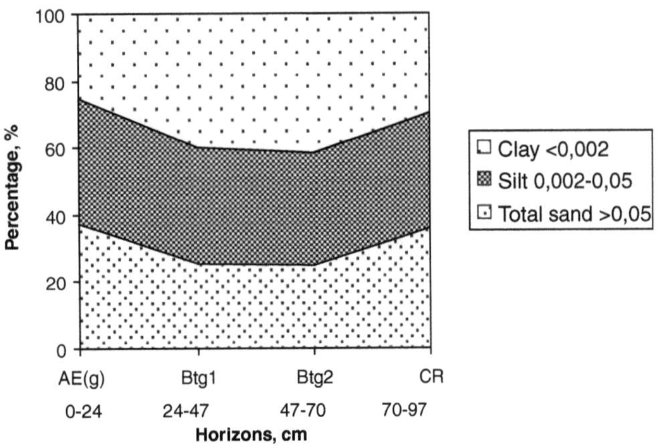

Fig. 5.15 Pattern of the particle size distribution (mm) in zheltozem podzolic (pseudopodzolic) forest soils (USDA)

about 20–30 % at the depth of 20–30 cm. Prevailing water stable aggregates with size over 0.25 mm are over 70 % at the surface and 40–45 % at the illuvial horizon.

In zheltozem podzolic (pseudopodzolic) soil 5–10 % of pore space is represented by macropores or the pores of aeration (with diameter > 300 μm); and 10–20 % are mesopores or with rapid moisture drainage (>30 μm diameter). Micropore space or water-filled is about 15–25 %, represented by slow drainage pores (diameter 0.2–30.0 μm) and 10–15 % of micropores with none remaining available for roots moisture (diameter <0.2 μm), (Dilkova 1985; Dilkova et al. 1998).

Total porosity is high at the surface (near 55 % at field capacity) and decreases with depth. Soils are characterized by low water permeability.

In spatial relationship zheltozem soils dominate in association consisting of Luvisols (Chromic), Umbric Leptosols, and Cambisols (Dystric) WRB (2006).

The organic matter content is 3–5 % below the litter and 1.0–1.5 % in the upper most of eluvial horizon, sharp decline is typical in woodlands (Fig. 5.16). The C/N ratio is 8–12. In humus composition fulvic acids (C_f) considerably prevail; at the lower part of the profile humic acids (C_h) are missing entirely. Fulvic humus type and prevailed clay of kaolinite composition determine low cation exchange capacity, which is less than 15 meq/100 g where exchangeable Al^{3+} and H^+ dominate. Soils have low base saturation of about 45–50 % at the surface, 45 % at the eluvial horizon, and 35 % at the CR horizon (Ganev and Arsova 1980) (Ganev et al. 1990) and with acid pH_{KCl} values 5.5 at the surface, 3.5 at the eluvial horizon, and below 3.0 down the depth.

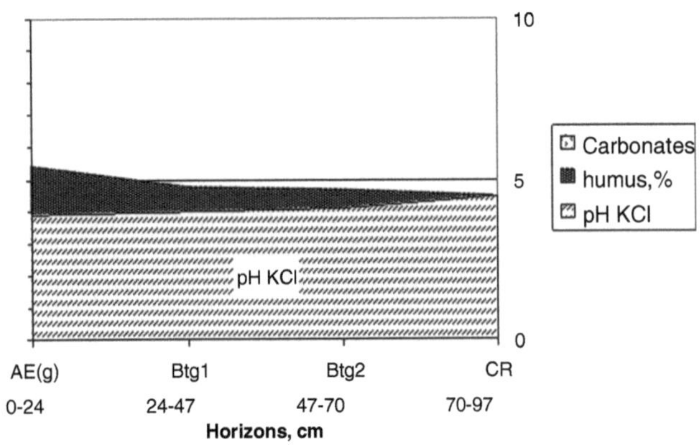

Fig. 5.16 Pattern of pH, humus, and carbonates distribution in zheltozem podzolic (pseudopodzolic) forest soils

The sand fraction is dominated by quartz, feldspar, mica, and small amounts of resistant minerals such as zircon, magnetite (ferrous-ferric oxide), and rutile. While the composition of the sand fraction is stable with depth of profile, the content significantly declines from the surface to the deeper horizons. Color varies from yellowish-red to dark brownish-red depending on the amount of iron.

5.4 Soil Properties and Soil Management

5.4.1 Soil Properties of the Zheltozem Soils

Chemical weathering in situ takes place over the whole year in wet conditions. Various combinations of the iron oxides haematite and goethite result in red-yellow color. Dominance of the non-silicate form of iron over its silicate form in all soil horizons is a typical feature and is evidence of advanced weathering. High content of the free forms of iron in all soil horizons is a typical feature and is also evidence of advanced weathering.

Zheltozem soils have low fertility due to their limited physical and chemical properties. They often have high water-holding capacity and negative physical properties such as low structural aggregation and low infiltration rates. Waterlogging may occur on the surface. A common feature is that the soils are sticky.

Loss of bases is by permanent leaching. In the composition of adsorbed cations H^+ and Al^{3+} prevail.

The soil reaction throughout the profile is strong acid at $pH_{KCl} < 4.0$.

5.4.2 Soil Management

These soils are of limited use due to the following physical properties: low water permeability, natural high water-holding capacity, low aeration, low structural aggregation, and slightly coherent soil particles in the surface layer. Erosion strongly affects the surface horizons in the dissected areas.

Zheltozem soils are used for forestry or rough grazing. Low nutrient status and low pH values make the utilization of these soils for agriculture difficult or impossible. For cultivation purposes more liming and the addition of nutrients are required.

References

Angelov E (1966) Zheltozem podzolic-like soils from Strandzha Mountain. J Soil Sci & Agroch, 1 (In Bulgarian)

Bondev I (1991) Vegetation in Bulgaria, map at scale 1:600 000 with explanatory text. St. Kliment Oxridski University Press, Sofia (In Bulgarian)

Dilkova R (1985) Structure and aeration in main soil varieties in Bulgaria. Theses, Institute of Soil Science Nikola Poushkarov, Sofia (In Bulgarian)

Dilkova R, Kerchev Zh, Kercheva M (1998) Evaluating and grouping of soils according to their susceptibility to anthropogenic degradation. In: Advances in Geoecology, vol 31. Catena Verlag, Reiskirchen, pp 125–131

Dokuchaev VV (1899) On study of nature's zones: horizontal and vertical soil zones. In: Municipal proceedings (In Russian)

FAO-UNESCO-ISRIC (1990) Revised legend of the soil map of the world. FAO World Soil Resources Report No 60, Rome

Ganev S, Arsova A (1980) Methods of determining the strongly acid and the slightly acid cation exchange in soil. Soil Sci Agrochem 15(3):22–33 (In Bulgarian)

Ganev S et al (1990) Physic-chemical properties of soils in Bulgaria and other world's regions. Department of chemistry, Institute of Soil Science Nikola Poushkarov, Sofia (In Bulgarian)

Gyurov G et al (1978) Some problems of genesis and classification of gray forest soils, brown forest and pseudopodzolic soils in Bulgaria. In: Proceedings of the Second scientific conference on soil science, Sofia (In Bulgarian)

Hadzhiyanakiev A (1989) Iron in Bulgarian soils. Zemizdat Press, Sofia (In Bulgarian)

Jokova M, Boyadzhiev T (1993) Forms of compounds of iron, aluminium and manganese in some Bulgarian soils. J For Sci 30(2):62–72 (In Bulgarian)

Jokova M (1994) Manganese of some Bulgarian soils—dependence of C Humic: C Fulvic acids on the content of manganese. In: Lal R, Kimble J, Lavine E (eds) Soil processes and greenhouse effect, vol 3. USDA, Soil Conservation Service, National Soil Survey Centre, Lincoln, pp 24–30

Jokova M, Boyadzhiev T (1998) Extractable compounds of Fe, Al, and Mn in Bulgarian textural differentiated soils. Bul J Agri Sci 4(1):17–24

Keys to Soil Taxonomy (2010) USDA United States Department of Agriculture, NRCS Natural Recourses Conservation Service, 11th edn. Soil Survey Staff, Washington, DC, p 338

Koynov V (1964) Peculiarities of the main Bulgarian soil types and their basic soil-genetic processes. In: Gerasimov IP et al. (eds) Soils of south–eastern Europe. Proceedings of the international symposium of soil science, June 1963, Sofia. Bulgarian Academy of Sciences Press, Sofia, pp 15–62 (In Russian)

Koynov V et al (1964) Classification and systematic of soils in Bulgaria. In: Gerasimov IP et al. (eds) Soils of south–eastern Europe. Proceedings of the international symposium of soil science, June 1963, Sofia. Bulgarian Academy of Sciences Press, Sofia, pp 231–246 (In Russian)

Penkov M et al (1992) Soil classification and diagnostic in Bulgaria in relation to the land reform fragmentation. Zemizat Press, Sofia (In Bulgarian)

Romashkevich IA (1979) Zheltozems: genetic soil types in sub-Kavkas. In: Gerasimov (ed) Science Press, Moscow, pp 66–81 (In Russian)

World Reference Base for Soil Resources (2006) A framework for international classification, correlation and communication. FAO-ISRIC, Rome

Yolevski M, Hadzhiyanakiev A (1976) Agro-grouping of soils in Bulgaria, Extended systematic list. Code of groups and soil varieties, Institute of Soil Science Nikola Poushkarov, Sofia (In Bulgarian)

Yolevski M, Ninov N, Krastanov S et al (1983) Brief diagnostic and criterion of the main taxonomic soil divisions in Bulgaria. Precise Classification Scheme, Institute of Soil Science Nikola Poushkarov, Sofia (In Bulgarian)

Yordanov D (1938) Vegetation relations in Bulgarian part of Strandzha mountain, vol 34. Sofia's University, Sofia (In Bulgarian)

Zacharov SA (1927) Achievements of Russian science in morphology of soils. Russian pedological investigations II, Academy of Science, Leningrad, pp 1–47

6 Smolnitsa

Derived from the Slavic word *smolyani,* meaning black, and from the Latin word *verto,* meaning a turn.

Undifferentiated black colored clayey soil; high shrinking and swelling activity produces specific gilgai microrelief.

6.1 Concept of the Type

Main concepts

1. Smolnitsa is a mineral soil characterized by high (>30 %) clay content;
2. Soil is base saturated (>90 %);
3. Characterized by intense dark color and low content of organic matter.

The main processes that have taken place in the development of smolnitsa are self-pedoturbation due to high shrinking-swelling activity and leaching.

6.2 Ecology of Soil Formation

6.2.1 Distribution

Black compacted soils are found in many parts of Asia (India), North America, Australia, North Africa, and Europe. They are known as black clayey soils, black tropical soils, tyrsi, smoniza, and smolnitsa. The world distribution of these soils is as follows: 60 % occurs in the tropics, 30 % occurs in the subtropics, and the rest occurs in other regions including Europe.

Smolnitsa is generally distributed in southern Bulgaria in regions bordered by cinnamonic forest soils. Smolnitsa occupies 595,000 ha or about 5.34 % of the entire country's territory.

The major areas of occurrence are in the Thracian Lowland, the southern Black Sea province, the Strandzha region, and in the western part of the country (with altitude from 550 to 700 m) (Fig. 6.1).

Climatic Conditions

Two climatic regions are associated with the distribution of smolnitsa in Bulgaria. The first is confined to the transitional continental climatic zone with Mediterranean influence in the Thracian Lowland. The second region is confined to the transitional continental zone in western Bulgaria.

In the central province of the Thracian Lowland the mean annual temperature is 12.0–12.5 °C. In January the mean annual temperature is −0.8 (1.7) °C. In July the mean annual temperature is 22.9–23.6 °C. The mean amount of precipitation varies from 473 to 667 mm.

In the eastern province of the Thracian Lowland in the region of Burgas valley the mean annual temperature is 11.4–12.7 °C. In January the mean annual temperature is 0.1–2.0 °C. In July the mean annual temperature is 22.1–23.1 °C. The mean amount of precipitation varies from 449 to 555 mm.

In the western province the mean annual temperature is 9.5–10.5 °C. In January the mean annual temperature is −1.7 (−2.6) °C. In July the mean annual temperature is 19.6–21.2 °C. The mean amount of precipitation varies from 589 to 688 mm.

Smolnitsa is formed under the alternation of wet and dry periods.

Topography

In other parts of the world black compacted soils are typically formed on flat slightly dissected plains and depressions. Commonly, they occur in river valleys and on lacustrine terraces.

In Bulgaria smolnitsa is located on the most stable parts of the landscape. These soils have developed on flat lowland or on land with gently sloping hilly relief like valley floors, level plains, and plateaus, usually on terraces and on depression terrain.

Only these soils form the specific topographic feature known as gilgai microrelief.

T. Shishkov and N. Kolev, *The Soils of Bulgaria*, World Soils Book Series, DOI: 10.1007/978-94-007-7784-2_6,

Fig. 6.1 Landscape with gilgai microrelief

If soils occur on elevated landforms or dissected plains, the location is usually associated with consolidated rock.

Parent Materials
Smolnitsa is confined to sedimentary basic and/or calcareous parent rock, containing a large amount of swelling clay usually in the form of smectites or illites. These materials produce a large amount of cations, which are favorable for the formation of new clay minerals.

Sometimes it is not possible to be certain about the origin of all these materials. Most materials are deposits of fine texture, usually of alluvium or lacustrine origin, but others appear to be colluvium derived, or are due to erosion with subsequent deposition.

Some soils have developed through chemical weathering of the underlying rock, which may be basalt, shale, limestone, or feldspars with a high content of plagioclase and ferromagnesian minerals.

It is a fact that the weathered products from the Pliocene Age are much finer than the weathered products of andesite rock, which themselves are much finer than the Quaternary Age deposits. The mineralogy of the andesite rock comprises hornblende, biotite, augite, hypersthene, orthoclase, plagioclase, and quartz.

The provincial differences of smolnitsa are defined by the character of these materials.

Thus in the central part of the Thracian Lowland deposits are on the Neogene Age continental sediments in the superimposed depressions of South Bulgaria (conglomerates, sands, and clays); on the Paleogen-Neogene (Oligocene–Miocene) Age clays, sands, and coals; on the Eocene Age limestones and marls; and on the Quaternary Age alluvial-drift and talus-drift sediments (boulders, pebbles, sands).

In the eastern part of the country these soils have developed on the Paleogene (Eocene–Oligocene) Age marls; and on the Upper Cretaceous Age volcanic trachyandesitobasalts and trachytes with packets of sediments (sandstones, siltstones, marls, clayey limestones).

In the western part of the country smolnitsa has formed on the continental sediments in the superimposed depressions of the Neogene Age terrigenous coal-bearing sediments (conglomerates, breccia-conglomerates, sandstones, siltstones, shales, sands, clays); on the Paleogene (Oligocene) Age continental-molasse coal-bearing sediments (conglomerate-sandstone, shale bituminous formations); on the Upper Cretaceous Age tuffs and tephroidal rocks interbedded with sedimentary rocks (andesites, conglomerates, sandstones, siltstones, clays, marls, organogenic limestones); and on the Quaternary Age alluvial-drift and talus-drift sediments (boulders, pebbles, sands).

Vegetation
Smolnitsa predominantly forms under grassland which may contain trees and shrubs. The grassland vegetation that is most common on the lowlands is bedstraw (*Gallium,*) feather grass (*Stipa stenophylla, S. joannis, S. capillata, S. Lessingiana,*) fescue (*Festuca* sp.,) oats (*Helictrotrichon desertorum, H. pubescence*), koeleria (*Koeleria cristata*), osoka (*Carex* sp.), clover (*Trifolium montanum, T.*

lupinaste), crested wheat grass (*Agropyron pectiniforme,*) sage *(Salvia nutans),* and bell-flower *(Campanula* sp.). The elevated lands are beneath oak forest (*Quercus virgiliana, Quercus cerris*), hawthorn (*Crataegus monogyna,*) dogwood (*Cornus mas*), and hornbeam (*Carpinus* sp.) (Bondev 1991).

Eastern and western parts of the country that were covered with forests are now agricultural or grassland.

Time

The age of smolnitsa is indicated by the material on which it was formed. While some varieties are comparatively young, they are probably still several thousands of years old.

In the areas where soils have formed from the underlying rock, they may date from the mid-Pleistocene Age or earlier. In the areas where soils have formed on transported old products or other sediments, they are dated later than the mid-Pleistocene Age.

Human influence

Historically smolnitsa was problematic for agricultural use due to drawbacks with regard to tillage. The soils become very hard during the dry period and are very plastic and sticky during the wet period. The major limitation is the short tillage period.

Repeated cultivation causes deterioration of soil structure. Irrigation practices provoke negative soil processes and secondary salinity at some sites.

6.2.2 Genesis

The most popular theory of the origin of Vertisols belongs to Singh (1956) and Kovda (1981) and posits a paleo-hydromorphic phase of soil development in which clay accumulation was followed by the secondary mineral formation of the smectites group.

Bulgarian smolnitsa can not be separated from similar soils that are found at the Balkan Peninsula in Serbia, in Romania, Albania, northern Greece, the European part of Turkey and Macedonia (Skopje). The principal variations within the type are thickness, humus content, and depth of carbonate accumulation.

Many Bulgarian researchers have studied the Smolnitsa soil type, namely Boykov and Doncheva (1928), Poushkarov (1928, 1931), Stranski (1935), Stranski and Koynov (1947), Koynov (1964), and Koynov et al. 1968, 1972). Primarily, the common understanding was that clayey soil has little or no textural spatial variation. However, the findings of the large soil survey proved that soil texture does have variety and differs in terms of textural classes. The hydromorphic origin of smolnitsa was rejected in 1963.

The main difference between chernozems and smolnitsa is clay mineralogy dominated by stable minerals of the smectites group. This fact indicates the physical properties of the soil and particularly shrinking and swelling activity.

Little is known about the high shrink swell behavior. The intrinsic feature of clay mineralogy and the link with organic matter is not clear.

Smolnitsa has cracks that open and close once each year. During the summer cracks remain open $\geq$60 days and during the winter they are closed $\geq$60 days. Seasonal wetting and drying cycles must occur in order to produce the cracks and initiate self-pedoturbation. Upon wetting when swelling is in force, the cracks close. The pressure deformation of peds during the swelling phase results in the sliding of some soil masses and the formation of a wedge structure and shiny ped faces named slickensides.

The slow gradual movement of the soil also causes an upward movement of the soil, producing the gilgai microrelief.

The main role of the swelling and shrinking is permanent movement of soil that minimizes the B horizon development in these soils. Annual repetition of this cycle produces the uniform soil profile. This is the second difference compared to the chernozems soil group.

The third difference with the chernozems is little fluctuation of soil solutions within the smolnitsa profile. The permanent lateral movement of the slightly alkali and neutral solutions contributes to the development of buffering conditions of smolnitsa. Commonly, these solutions are enriched with Ca^{2+}, Mg^{2+}, iron, soluble salts, etc.

The content of organic matter is low and decreases gradually with increasing soil depth. A change in organic matter content does not result in a change of color. Singh (1956) suggested that it might be due to a dark colored complex of organic matter and montmorillonite which forms in the wet environment.

The fourth difference with the chernozems soil group is unfavorable physical properties.

The fifth difference with the chernozems soil group is low biological activity.

The sixth difference with the chernozems is decreasing of the water stable aggregates (with size 1–3 mm) at the depth of the prevailing >70 % in the surface horizon.

The internal porosity of aggregates is low.

6.3 Soil Diagnostic and Classification

6.3.1 Morphology

A typical feature of smolnitsa is variability at short distance of the thickness of the humus horizon due to gilgai relief. The soil profile is uniform in texture and throughout has a very dark color to a considerable depth (Fig. 6.2). Shallow soils are, however, common (Fig. 6.3). Horizon differentiation is only on the basis of compaction pronounce and carbonate distribution.

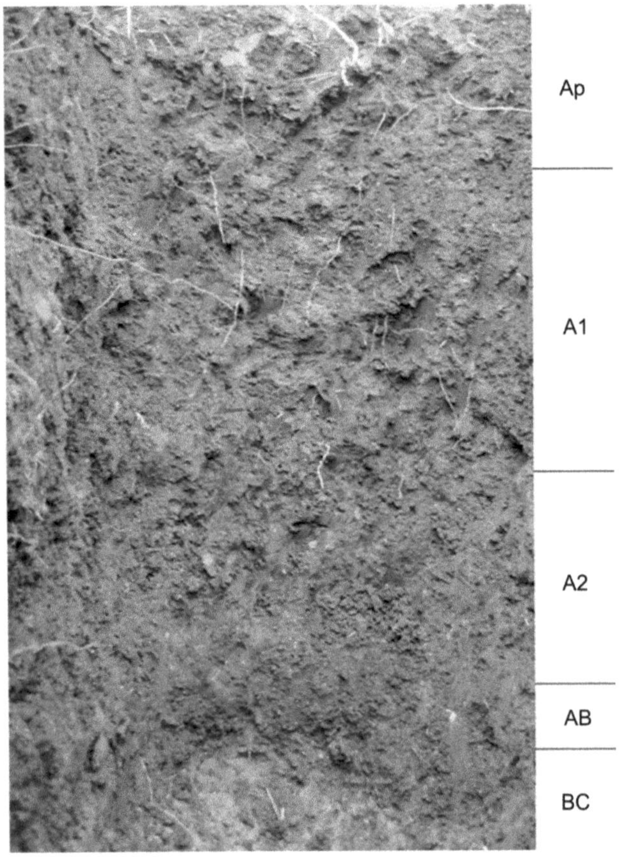

Fig. 6.2 Strongly leached smolnitsa, Bulgarian classification Yolevski and Hadzhiyanakiev (1976), *Haplic Vertisol* (*Eutric, Pellic,*) WRB (2006), *Typic Haploxererts,* Keys to Soil Taxonomy (2010), *Haplic Vertisol*, Revised Legend FAO-UNESCO-ISRIC (1990)

Most smolnitsa soils have A(p)-AB-BC(k)-C(k) pattern of horizon sequence.

Ah—The exposed upper mineral humus horizon is <2 cm thick, very dark brown or black (moist) and very dark grayish (dry) colored; naturally occurs in pristine lands. Structure is granular less than very fine (<1 mm diameter), consistence is single grain (individual grains). Beneath is clear transition to the lower

Ai(k), (Ap)—mineral humus accumulative horizon that can be subdivided into prime, second, etc., and homogeneous, without color change with hue 10YR dark brown. Structure is subangular blocky (blocks having mixed rounded and plane faces of aggregates), with coarse size (5–10 mm diameter), strong pedality grade, and consistence is firm (moist) and very hard (very resistant to pressure) dry. Large aggregates comprise many small ones. Downward soil mass is compacted and forms rectangular wedged shapes. Cracks are 0.5–1.0 cm wide stretching to a depth of over 1 m. Intersecting slickensides are commonly observed on ped faces, particularly in the lower part of (A) horizon (Fig. 6.4).

Secondary calcium carbonate is deposited as pseudomycelia (threads) in case of presence. Concretions of calcium carbonate are often distributed through the soil. Porosity is high. Many fine roots with size 0.5–2 mm; gradually grade beneath to

ABi(k)—a transitional mineral horizon, dark brown colored; structure is with coarse size (20–50 mm) subangular blocky structure (blocks having mixed rounded and plane faces of aggregates), where compaction is pronounced, with strong pedality grade, consistence is very firm (moist) and hard (dry). Intersecting slickensides are common on ped surfaces. Cracks are common. Secondary calcium carbonate threads are deposited in case of presence. Many fine roots with size 0.5–2 mm and (medium) 2–5 mm are in the soil mass; gradually grade beneath to

BC(k)—a transitional horizon to the upper part of the parent materials; with brown color, clay enriched, commonly with carbonates like soft spots and concretions. Structure is angular blocky medium size (10–20 mm) and very fine (1–5 mm), moderate pedality grade, consistence is firm (moist) and hard (dry), porosity is medium. There are many iron-manganese concretions in the soil mass. Large cracks are filled with upper material.

There is clear transition to the parent materials. Slickensides are

Ck—parent materials show little change, and are yellowish brown colored. The structure is massive (structureless). There is prominent accumulation of the secondary carbonates, mostly soft spots and concretions (Fig. 6.5).

The main process that takes place in the smolnitsa is self-pedoturbation due to high activity of swelling and shrinking. Seasonal leaching washes out soluble salts and carbonates. Iterative periods of wetting and drying fix oxidation and reduction.

Beneath is gradual change with mixed pattern down to the parent materials.

6.3.2 Classification

Despite the fact that black compacted soils have been well studied, the question of their classification is open to discussion. The soil group of Grumosols was defined in 1950. In Russia Glazovskaya (1971) defined the soil group of Slitozem.

Since 1982, the group of Vertisols was confined to those soils with features of gilgai relief and rectangular structure with oblique shape.

In the Revised Legend of the World Soils (FAO-UNESCO-ISRIC 1990) the group of Vertisols was subdivided into the dark and colored soils.

The Bulgarian classification (Koynov et al. 1964) and (Yolevski and Hadzhiyanakiev 1976) subdivided the type of

Fig. 6.3 Shallow smoltiza developed on andesites, *Leptic Haploxererts,* Keys to Soil Taxonomy (2010)

Fig. 6.4 Slickenside

smolnitsa into subtypes based on the degree of carbonate leaching. However, the depth of carbonate leaching is hard to equate with contemporary rainfall distribution.

The soil classification of Yolevski et al. (1983) rejected the subtype of podzolized (degraded) smolnitsa and included chernozem-like and cinnamonic-like subtypes. There was classification in Penkov et al. (1992) which at the theoretic level named subtypes of Smolnitsa (chromic, gleyic, haplic) which has not been applied so far.

Within the subtype soil differs in the thickness of the humus horizon (slightly thick <40 cm, medium 40–80 cm and thick >80 cm) and in rates of erosion.

Fig. 6.5 Accumulation of carbonate concretions in Ck horizon

Cinnamonic-like and chernozem-like smolnitsa varieties are subordinated on the level "group" and implies difference in display of a vertic feature and color.

Classification after WRB (2006) required *Mollic* diagnostic horizon in the reference group of Vertisols.

Keys to Soil Taxonomy (2010) required a *Mollic* or *Ochric* epipedon for Vertisols order classification. In Soil Taxonomy there are demands for subsurface horizons allowing a*rgillic* horizon with high base saturation (>35 %); *calcic, gipsic, salic, or duripan.*

In this case *Mollic* epipedon may occur in other orders without Mollisol classification. Thus subsurface horizon is *vertic* in Vertisols and takes precedence over the Mollic horizon.

According to the USDA content of particles <0.002 mm has to be no less than 35 %.

6.3.3 Smolnitsa Soil Subtypes in Bulgaria

6.3.3.1 Calcareous and typical smolnitsa

Calcareous and **typical** (*subtype*) **Smolnitsa** (*type*) is named after the Bulgarian classification of Koynov et al. (1964), and Yolevski and Hadzhiyanakiev (1976).

Epicalcic Mollic Vertisol (Calcaric) is named according to WRB (2006) and *Typic Calcixererts* according to Keys to Soil Taxonomy (2010).

Calcic Vertisol: Revised Legend of the World Soils (FAO-UNESCO-ISRIC 1990).

Calcareous and typical smolnitsa occupy 40,000 ha or about 0.36 % of the territory of the entire country.

They are distinguished by the appearance of secondary carbonate at the surface (calcareous smolnitsa) or beneath 10 cm (typical smolnitsa). The carbonate accumulation is in the form of pseudomycelium, commencing in the upper part of the humus horizon. Carbonates comprise calcite and powder calcite in segregations.

The particle-size distribution shows that the clay fraction (<0.001 mm) dominates (Table 6.1 and Fig. 6.6). Texture is homogeneous at the depth of profile (Table 6.2 and Fig. 6.7).

The profile of calcareous and typical smolnitsa is not differentiated in terms of clay distribution. Commonly, the amount of clay decreases in the lowest soil layer and parent materials.

In the clay fraction the content of humus colloids is low and plays an insignificant role in the high rates of sorption capacity.

The thickness of the humus horizon (A + AC) is 40–80 cm. Content of humus varies from 2–4 %. Stock of organic matter is 200–350 tons per hectare to a depth of 100 cm.

Soil reaction is neutral to slightly alkali. Cation exchange capacity is 35–45 meq/100 g soil. Exchangeable cations are Ca^{2+} and Mg^{2+}.

Calcareous and typical smolnitsa have very high water-holding capacity. The soils are characterized by high rates of water percolation (generally through the cracks) when

Table 6.1 Variables of the particle-size distribution in 17 referenced profiles of calcareous smolnitsa in Bulgaria

Profile count 17	Horizon	Upper depth	Lower depth	Loss HCl	Particle-size distribution (mm), % (according method of Kachinskiy revised)							
					>1 mm	1–0.25	0.25–0.05	0.05–0.01	0.01–0.005	0.005–0.001	<0.001	<0.001[a]
Mean	A′kp	0	29	6.9	0.2	3.0	8.0	15.6	8.1	10.0	48.1	66.1
	A″k	29	56	7.0	0.1	3.0	7.9	15.1	7.5	8.7	50.8	67.0
	A‴k	56	84	10.7	0.1	2.7	6.9	14.9	7.2	8.1	49.4	64.7
	ACk	84	110	13.6	0.2	2.5	7.7	13.6	8.3	10.4	43.7	62.3
	Ck	110	141	23.2	0.1	4.1	8.5	12.5	6.1	8.7	36.8	51.6
Stdev	A′kp	0	4	3.7	0.5	2.1	2.6	2.1	1.5	2.7	7.0	5.4
	A″k	4	6	4.1	0.3	1.9	3.0	2.5	1.2	1.8	6.4	5.9
	A‴k	5	8	7.8	0.3	1.7	2.3	2.9	2.6	3.2	8.5	9.5
	ACk	7	12	5.2	0.5	1.5	4.3	3.2	4.1	4.1	12.3	6.9
	Ck	9	13	7.4	0.5	4.8	4.2	5.6	2.3	2.1	12.2	10.7

[a] Physical clay, the parameter characterized sum of the particle-size distribution, according to Kachinskiy method (mm) $\sum < 0.01 = <0.001 + (0.005 - 0.001) + (0.01 - 0.005)$

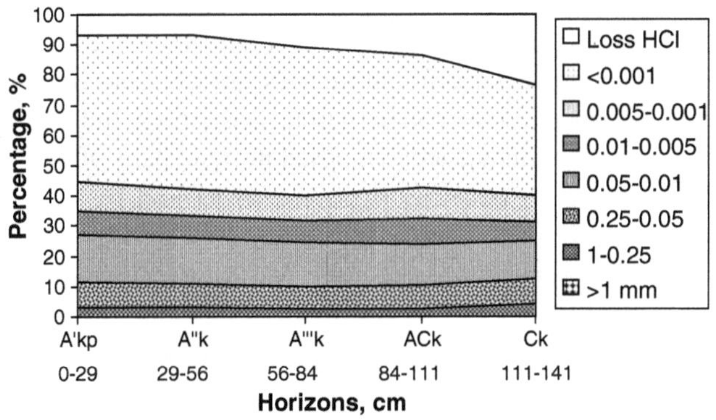

Fig. 6.6 Pattern of the particle-size distribution (mm) in calcareous smolnitsa soil (according to revised Kachinskiy method)

dry and very low rates or none when wet. Soil porosity varies according to soil water saturation. Presence of carbonates implies a lower rates of shrinkage of soil mass.

These soils are poor in biological activity.

Naturally, the spatial relationship is with slightly leached smolnitsa and with calcareous regosols.

Leached smolnitsa

Leached (*subtype*) **Smolnitsa** (*type*) named after the Bulgarian classification of Koynov et al. (1964), and Yolevski and Hadzhiyanakiev (1976).

Distinguishing features are high rates of compaction, lack of clay differentiation, and leaching of carbonates.

Bulgarian classification Koynov et al. (1964), Yolevski and Hadzhiyanakiev (1976) defined three varieties of leached smolnitsa (Table 6.3).

Table 6.2 Variables of humus, pH, carbonates, and particle-size distribution (mm) according to USDA in 17 referenced profiles of calcareous smolnitsa in Bulgaria

Profile count	Horizon	Upper depth	Lower depth	Humus (%)	pH_{KCl}	Bulk density g.cm^3 calc	Clay <0.002	Silt 0.002–0.05	Fine sand 0.05–0.2	Total sand 0.05–2.0
17										
Mean	A′kp	0	29	3.31	6.8	1.21	56.9	31.4	8.6	11.7
	A″k	30	56	2.87	6.8	1.25	59.0	29.6	8.4	11.5
	A‴k	57	84	2.23	6.9	1.31	59.2	30.3	7.7	10.4
	ACk	84	111	1.41	7.0	1.39	56.3	32.3	8.8	11.4
	Ck	110	141	0.65	7.1	–	53.7	32.0	10.3	14.4
Stdev	A′kp	0	4	0.52	0.1	0.05	5.3	3.5	2.7	3.7
	A″k	4	6	0.41	0.2	0.04	4.9	3.7	3.1	4.5
	A‴k	5	8	0.48	0.3	0.04	5.6	3.3	2.3	3.5
	ACk	7	12	0.45	0.2	0.09	10.7	6.9	4.1	4.7
	Ck	9	13	0.19	0.2	–	12.0	7.4	4.1	7.8

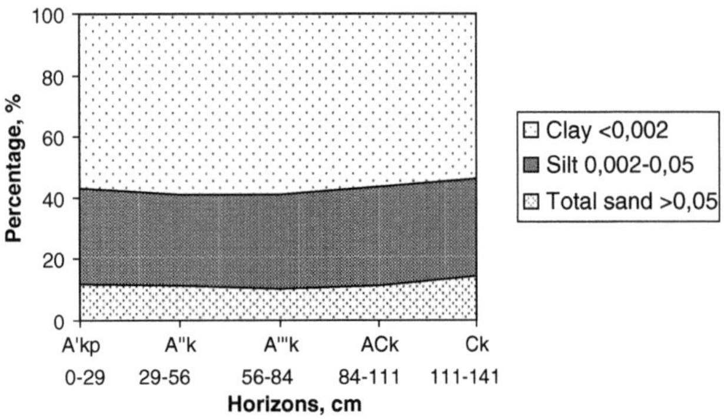

Fig. 6.7 Pattern of the particle-size distribution (mm) in calcareous smolnitsa soil (USDA)

Carbonates are leached out of the humus (A) horizon in the slightly leached smolnitsa; carbonates are leached to the transitional (BC) horizon in the leached smolnitsa. Carbonates are removed out to the (C) horizon without accumulation in the profile in the strongly leached variety.

The particle-size distribution shows that the clay fraction (<0.001 mm) dominates (Table 6.4 and Fig. 6.8). Texture is uniform at the depth (Table 6.5 and Figs. 6.9, 6.10). The amount of clay decreases in the parent materials.

Prevailing water stable aggregates with the size over 0.25 mm are 85 % and with size over 1 mm are 65–70 %. The average particle density is 2.45–2.55 g/cm^3 and depends on the humus content. Porosity is high (60–65 % at field capacity) in the humus horizon. Field water capacity is 65–75 %. Aeration is more than 30 % of soil volume.

Development of structure defines the pore spaces. In the leached smolnitsa 5–15 % of pore space is comprised of macropores or the pores of aeration (with diameter >300 μm); and 15–25 % is the mesopores or pores with rapid moisture drainage (>30 μm diameter). Micropore space or water-filled is about 10–30 %, comprised of the slow drainage pores (diameter 0.2–30.0 μm) and 20–25 % of micropores is with none remaining available for roots moisture (diameter <0.2 μm) (Dilkova 1985; Dilkova et al. 1998).

Commonly the thickness of leached smolnitsa is 80–120 cm, where the thickness of the humus horizon (A + AC) is about 40–80 cm. The content of organic matter can be 3.5–5.0 % at the upper part of the humus horizon in pristine land and gradually decreases downward (Fig. 6.10), but is usually below 2.5 % with C/N ratio 10–14. Bounded with calcium, the humic substances are characterized by high polymerization, condensation, and strong links with clay. Forms of humus are composed in equal parts of humin

Table 6.3 Varieties of leached smolnitsa in Bulgaria

Bulgarian soil classifications (1964, 1976)	WRB (2006)	Keys to Soil Taxonomy (2010)	Revised FAO Legend (1990)
Slightly leached smolnitsa	Epicalcic Mollic Vertisol (Calcaric)	Typic Calcixererts	Calcic Vertisol
Leached smolnitsa	Endocalcic Mollic Vertisol	Typic Haploxererts Typic Haplusterts	Haplic Vertisol
Strongly leached smolnitsa	Bathycalcic Vertisol Eutric Pellic	Typic Haploxererts Typic Haplusterts	Haplic Vertisol

Table 6.4 Variables of the particle-size distribution in 66 referenced profiles of leached smolnitsa in Bulgaria

Profile count 66	Horizon	Upper depth	Lower depth	Loss HCl	Particle-size distribution (mm), % (according method of Kachinskiy revised)							
					>1 mm	1–0.25	0.25–0.05	0.05–0.01	0.01–0.005	0.005–0.001	<0.001	<0.01[a]
Mean	A′p	0	29	2.5	0.2	3.7	11.2	17.3	8.0	9.6	47.6	65.1
	Ai″	29	56	2.4	0.1	3.7	10.8	16.9	7.4	9.1	49.7	66.1
	Ai‴	56	84	2.4	0.1	3.7	10.6	16.6	7.3	8.9	50.4	66.6
	ACk	84	112	4.5	0.1	3.6	10.6	16.0	7.4	9.3	48.5	65.3
	Ck	112	139	13.4	0.2	3.4	10.7	15.5	7.0	8.6	41.2	56.7
Stdev	A′p	0	2	1.0	0.6	3.3	4.5	3.2	2.2	3.4	5.2	5.3
	Ai″	2	4	0.9	0.7	3.1	4.2	3.0	1.6	3.9	5.9	5.3
	Ai‴	4	7	0.9	0.6	3.6	4.1	3.0	1.6	4.1	5.9	5.3
	ACk	7	11	4.5	0.3	3.9	5.1	3.9	1.4	5.6	7.3	5.6
	Ck	11	10	7.2	0.8	4.8	5.0	4.5	1.8	4.5	9.5	9.5

[a] Physical clay, the parameter characterized sum of the particle-size distribution, according to Kachinskiy method (mm) $\sum < 0.01 = <0.001 + (0.005 - 0.001) + (0.01 - 0.005)$

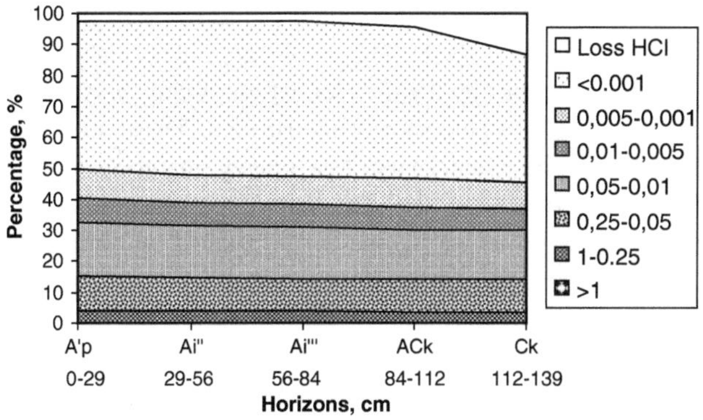

Fig. 6.8 Pattern of the particle-size distribution (mm) in leached smolnitsa soil (according to revised Kachinskiy method)

and humic acids (Petrova 1980). The stock of organic matter is 200–400 tons per ha to a depth of 100 cm.

The qualitative composition of clay minerals, determined by X-ray diffraction analysis and transitional electronic microscopy, has shown the dominance of montmorillonite at rates of 50 % in the clay (<0.001 mm) and 22 % in the soil at the surface (A) horizon, gradually increasing at the depth, where in the (C) horizon it is about 80 % in the clay and 25 % in the soil. There is a significant content of vermiculite at rates of 22 % in the clay and 9 % in the soil at the surface (A) horizon and (C) horizon, and at the transitional (AB) horizon there is a slight increase to 34 % in the clay and 15 % in the soil. Leached smolnitsa shows a high content of 70–80 % in the clay and 30–35 % in the soil in the association dominated by montmorillonite and vermiculite. This mineral composition determines many soil properties. Hydrated micas content is at rates of 17 % in the clay and 7 % in the soil at the surface (A) horizon, gradually decreasing at the depth, where in the (C) horizon it is about 12 % in the clay and 4 % in the soil. Kaolinite is about 6 % in the clay (<0.001 mm) and 3 % in the soil (Halil et al. 1990). Clay minerals distribution in the profile manifests segmentation.

The amount of total iron (Fe_2O_3) (Arinushkina 1970) is uniformly distributed in the profile. In all horizons the silicate form of iron (75–88 % of the total iron Fe_2O_3) greatly dominates and the highest values are in the parent materials. This means that most of the iron is bounded with silicate minerals and the weathering is near the surface (Hadzhiyanakiev 1989).

The non-silicate iron (Mehra and Jackson 1960) is about 12–25 % (of the total iron Fe_2O_3). Almost all non-silicate iron appears in crystallized form (iron oxides and hydroxides). The amount of the crystallized form of iron is less than that of leached chernozems (70–80 % of the amount of non-silicate iron). The conditions for non-silicate iron crystallization are less favorable compared to the leached chernozems.

The hydroxide amorphous form of iron (Tamm 1922) is only 2.5–9.5 % (of the total iron) and without differentiation in the soil.

The free amorphous form of iron is about 19–37 % (of the total non-silicate amorphous iron). This amount is considerably higher than that found in leached chernozems and is evidence of wetter conditions in the soil.

The amorphous form of iron bounded with organic matter (Bascomb 1968) is insignificant and only 0.6–2.6 % (of the total non-silicate amorphous iron). This is due to the fact that humic acids prevail in the humus composition and are characterized by a low ability to make complex formations with iron (Hadzhiyanakiev 1989).

In the spatial relationship leached smolnitsa is bordered by the cinnamonic forest soil and by Eutric Regosols.

The Eutric Regosols are characterized by high clay content in the profile and parent materials (Tables 6.6, 6.7). Texture is homogeneous at depth (Figs. 6.11, 6.12).

Table 6.5 Variables of humus, pH, carbonates, and particle-size distribution (mm). according to USDA in 66 referenced profiles of leached smolnitsa in Bulgaria

Profile count	Horizon	Upper depth	Lower depth	Humus (%)	pH_{KCl}	Bulk density g.cm^3 calc	Clay <0.002	Silt 0.002–0.05	Fine sand 0.05–0.2	Total sand 0.05–2.0
66										
Mean	A′p	0	29	3.0	6.1	1.24	53.75	31.04	11.40	15.20
	Ai″	29	56	2.4	6.1	1.29	55.52	29.70	11.01	14.78
	Ai‴	56	84	1.9	6.1	1.34	56.19	29.28	10.75	14.53
	ACk	84	112	1.4	6.5	1.35	55.63	29.84	10.93	14.53
	Ck	112	139	1.1	7.0	–	52.76	31.97	11.76	15.27
Stdev	A′p	0	2	0.8	0.5	0.07	4.80	4.25	4.47	5.99
	A″	2	4	0.5	0.4	0.04	5.00	4.75	4.25	5.73
	A‴	4	7	0.4	0.4	0.04	4.89	4.33	4.07	6.06
	ACk	7	11	0.3	0.5	0.06	5.82	4.64	5.01	6.88
	Ck	11	10	0.3	0.3	–	8.54	5.77	4.88	7.83

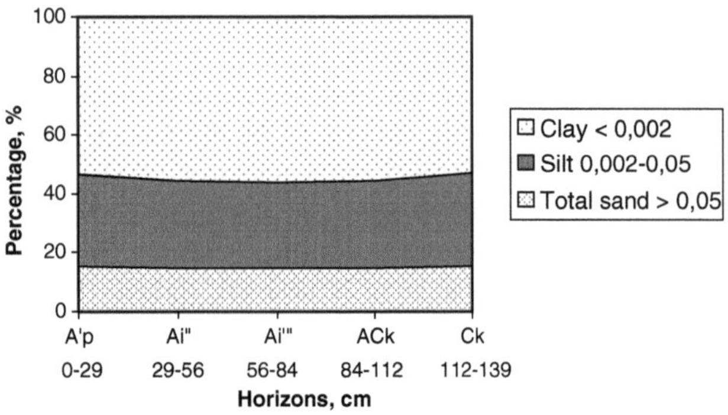

Fig. 6.9 Pattern of the particle-size distribution (mm) in leached smolnitsa soil (USDA)

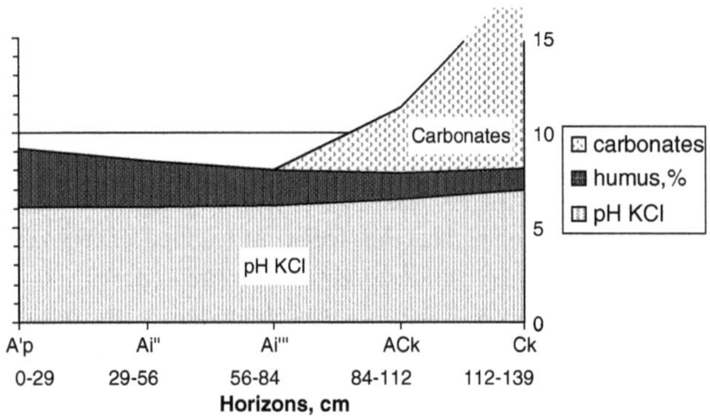

Fig. 6.10 Pattern of pH, humus, and carbonates distribution in leached smolnitsa

6.3.3.2 Degraded (Podzolized) Smolnitsa

Degraded (Podzolized) (*subtype*) **Smolnitsa** (*type*) is named after the Bulgarian classification of Koynov et al. (1964), and Yolevski and Hadzhiyanakiev (1976).

Stagnic Endogleic Vertisol is named according to WRB (2006) and *Aquic Hapluderts* according to Keys to Soil Taxonomy (2010).

Haplic Vertisol: Revised Legend of the World Soils (FAO-UNESCO-ISRIC 1990).

Soil diagnostic was problematic due to negligible distribution (Koynov 1964). These soils are distinguished from the other subtypes by the differentiation of clay, compaction pronounce, and light color at the surface. These soils have formed on the Quaternary Age clayey materials of acid rocks that occur on the Thracian plain. Degradation is indicated in terms of change in the typical smolnitsa features. Degraded (podzolized) smolnitsa is free of carbonates.

Research has proved that these soils are not typically podzolized, but are characterized by advanced weathering in situ that has resulted in differentiation. There is no destruction of the mineral part. Seasonal waterlogging promotes bleaching of the upper layer with grayish brown color.

In the spatial distribution podzolized (degraded) smolnitsa is bordered by strongly leached smolnitsa and strongly leached to slightly podzolize cinnamonic forest soils.

6.4 Soil Properties and Soil Management

6.4.1 Soil Properties

Smolnitsa is poor in biological activity. The intense black color is not associated with humus content, but with the specific quality of the humus.

Soil reaction is generally neutral; it is slightly alkaline in the upper part of the calcareous and typical smolnitsa varieties and is slightly acid in the strongly leached and degraded varieties.

Soils are base saturated 90–95 % with Ca^{2+} and Mg^{2+} (Ganev and Arsova 1980; Ganev et al. 1990). Cation exchange capacity is over 35–45 meq/100 g soil. High values of cation exchange capacity are determined by the high content of smectite clay minerals.

The leached smolnitsa is not differentiated in terms of SiO_2, Al_2O_3, Fe_2O_3 compounds; there is no evidence of MgO and K_2O migration, only CaO.

Mineralogy composition is uniform in the soil profile, consisting of montmorillonite (nontronite) and amorphous

Table 6.6 Variables of the particle size distribution in 18 referenced profiles of Eutric Regosol in Bulgaria

Profile count 18	Horizon	Upper depth	Lower depth	Loss HCl	Particle-size distribution (mm), % (according method of Kachinskiy revised)							
					>1 mm	1–0.25	0.25–0.05	0.05–0.01	0.01–0.005	0.005–0.001	<0.001	<0.01[a]
Mean	Ap	0	26	4.0	1.8	13.3	15.3	12.6	6.0	7.4	39.7	53.6
	AC	26	51	3.1	1.5	10.9	14.9	13.4	6.6	7.9	41.7	56.4
	Ck	51	80	14.5	3.5	11.4	14.1	11.4	6.4	8.6	30.2	46.9
Stdev	Ap	0	4	3.5	3.4	13.5	6.9	4.2	2.0	3.3	12.5	17.0
	AC	4	5	2.1	2.8	8.2	5.1	2.5	1.8	2.6	9.6	12.5
	Ck	5	8	10.1	8.5	9.0	7.5	2.5	3.4	6.6	11.2	11.2

[a] Physical clay, the parameter characterized sum of the particle-size distribution, according to Kachinskiy method (mm) $\sum < 0.01 = <0.001 + (0.005 - 0.001) + (0.01 - 0.005)$

Table 6.7 Variables of humus, pH, carbonates, and particle size distribution (mm) according to USDA in 18 referenced profiles of Eutric Regosols in Bulgaria

Profile count	Horizon	Upper depth	Lower depth	Humus (%)	pH_{KCl}	Bulk density g.cm^3 calc	Clay <0.002	Silt 0.002–0.05	Fine sand 0.05–0.2	Total sand 0.05–2.0
18										
Mean	Ap	0	26	2.3	5.8	1.3	46.0	24.0	15.7	30.0
	AC	26	51	1.6	5.9	1.4	47.7	25.4	15.2	26.9
	Ck	51	80	0.8	6.6	–	43.1	28.2	15.5	28.7
Stdev	Ap	0	4	0.8	0.9	0.1	12.9	6.1	7.0	17.7
	AC	4	5	0.7	0.9	0.1	10.3	3.7	5.1	12.5
	Ck	5	8	0.3	0.8	–	10.1	7.3	7.2	12.9

substances. In soils developed on Pliocene Age sediments some kaolinite occurs.

In the soil colloids content of Fe_2O_3 is twice that of Al_2O_3. SiO_2 content is uniform in all horizons. Ratio $SiO_2 : R_2O_3$ shows the presence of nontronite clay mineral in stable form and montmorillonite.

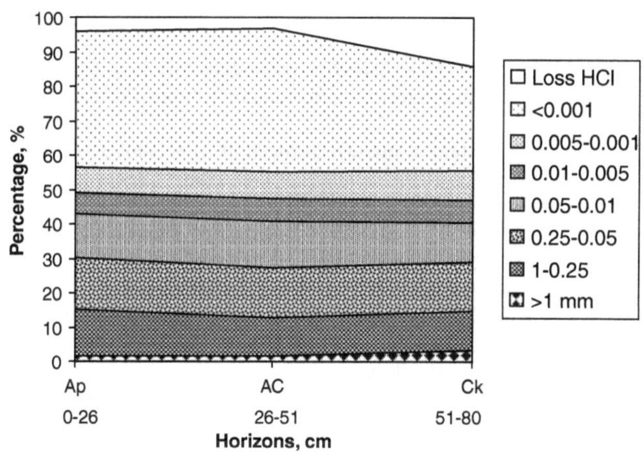

Fig. 6.11 Pattern of the particle-size distribution (mm) in the Eutric Regosol (according to revised Kachinskiy method)

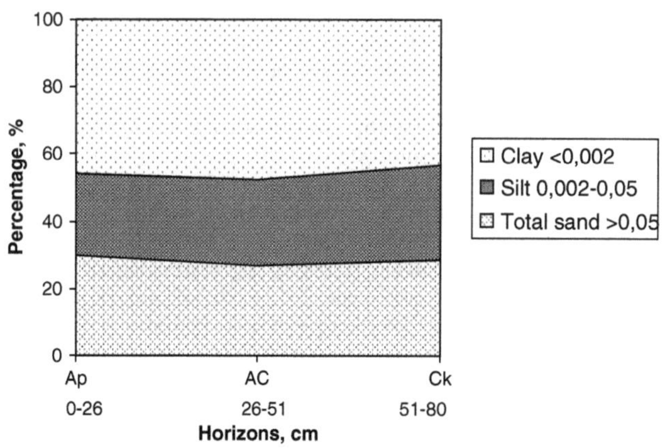

Fig. 6.12 Pattern of the particle-size distribution (mm) in the Eutric Regosol (USDA)

Smolnitsa has very high water-holding capacity, and high rates of the none available for plant roots moisture. The wilting point is 60–70 % (at field capacity). Soils are characterized by high rates of water percolation (generally through the cracks) when dry and very low rates or none when wet. Soil porosity varies according to soil water saturation.

Deep cracks formation promotes dryness of the soil mass at the depth. Consequently, the aeration of the soil then dry affects the deeper layers, but because of high compaction the internal aeration of aggregates is low.

Despite the clayey texture, these soils tend to be loosely structured at the exposed surface horizon when dry. This is due to very fine granular structure and explains why the soils do not form a crust.

Contrary to the chernozems soil group, smolnitsa soil is characterized by poor physical properties. Soils are hard when dry, firm when moist and plastic, and sticky when wet.

6.4.2 Soil Management

Smolnitsa has many land-use restrictions. The high shrink and swell activity causes a shifting of foundations, road-beds, buried pipes and wires, septic systems, etc.

Smolnitsa has natural high fertility, good structure, and high water-holding capacity which makes it suitable for agriculture. Since rainfalls is relatively low and summer drought is sometimes a hazard, moisture conservation is important. The nutrients stored by the soil, especially organic nitrogen and phosphorous, are released for use by plants as organic matter decomposes.

Cultivated soils differ from those in pristine lands in the fact that nutrients uptake by crops prevails and is not compensated for by residues that enter back into the soil. The conversion of abandoned lands with leached smolnitsa into virgin land is not associated with a trend increase in the content of humic substances, at least in the initial period (Petrova and Shishkov 2005).

Smolnitsa occupies flat terrains, its occurrence on pristine land is negligible/almost nil.

Conditions for nitrogen mineralization are not favorable due to the high clay content and poor aeration. The amount of available nitrogen is not sufficient for producing high yields.

These soils often give rise to serious management problems due to negative/poor physical properties that make them difficult to wet and till. Naturally, clayey soils are vulnerable to compaction and poor aeration. The tillage period is short. Water stable aggregates are dramatically decreased after two decades of use.

Smolnitsa is susceptible to all forms of erosion. Wind erosion affects the non-coherent soil particles.

Large areas of smolnitsa terrain are used for growing cotton, sugar beets, cereals, maize, sunflower, ground nuts, beans, and alfalfa. Vegetables can be grown under irrigation.

The retention of water for plant roots is a favorable in case of drought stress for crops with prolonged growing season such as maize, sunflower, cotton, and beet.

In the western region spring comes late due to the climatic conditions and higher altitude location. Due to the water saturation, poor aeration, and low temperatures the region is not suitable for growing cotton, tobacco, rice, or for the cultivation of vineyards.

References

Arinushkina E (1970) Manual on chemical soil analysis. Moscow State University Press, Moscow (In Russian)

Bascomb CL (1968) Distribution of pyrophosphate extractable iron and organic carbon in soils of various groups. J Soil Sci 19(1):251–268

Bondev I (1991) Vegetation in Bulgaria, map at scale 1:600,000 with explanatory text. St. Kliment Oxridski University Press, Sofia (In Bulgarian)

Boykov P, Doncheva L (1928) Annual report of the state agricultural experimental and control station for 1920, Sofia (In Bulgarian)

Dilkova R (1985) Structure and aeration in main soil varieties in Bulgaria. Theses, Institute of Soil Science Nikola Poushkarov, Sofia (In Bulgarian)

Dilkova R, Kerchev Zh, Kercheva M (1998) Evaluating and grouping of soils according to their susceptibility to anthropogenic degradation, advances in geoecology. Catena Verlag, Reiskirchen, pp 125–131

FAO-UNESCO-ISRIC (1990) Revised legend of the soil map of the world. FAO world soil resources report no 60, Rome

Ganev S, Arsova A (1980) Methods of determining the strongly acid and the slightly acid cation exchange in soil. Soil sci Agrochem J 15(3):22–33, Sofia (In Bulgarian)

Ganev S et al (1990) Physic-chemical properties of soils in Bulgaria and other world's regions. Department of Chemistry, Institute of Soil Science Nikola Poushkarov, Sofia (In Bulgarian)

Glazovskaya MA (1971) Landscape-geochemical relicts of Australia, Vestnik MGU, Series Geography, pp 45–52

Hadzhiyanakiev As (1989) Iron in Bulgarian soils. Zemizdat Press, Sofia, p 2345 (In Bulgarian)

Halil R, Boneva K, Houbenov G (1990) Investigations on the composition of clayic minerals in some Bulgarian soils. Soil Sci Agrochem J 25(4):13–22, Sofia (In Bulgarian)

Keys to Soil Taxonomy (2010) USDA United States Department of Agriculture, NRCS Natural Recourses Conservation Service, 11th edn. Soil Survey Staff, Washington, DC, p 338

Kovda VA (1981) Soil cover, its improvement, use and protection. Nauka Press, Moscow (In Russian)

Koynov V (1964) Smolnitsas in Bulgaria and their peculiarities. In: Gerasimov IP et al. (eds) Soils of south–eastern Europe. Proceedings of the international symposium of soil science, June 1963, Sofia. Bulgarian Academy of Sciences Press, Sofia: 95–112 (In Russian)

Koynov V, Trashliev Ch, Yolevski M et al (1968) Soil map of Bulgaria at scale 1:400,000 with description, Institute of soil science Nikola Poushkarov, general office of geography and mapping. Bulgarian Academy of Sciences Press, Sofia (In Bulgarian and English)

Koynov V, Trashliev Ch, Yolevski M et al (1972) Soil resources of Bulgaria and their use. First national soil science congress, 23–25 Sept 1969. Bulgarian Academy of Sciences Press, Sofia (In Bulgarian)

Mehra OP, Jackson ML (1960) Iron oxide removal from soils and clays by dithionite—citrate—system buffered with sodium bicarbonate. Clays Clay Miner 7:317–327

Penkov M et al (1992) Soil classification and diagnostic in Bulgaria in relation to the land reform fragmentation. Zemizat Press, Sofia (In Bulgarian)

Petrova L (1980) State of organic nitrogen in some soils in Bulgaria. Theses, Institute of Soil Science N Poushkarov, Sofia, p 138 (In Bulgarian)

Petrova L, Shishkov T (2005) Changing of humus content in abandoned arable soils, ecology and industry, vol 7(1). Balcanereco, Sofia, 8–10 June 2005, pp 83–85 (In Bulgarian)

Poushkarov N (1928) The experiment of improvement of Sofia's clay chernozem. Annual report of the Agricultural experimental station 1927–1928, Sofia (In Bulgarian)

Poushkarov N (1931) Soil map of Bulgaria at scale 1:500,000, Sofia

Singh S (1956) The formation of dark colored clay-organic complexes in black soils. J Soil Sci 7:43–58

Stranski I (1935) Black soils of Sofia town region. Annual report of Sofia's University, Faculty of Agronomy, vol 14, Sofia (In Bulgarian)

Stranski I, Koynov V (1947) Smolnitsa in Chirpan region. In: Proceedings of the Institute of Soil Science, No 1, Bulgarian Academy of Sciences Press, Sofia (In Bulgarian) 6

Tamm O (1922) Eine methode zur bestimung der anorganischen komponente ds gelcomplexes im boden. Statens Skogsforsoksanstalt, Medd 19:387–404 (by Arinoushkina E (1972) Manuel of chemical analysis of soils, Moscow)

World Reference Base for Soil Resources (2006) A framework for international classification, correlation and communication. FAO-ISRIC, Rome

Yolevski M, Hadzhiyanakiev A (1976) Agro-grouping of soils in Bulgaria, Extended systematic list. Code of groups and soil varieties. Institute of Soil Science Nikola Poushkarov, Sofia (In Bulgarian)

Yolevski M, Ninov N, Krastanov S et al (1983) Brief diagnostic and criterion of the main taxonomic soil divisions in Bulgaria. Precise Classification Scheme, Institute of Soil Science Nikola Poushkarov, Sofia (In Bulgarian)

7 Cinnamonic Forest Soils

Derived from the Greek word *canella,* meaning cinnamon and from the Latin word *eluere,* meaning to wash out.

Cinnamonic forest soils are reddish colored texturally differentiated soils.

7.1 Concept of the Type

Main concepts

1. Cinnamonic forest soils are mineral soils with reddish color and clay enrichment as a result of their evolution;
2. Sesquioxides (oxides, hydrous oxides, hydroxides) have accumulated in the soil profile;
3. Developed on old parent materials of different composition and age;
4. Characterized by high base saturation (>80 %).

The main processes that have taken place in the development of cinnamonic forest soils are weathering in situ, lessivage, leaching, and humification.

7.2 Ecology of Soil Formation

7.2.1 Distribution

Cinnamonic forest soil is widely distributed in southern Bulgaria, located on plateaus, in valleys, on medium gradient hills, and mountainous highlands (Fig. 7.1). These soils occupy approximately 21.91 % of the entire country's territory. Most of the soils are shallow.

These soils are mostly distributed from east to west on the foot slopes of the sub-Balkan province and in the southern part of Fore Balkan. They occupy the low elevated slopes of Sredna Gora Mountain, the western and eastern parts of Rhodope Mountains, Sakar Mountain, the northeastern part of Strandzha Mountain, and the western Sofia-Kraishte province. Soils are located on the Upper Thracian Lowland, and in the southwestern Ossogovo province in the region of Struma and Mesta River valleys.

In the geographic distribution, cinnamonic forest soils are found in elevated areas, where the level of ground water is of considerable depth. Meadow soil varieties occur near the river valleys and old terraces.

In the mountainous regions, the variety is known as shallow cinnamonic forest soil and overlies hard rock. Due to the permanent process of erosion the soil has low thickness and is somewhat skeletal.

Climatic conditions

Development of the cinnamonic forest soils implies that they formed in various climatic conditions since the Pliocene Age. Nowadays cinnamonic forest soils are found in regions with transitional continental climate influenced by the Mediterranean. The southernmost part of the country does in fact have sub-Mediterranean climatic conditions.

The climate is characterized by warm winters and hot dry summers. Autumn is warm and prolonged. The temperature regime and total precipitation are variable from one region to another. In the different regions the amount of precipitation consists of two maximum falls during summer and winter. Dryness usually occurs from June to September.

In the central province of the Thracian Lowland in the region of Maritsa and Tundzha River valleys the mean annual temperature is 10.1–12.9 °C. In January the mean annual temperature is −1.0 (1.3) °C. In July the mean annual temperature is 20.3–24.0 °C. The mean amount of precipitation varies from 577 to 656 mm.

In the western province the mean annual temperature is 8.0–10.1 °C. In January the mean annual temperature is −1.7 (−3.9) °C. In July the mean annual temperature is 18.1–20.3 °C. The mean amount of precipitation varies from 585 to 688 mm.

In the southwestern province in the region of Ossogovo and Struma and Mesta River valleys the mean annual

T. Shishkov and N. Kolev, *The Soils of Bulgaria*, World Soils Book Series, DOI: 10.1007/978-94-007-7784-2_7,

Fig. 7.1 The landscape of leached cinnamonic forest soils

temperature varies from 9.0 to 13.9 °C. In January the mean annual temperature is −0.8 (2.1) °C. In July the mean annual temperature varies from 18.6 to 24.9 °C. The mean amount of precipitation varies from 533 to 695 mm.

In the southeastern Rhodopes province the mean annual temperature is 12.2–12.8 °C. In January, the mean annual temperature is 0.8 (1.4) °C. In July the mean annual temperature is 23.0–23.5 °C. The mean amount of precipitation varies from 609 to 761 mm.

Topography

Cinnamonic forest soils are found in valleys and on hilly and low mountain territories of up to 700 m altitude. The plains and valleys have undulating relief and are more or less dissected. The foot slopes of the mountains have high gradient rolling relief and are rather broken. Most of the regions are strongly dissected and affected by erosion.

Cinnamonic forest soils usually occupy uplands on the plains, elevated sites in the valleys, and the foot slopes of mountains with elevation below 700 m.

Parent materials

Cinnamonic forest soils occur on a broad variety of parent weathered materials and rocks.

In the western part of the country, cinnamonic forest soils have formed on the Quaternary Age alluvial-drift and talus-drift sediments (boulders, pebbles, and sands); on the Paleogene/Oligocene Age continental-molasses coal-bearing sediments (conglomerate-sandstone, shale bituminous formations); on the Paleogene/Eocene–Oligocene Age continental volcanogenic-sedimentary rocks (breccia, conglomerates, sandstones, shales, tuffs, tuffites); on the Upper Cretaceous (Maastrichtian) Age (Mesozoic Era) terigenous-carbonate formations (conglomerates, sandstones, clays, marls, organogenic limestones); on the eluvium of the Ordovician Age (Paleozoic Era) diorites, gabrodiorites, and granodiorites; and on the Precambrian Age (Paleozoic Era) amphibolites, muscovite to two mica gneiss-schists, biotite gneisses, schists, graphite-bearing quartzites, leptinites, amphibolites, and metaconglomerates.

In the central Balkan and Sredna Gora Mountain regions, the soils have formed on the eluvium of the Upper Cretaceous (Turonian-Campanian) Age (Mesozoic Era) gray and red limestones, limestones, with flint, trachybasalts, trachyandesitobasalts, trachytes, with packets of sediment sandstones, siltstones, marls, and clayey limestones; and on the Carboniferous Age (Paleozoic Era) granites and granodiorites.

In the central Thracian Lowland, the soils have formed on the Quaternary Age alluvial-drift and talus-drift sediments (boulders, pebbles, and sands); on the Neogene Age continental sediments in the superimposed depressions of South Bulgaria (conglomerates, sands, and clays); on the Paleogene-Neogene (Oligocene–Miocene) Age clays, sands, and coals; and on the Paleogene/Eocene Age limestones and marls.

In the southwestern part, cinnamonic forest soils have formed on the Quaternary Age alluvial-drift and talus-drift sediments (boulders, pebbles, and sands); on the Neogene Age terrigenous continental sediments in the superimposed depressions (conglomerates, breccia-conglomerates, sandstones, siltstones, shales, sands, clays); on the Paleogene/Oligocene Age continental-molasses coal-bearing sediments (conglomerate-sandstone, shale bituminous formations); on the Paleogene/Eocene–Oligocene Age continental volcanogenic-sedimentary rocks (breccias, conglomerates, sandstones, shales, tuffs, tuffites); and on the eluvium of the Precambrian Age (Paleozoic Era) granitized biotite, two-mica gneisses, migmatites, granite-gneisses, gneisses, amphibolites, silimanite-garnet schists, schists, graphite-bearing quartzites, leptinites, and metaconglomerates.

In the southwestern region of Struma and Mesta Rivers, the cinnamonic forest soils have formed on the Neogene Age terrigenous continental sediments in the superimposed depressions (conglomerates, clays); on the Paleogene-Neogene/Oligocene–Miocene Age sands, sandstones, and conglomerates; on the eluvium of the Paleogene/Oligocene Age rhyolites, rhyodacites, trachyrhyolites, delenites, and acid pyroclastics tuffs, tuffites, tuffaceous sandstones, and reef limestones; on the Paleogene/Upper Eocene Age volcanic rocks (latites, andesites, shoshonites, absarokotes) and on volcanogenic-sedimentary rocks (tuffs, tuffites, sandstones, limestones, olistostromes, terrigenous coal-bearing sediment conglomerates, sandstones, siltstones, marls, coal shists, coal, breccia conglomerates, and gritstones); on the Lower Triassic Age (Mesozoic Era) metasandstones, metapelites, schists, calcite, and dolomite marbles; on the eluvium of the Carboniferous Age (Paleozoic Era) granites and granodiorites; on the Precambrian (Paleozoic Era) granitized biotite, two mica gneisses, migmatites, silimanite-garnet schists, metaconglomerates, porphyroblastic gneisses, granite-gneisses, leptinites, amphibolites, gneiss-schists, schists, migmatized gneisses, quartzites, and marbles.

In the eastern Strandzha region, soils have formed on the Paleogene marine sedimentary rocks of Paleocene-Middle Eocene Age flysch alternation of sandstones, siltstones, sandy limestones, clayey marls, clay packets of breccia-conglomerates; on the Upper Cretaceous (Maastrichtian) Age (Mesozoic Era) flysch formation interbedding of sandstones, siltstones, limestones, and marls; on the Upper Cretaceous (Turonian-Coniacian) Age fine to block breccia with packets of sandstones, siltstones, schales, marls, clayey limestones, and volcanics basalts, trachybasalts, and trachyandesite basalts; on the Lower Triassic Age (Mesozoic Ea) metasandstones, metapelites, and marbles; on the eluvium of the Carboniferous Age (Paleozoic Era) granites and granodiorites; and on the Precambrian Age (Paleozoic Era) migmatites, biotites, two micas gneisses, gneiss-schists, amphibolites, and interbedded marbles.

Vegetation

Cinnamonic forest soils have a mixed cover of xerophytic, mesophytic, and mesothermic vegetation dominated by xerothermic oak forest and shrubs. There may have been various sequences of vegetation over a long period of time. Indigenous forest formations that occupy the dry lands of the hilly plains are: *Quercus cerris, Quercus frainetti, Quercus pubesccens, Quercus virgilianae, Fraxineta orni, Acereta monspessulani,* and *Junipereta excelsae.* Areas of destructed woodlands are occupied by *Carpineta orientalis,* and the shrub formations of *Paliureta spina-christi, Genisteta rumelicae, Genisteta lidiae, Tragacanthus spinose, Astagaleta augustifoliae, Astragaleta aitosensis,* and *Astragaleta thracicae.* Nowadays the most common plant community is oak forest, but it be could be shrubs and open grassland. Species vary from place to place and can include common beech (*Fagus orientalis*), oak including *Quercus sesiliflora, Quercus conferta, Quercus pedunculata, Quercus pedunculiflora, Quercus stranjensis,* and *Quercus orientalis,* hornbeam (*Carpinus betulus*), aspen (*Populus tremula*), almond trees (*Amygdalus nana*), fig trees (*Ficus carica*), and other species. Steppe vegetation includes the following formations: *Amygdaleta nanae, Artemisieta albae, Agropireta pectiniforme, Agropireta brandse, Brometa riparii* and others such as *Dichantieta ischaemi, Poaeta bulbosae, Chrysopogoneta grylli, Cynodonteta dactyloni, Lolieta perenne,* and *Ephemereta* (Bondev 1991).

In southern regions of the Struma valley, the evergreen shrub formations of *Phyllireeta latifoliae* and *Querceta cocciferae* grow in shallow soils in rocky areas. In the central Balkan and Sredna Gora Mountain regions oak forest is preserved on the sloping areas. The eastern part of the Balkan Mountains is covered with oak and hornbeam forest.

In the past, the low elevated slopes of Rhodope Mountains were covered with oak forest; however, today they are dominated by the cultivation of tobacco and to a lesser extent potatoes.

Two centuries ago the region of Maritsa River in the Thracian Lowland was covered with dense oak forest; today the area is agricultural.

Time

Most cinnamonic forest soils are quite old (hundreds of thousands of years) and retain features dating back to Pleistocene.

Human influence

Over the past two centuries forest areas have been reduced and/or converted into arable lands. Cultivated lands are 41 % of the total area occupied by the cinnamonic forest soils. In the forest areas, the species composition has now changed.

Today the endemic oak forest is preserved at the Balkan Mountains and Strandzha Mountain.

Converting the woodlands into arable land has led to enhanced mineralization of organic matter and loss of structure aggregation.

Texturally, differentiated subsoil is more vulnerable to compaction.

7.2.2 Genesis

The genesis of soils with differentiated profile is still open to discussion. The precise time of *argic* formation is not clear. Some hypotheses have considered the possibility that it evolved since Pleistocene. In many soils the clay-rich horizon hold relict features of previously existed conditions.

Another theory is that the soils were formed in the conditions that prevailed during the long migration downwards. The precise mechanism of clay deposition is not understood. Soluble salts and primary calcium carbonates were leached out salts and carbonates were leached by precipitation; this was followed by the gradual translocation of clay by gravity. Textural differentiation was enhanced by weathering in situ.

It is supposed that the processes of leaching, lessivage, and weathering in situ took place in more humid conditions, while the rubification (dehydration of iron oxides) took place during hot dry periods. As regards the leaching of carbonates, it is also open to discussion whether the depth of accumulation was influenced by soil formation.

Concerning soil evolution, these soils are of old origin and there are relict features in the soil profile. Intensive weathering during the winter formed clay-rich at the middle part of the horizon. Advanced hydrolysis of the silicates results in the release of iron oxides that give the soils a reddish color.

Parent materials are often represented by old washed out soils and crusts that result from weathering.

7.3 Soil Diagnostic and Classification

7.3.1 Morphology

A typical distinguishing feature is a well-developed textural horizon (Figs. 7.2, 7.3, 7.4, 7.5).

Morphologically these features are red colored, clay-rich and with compacted horizon. Usually, the illuvial horizon commences at the lower part of the eluvial AE horizon.

Fig. 7.2 Cinnamonic forest soil, Bulgarian classification Yolevski and Hadzhiyanakiev (1976), *Cutanic Luvisol* (*Skeletic, Chromic*), WRB (2006), *Typic Haplxeralfs*, Keys to Soil Taxonomy (2010), *Chromic Luvisol*, Revised Legend of the world soils FAO-UNESCO-ISRIC (1990)

The pattern of horizon sequence in the profile is O–AE–Bt1–Bt2–B–C(k).

O—A thin (<3 cm) leafy litter, consisting of partly decomposed leaves and grass remnants, naturally occurs in woodlands; below is clear transition into

***A*(*Ap*)**—A thin humus accumulative horizon 5–10 cm thick, color varies from dark grayish brown to dark reddish brown (moist) and brown or reddish brown (dry) colored, homogeneous, with fine crumb size (1–2 mm) structure (spheroids or polyhedrons), strong pedality grade, consistence is firm (moist), and slightly hard (dry). Abundance of roots with fine size (0.5–2.0 mm) and medium size (2.0–5.0 mm), presence of earthworm casts and vertebrate

Fig. 7.3 Leached cinnamonic forest soil, Bulgarian classification Yolevski and Hadzhiyanakiev (1976), *Haplic Luvisol* (*Chromic*), WRB (2006), *Typic Haplustalfs*, Keys to Soil Taxonomy (2010), *Chromic Luvisol,* Revised Legend of the world soils FAO-UNESCO-ISRIC (1990)

Fig. 7.4 Leached Cinnamonic forest soil, Bulgarian classification Yolevski and Hadzhiyanakiev (1976), *Cutanic Luvisol* (*Chromic*), WRB (2006), *Typic Paleustalfs*, Keys to Soil Taxonomy (2010), *Chromic Luvisol,* Revised Legend of the world soils FAO-UNESCO-ISRIC (1990)

passages; naturally occurs in the pristine lands or it is plough; clear grades into

AE(p)—a moderately thick humus-eluvial horizon, reddish dark brown or brown colored (moist), no visible evidence of bleaching, structure is subangular blocky with very fine size (<5 mm), strong pedality grade, consistence is firm (moist) and hard (dry), porosity is high; sharp change beneath into

Bt(g)—a thick illuvial horizon that can be subdivided into prime, second etc. Dark reddish brown, reddish brown or strong brown colored with hue 7.5YR or 5YR or less; usually the reddish color is due to the presence of iron oxides. Shows an increase in clay with coarse (20–50 mm) angular blocky structure (blocks having mixed, rounded, and sharp angled faces of aggregates) and compaction pronounce, strong pedality grade, consistence is very firm (moist) and very hard (very resistant to pressure) dry. Porosity is medium consisting mainly of fine pores and cracks. Many skins are on ped surfaces or pore spaces and cracks. Commonly there are coarse grains in the soil mass; many rusty spots and soft black manganese concretions occur; grade into

BC—a transitional mineral illuvial-metamorphic horizon, with a lower content of clay and where cutans are few or absent; strong brown, reddish brown or yellowish red colored, structure is medium size (10–20 mm) subangular blocky, strong pedality grade, consistence is very firm (moist) and very hard (very resistant to pressure) dry; sharp change beneath into

C(k)—parent materials which are reddish brown, reddish yellow, yellowish red or brownish yellow colored, structure is massive (no structure), carbonates can occur lower down.

In the humus-eluvial layer (A + AE) processes are related to the mineralization of organic remains and accumulation of organic matter. Leaching of soluble salts and carbonates contributes to further lessivage development. Oriented skins of clays require dispersal of the clay in the

Fig. 7.5 Leached cinnamonic smolnitsa-like soil, Bulgarian classification Yolevski and Hadzhiyanakiev (1976), *Vertic Cutanic Luvisol* (*Chromic*) or may be *Mollic Vertisol* (*Eutric*), WRB (2006), *Vertic Paleustalfs* may be *Typic Haplusterts*, Keys to Soil Taxonomy (2010), *Verti-Chromic Luvisol* may be *Eutric Vertisol*, Revised Legend of the world soils FAO-UNESCO-ISRIC (1990)

eluvial horizon, translocation of the clay to the illuvial horizon and physical precipitation on ped faces.

Oxidation and reduction of iron oxides is apparent throughout the profile in the form of rusty mottles. Chemical weathering in situ takes place over the whole year.

7.3.2 Classification

Poushkarov (1931) classified the soil as a podzolized unit in the group of brown forest soils. The Bulgarian classification of (Antipov-Karataev and Gerasimov 1948) defined the type of cinnamonic forest soil and the subtypes of typical, leached, and podzolized. The classification of Koynov et al. (1964) shifted the subtype podzolic to the single type. On the legend of the Soil map 1:400,000, Koynov et al. (1968) defined four subtypes: typical, leached, smolnitza-like, and strongly leached (lessive). Gyurov and Ninov (1964), Gyurov et al. (1978) proposed two soil subtypes—typical and strongly leached (lessive). In the soil classifications of Yolevski et al. (1983) the subtype of typical was renamed as calcareous. There was classification in (Penkov et al. 1992) which at the theoretic level named subtypes of the cinnamonic forest soil (mollic, albic, calcic, vertic, rodic, haplic) but this has not been applied in practice.

The distinguishing feature is pronounced textural differentiation in the profile on the subtype level. Subtype varieties differ in the degree of leaching (slight, moderate, and strong), in the thickness of the humus horizon (non-eroded, slightly eroded, and accumulated), in the degree of erosion (slightly, moderately, or not eroded), and in the occurrence of gleyic or vertic properties in the profile.

WRB (2006) allowed that soils that respond to the demands of *argic* horizon can be referred to as Luvisols.

Keys to Soil Taxonomy (2010) required the *ochric* epipedon and the *argillic* horizon for Alfisols order classification. In Soil Taxonomy there are demands for subsurface horizons allowing *argillic* horizon with high base saturation (>35 %) expressing mottling, gleying, and coarse structure development; *or kandic* horizon with low (<35 %) base saturation in Ultisols; *calcic or petrocalcic* horizon; and*albic, natric, or none*.

Ochric epipedon is most commonly a thin A horizon showing some incorporation of organic matter.

7.3.3 Cinnamonic Forest Soils Subtypes in Bulgaria

7.3.3.1 Typical Cinnamonic Forest Soils

Typical meaning calcareous (*subtype*) **cinnamonic forest soil** (*type*) is named after the Bulgarian classification of Koynov et al. (1964), Yolevski and Hadzhiyanakiev (1976).

Epicalcic Luvisol (*Chromic*) or *Vertic Cambisol* (*Chromic*) is named according to WRB (2006) and *Calcic Haplexeralfs* or *Calcic Haploxerepts* according to Keys to Soil Taxonomy (2010).

Chromo-Calcic Luvisol: Revised Legend of the World Soils (FAO-UNESCO-ISRIC 1990).

Distinguishing features are carbonate occurrence and little or no textural differentiation. Morphologically, these features show homogeneous texture and the occurrence of soft carbonates segregation. Most soils are usually shallow (50 cm profile depth) or moderately shallow (less than 100 cm deep) and developed on content-rich calcareous parent materials.

The pattern of the horizon sequence is A(Ap)–AB(k)–Bk–BCk–Ck. The average thickness of the humus-eluvial horizon (A + AE) is 20–30 cm. The surface humus (A) horizon is thin and dark brownish-gray, and humus content varies from 2 to 4 %. Stock of organic matter in a 100 cm layer is about 150 ton per ha in pristine lands and is less in arable lands. In humus composition humic acids bound with calcium prevail. Soil reaction is neutral to slightly acid at

Table 7.1 Variables of the particle size distribution in 30 referenced profiles of strongly leached cinnamonic forest soils in Bulgaria (according to revised Kachinskiy method)

Profile count 30	Horizon	Upper depth	Lower depth	Loss HCl	Particle-size distribution (mm), (%) (according method of Kachinskiy revised)							
					>1 mm	1–0.25	0.25–0.05	0.05–0.01	0.01–0.005	0.005–0.001	<0.001	<0.01[a]
Mean	AEp	0	26	2.0	0.7	6.9	15.8	19.4	7.9	9.2	38.1	55.2
	Bt1	26	51	1.9	0.7	5.6	12.9	14.9	6.8	7.9	49.3	64.0
	Bt2	51	76	2.0	0.6	5.4	13.3	14.6	6.5	8.0	49.5	64.1
	B	76	101	4.3	0.7	5.8	12.6	15.2	6.9	8.5	46.0	61.3
	Ck	101	126	13.8	1.0	5.5	14.3	13.7	6.8	8.6	36.3	51.6
Stdev	AEp	0	4	1.7	1.1	5.9	5.2	5.2	2.4	2.6	8.3	8.7
	Bt1	4	5	0.8	1.4	5.0	4.9	4.2	2.0	2.0	6.6	7.8
	Bt2	5	9	1.1	1.5	5.5	4.3	4.0	1.9	2.1	7.2	8.0
	B	9	13	6.7	2.1	6.0	5.4	3.6	2.3	4.3	9.6	10.5
	Ck	13	16	9.7	1.9	6.1	9.6	3.1	2.2	4.6	9.3	11.3

[a] Physical clay, the parameter characterized sum of the particle-size distribution, according to Kachinskiy method (mm) $\sum <0.01 = <0.001 + (0.005-0.001) + (0.01-0.005)$

Table 7.2 Variables of the particle size distribution in 20 referenced profiles of strongly leached cinnamonic forest soils in Bulgaria (according to revised Kachinskiy method)

Profile count 20	Horizon	Upper depth	Lower depth	Loss HCl	Particle-size distribution (mm), (%) (according to method of Kachinskiy revised)							
					>1 mm	1–0.25	0.25–0.05	0.05–0.01	0.01–0.005	0.005–0.001	<0.001	<0.01[a]
Mean	AEp	0	26	1.5	3.8	14.3	17.9	16.7	8.1	7.5	30.0	45.7
	Bt1	26	56	1.4	2.2	10.9	14.4	14.6	6.9	7.4	42.3	56.6
	Bt2	56	87	1.4	2.1	11.2	16.0	14.8	6.5	7.6	40.4	54.5
	Ck	87	112	7.0	1.6	19.2	18.5	14.0	6.6	6.7	26.5	39.8
Stdev	AEp	0	4	1.3	5.3	11.2	6.4	3.5	2.9	2.9	14.4	17.5
	Bt1	4	10	0.7	3.9	8.8	4.7	3.2	2.0	2.1	11.7	13.3
	Bt2	10	19	0.7	4.1	8.9	5.6	3.6	2.1	2.2	11.1	12.7
	Ck	19	20	6.7	2.7	10.3	8.3	4.0	1.7	2.7	11.5	11.3

[a] Physical clay, the parameter characterized sum of the particle-size distribution, according to Kachinskiy method (mm) $\sum <0.01 = <0.001 + (0.005-0.001) + (0.01-0.005)$

the surface, changing to alkali in the B horizon. Exchangeable cations Ca^{2+}, Mg^{2+}, and K^{+} prevail at the cation exchange capacity. Base saturation is over 90 % (Ganev and Arsova 1980) and the cation exchange capacity is about 35–45 meq/100 g, and significantly higher in the lower part of the soil.

7.3.3.2 Leached Cinnamonic Forest Soils

Leached (*subtype*) **cinnamonic forest soil** (*type*) is named after the Bulgarian classification of Koynov et al. (1964), Yolevski and Hadzhiyanakiev (1976).

Cutanic Luvisol (*Chromic*) is named according to WRB (2006) and *Typic Palexeralfs or Haplxeralfs* according to Keys to Soil Taxonomy (2010).

Verti-Chromic Luvisol or *Chromic Luvisol*: Revised Legend of the World Soils (FAO-UNESCO-ISRIC 1990).

The eluvial (A + AE) horizon has a thickness of 25–50 cm. A typical feature of soils formed under forests is a sharp decrease in organic matter with the depth. The illuvial (Bt) horizon has a thickness of 60–120 cm. The translocation of fine material without further destruction is typical. The mineral composition of the clay fraction consists of hydromicas, with less montmorillonite (or smectite group) and kaolinite.

Thin section analysis showed a prevalence of illuviated cutans and less diffuse cutans and papules occurrence in the upper part of the Bt horizon, which is evidence of the dominance of lessivage over weathering in situ.

The particle size distribution shows that the clay fraction (<0.001 mm) prevails followed by the coarse silt (0.05–0.01 mm) or fine sand (0.25–0.05 mm) (Tables 7.1, 7.2 and Figs. 7.6, 7.7).

Table 7.3 Variables of humus, pH, carbonates, and particle-size distribution (mm) according to USDA in 30 referenced profiles of strongly leached cinnamonic forest soils in Bulgaria

Profile count 30	Horizon	Upper depth	Lower depth	Humus (%)	pH_{KCl}	Carbonates	Bulk density g.cm³ calc	Clay <0.002	Silt 0.002–0.05	Fine sand 0.05–0.2	Total sand 0.05–2.0
Mean	AEp	0	26	2.61	5.1	0.0	1.28	43.9	32.8	16.0	23.3
	Bt1	26	51	1.42	5.1	0.0	1.34	54.5	26.5	13.1	19.1
	Bt2	51	76	1.06	5.3	0.0	1.37	54.8	26.0	13.5	19.2
	B	76	101	0.89	5.8	0.3	1.41	52.7	28.0	13.0	19.2
	Ck	101	126	0.77	6.8	7.1	–	46.5	29.1	14.9	20.7
Stdev	AEp	0	4	1.87	0.7	0.0	0.17	8.7	6.7	5.3	9.4
	Bt1	4	5	0.48	0.7	0.0	0.07	6.9	5.7	5.0	8.5
	Bt2	5	9	0.29	0.8	0.0	0.06	7.1	5.1	4.3	7.9
	B	9	13	0.26	0.9	0.9	0.07	8.3	6.0	5.5	9.8
	Ck	13	16	0.23	0.7	5.4	–	13.9	8.3	9.5	14.7

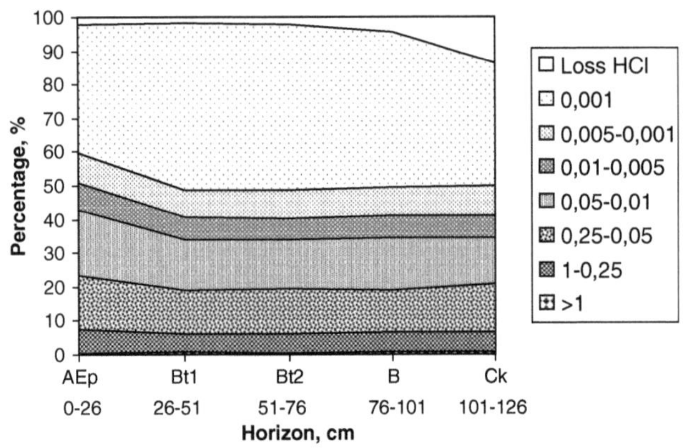

Fig. 7.6 Pattern of the particle size distribution (mm) in strongly leached cinnamonic forest soil (according to revised Kachinskiy method)

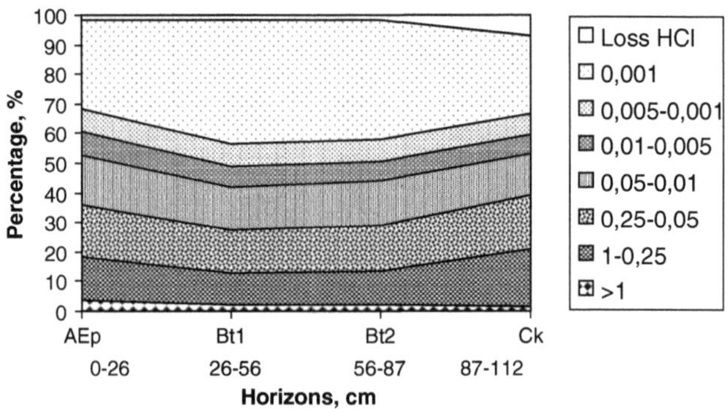

Fig. 7.7 Pattern of the particle size distribution (mm) in strongly leached cinnamonic forest soil (according to revised Kachinskiy method)

There is evidence of clay differentiation in the profile (Tables 7.3, 7.4 and Figs. 7.8, 7.9, 7.10). The amount of clay is less in the eluvial AE horizon and increases in the illuvial Bt horizon where cutans are often observed on ped surfaces. Soils are characterized by evidence of natural compaction. When dried the fine textural horizon turns very hard in consistence.

The average particle density is 2.58–2.65 g/cm³ at the upper part and is 2.65–2.70 g/cm³ at the depth. Bulk density is about 1.25–1.4 g/cm³ at the upper part and is about 1.5 g/cm³ in the middle part of the soil.

Water stable aggregates with size >1 mm are about 55 % at the surface (0–15 cm) and sharply decrease to about 25–35 % at the depth of 20–30 cm in pristine land. Prevailing water stable aggregates with size over 0.25 mm are over 70 % in the upper part and decrease at depth.

In leached cinnamonic forest soils nearly 5–15 % of pore space is represented by macropores or the pores of aeration (with diameter >300 μm); and 10–20 % is mesopores or pores with rapid moisture drainage (>30 μm diameter). Micropore space or water-filled is about 20 %, represented by slow drainage pores (diameter 0.2–30.0 μm), and 10–20 % of micropores is with none remaining available for roots moisture (diameter <0.2 μm) (Dilkova 1985; Dilkova et al. 1998).

Total porosity is 50–57 % (at field capacity) at the surface and decreases with depth. Soils are low water permeable.

In spatial distribution leached cinnamonic forest soils are bordered by leached smolnitsa, podzolic (pseudopodzolic) soils, by brown forest soils, and by Eutric Regosols.

Clay minerals commonly occur as mixed layers of two or more different minerals in the soil. The qualitative composition of clay minerals, determined by X-ray diffraction

Table 7.4 Variables of humus, pH, carbonates, and particle-size distribution (mm) according to USDA in 20 referenced profiles of strongly leached cinnamonic forest soils in Bulgaria

Profile count 20	Horizon	Upper depth	Lower depth	Humus (%)	pH_{KCl}	Bulk density $g.cm^3$ calc	Clay <0.002	Silt 0.002–0.05	Fine sand 0.05–0.2	Total sand 0.05–2.0
Mean	AEp	0	26	1.74	4.7	1.38	35.2	29.7	18.5	35.1
	Bt1	26	56	1.01	4.9	1.39	47.2	26.0	14.7	26.8
	Bt2	56	87	0.79	5.1	1.43	45.5	25.9	16.4	28.6
	Ck	87	112	0.60	6.0	–	34.0	26.8	19.2	39.2
Stdev	AEp	0	4	0.69	0.9	0.10	15.1	6.6	6.7	18.0
	Bt1	4	10	0.44	0.8	0.11	11.9	5.2	4.8	13.8
	Bt2	10	19	0.30	0.7	0.07	11.3	5.2	5.7	13.1
	Ck	19	20	0.22	1.1	–	13.0	5.9	8.2	13.8

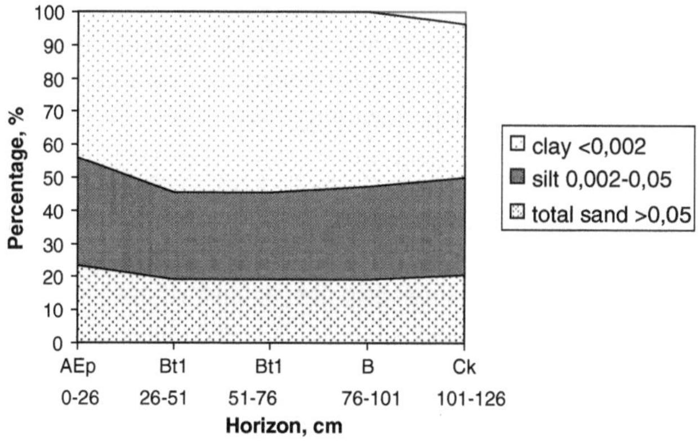

Fig. 7.8 Pattern of the particle-size distribution (mm) in strongly leached cinnamonic forest soil (USDA)

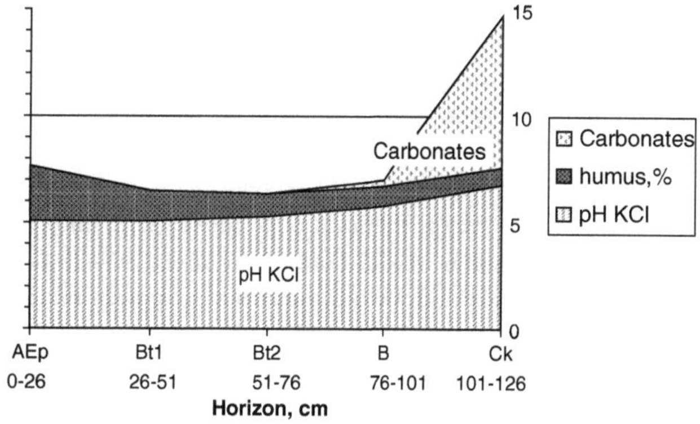

Fig. 7.10 Pattern of pH, humus, and carbonates distribution in strongly leached cinnamonic forest soil

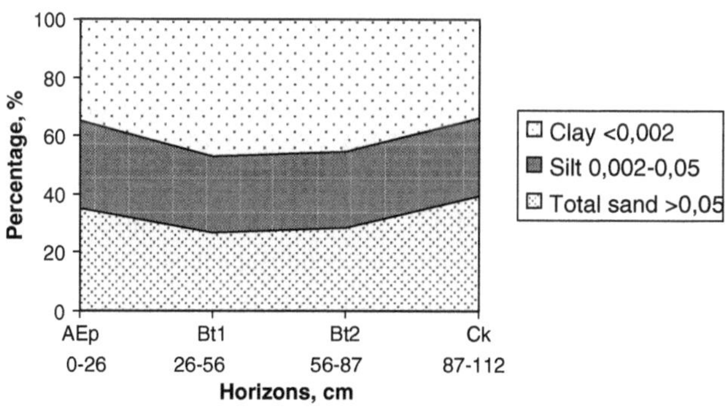

Fig. 7.9 Pattern of the particle-size distribution (mm) in strongly leached cinnamonic forest soil (USDA)

analysis and transitional electronic microscopy, has shown a prevalence of montmorillonite at rates of 49 % in the clay (<0.001 mm) and 15 % in the soil at the surface (AE) horizon, gradually increasing at depth, where at the illuvial (Bt) horizon it is 53 % in the clay and 23 % in the soil, and in the (C) horizon it is about 57 % in the clay and 12 % in the soil. The content of montmorillonite at the quantity over 20 % in the soil resembles that which occurs in the Vertisols group of soils. The illite content is at rates of 18 % in the clay and 6 % in the soil at the surface (AE) horizon, gradually decreasing at depth, where in the (C) horizon it is about 12 % in the clay and 3 % in the soil. Vermiculite (or hydrated micas) is 10 % in the clay and 3–4 % in the soil, and slightly increases to 13 % in the clay and 4 % in the soil at the (C) horizon. Kaolinite is about 15 % in the clay (<0.001 mm) and 5–7 % in the soil, and slightly decreases to 11 % in the clay and 3 % in the soil at the (C) horizon (Boneva 1998).

The amount of total iron (Fe_2O_3) (Arinushkina 1970) distribution in the profile is differentiated. The amount of total iron in the eluvial horizon is about 2.5–4.5 %; it is 4.5–6.8 % in the illuvial Bt horizon and 4.0–5.0 % in the parent materials. The distribution of total iron in the soil is similar to the distribution of the clay fraction (Hadzhiyanakiev 1989).

The silicate form of iron which is 60–80 % (of the total iron Fe_2O_3) dominates in all horizons. Values of silicate iron are lowest at the eluvial horizon and higher further down.

In the strongly leached cinnamonic forest soils, the amount of silicate iron is twice that of non-silicate iron.

Bulk chemical composition revealed a decrease in silica (2–6 % SiO_2) and aluminum (4–8 % Al_2O_3) as a result of weathering and significant enrichment with iron (Fe_2O_3).

The non-silicate iron (Mehra and Jackson 1960) is about 31–41 % (of the total iron Fe_2O_3) in the upper eluvial horizon and is 30–38 % in the Bt horizon. Most of the non-silicate iron appears in slightly crystallized form. Some non-silicate iron released from primary silicates may be involved in the newest clay formation.

The amorphous form of iron (Tamm 1922) has its highest values (about 10–20 % of the total iron) in the eluvial horizon, and its lowest values (about 5.5–15.5 % of the total iron) in the illuvial Bt horizon.

The free amorphous form of iron shows differentiated distribution; in the eluvial horizon it is about 25–45 % (of the total non-silicate iron), gradually decreasing in the illuvial Bt horizon to 18–40 % (of the total non-silicate iron) and then sharply decreasing in the C horizon to about 1.8–26.8 %. This fact is explained by periodic water saturation and weathering at the surface layer (Hadzhiyanakiev 1989).

The distribution of the amorphous form of iron bounded with organic matter (Bascomb 1968) varies significantly. The highest values are in the upper part of the humus eluvial horizon 3–5 % (of the total non-silicate amorphous iron) and sharply decrease with the depth to about 0.5 % in the lower part of the Bt horizon. This is due to the fact that fulvic acids prevail in the humus composition and are characterized by a higher ability of complex formation with iron. The amorphous form of iron bounded with organic matter showed luck of mobility down the profile.

7.4 Soil Properties and Soil Management

7.4.1 Soil Properties of Cinnamonic Forest Soils

Cinnamonic forest soils range from place to place. Relief and parent materials strongly influence the soil properties. Commonly soils are vulnerable to erosion. Chemical weathering in situ is pronounced where carbonates are leached.

Annual organic decay of vegetation during the biological cycle promotes the supply of considerable amounts of nitrogen, bases, and calcium back into the soil.

The accumulation of humus takes place in the upper 20–30 cm. The content of humus in the upper part varies from 1.5 to 4.5 % where the higher values are in forest land. The upper 20 cm of the soil stores 60 % of the organic matter in a 1m by 1 m layer. The stock of organic matter in a 100 cm layer is 100–200 tons per ha (Filcheva et al. 2002). Forms of humus are composed of humin and fulvic acids. The process of transformation of organic residues is accompanied by nitrogen enrichment of humic substances shown by the ratio C/N 10–12 which means high rates (Petrova 1980).

Base saturation is variable from 60 to 85 % in the (A + AE) horizon to over 80–90 % in the (Bt) horizon. Cation exchange capacity is about 15–25 meq/100 g in the upper part and is much higher in the middle part. Exchangeable Ca^{2+}, Mg^{2+} and K^+ prevails in the cation exchange capacity over H^+ or Al^{3+}. Soil reaction is acid or neutral.

These soils have high base buffering capacity and high water-holding capacity. Porosity decreases with depth. The soils have low water permeability.

7.4.2 Soil Management

Cinnamonic forest soils are fertile, and despite the low organic matter content (1.5–3.5 %) they have favorable properties for agricultural practices.

Cinnamonic forest soils are the major cultivated soil variety in southern Bulgaria. As well as agriculture, they are commonly used for forestry. The soils are characterized by natural low amounts of nitrogen, phosphorous, and potassium.

In the past, large areas of woodland were destroyed and replaced with vineyards, tobacco and peanut plantations, orchards, and land for growing crops and vegetables.

Almost all known cultivated plants in Bulgaria can be grown in the lowlands: winter wheat, barley, maize, sunflower, beet, cotton and rice, vegetables (under irrigation), beans, vetch, peas, alfalfa, and peanuts; viticulture can also be established. Warmer winter conditions benefit the growth of some specific plant species such as (*F. carica*), pomegranate (*Punica granatum*), persimmon (*Diospyros kaki*), almond (*Amygdalus communis*), olive (*Olea europea*), and lemon trees (*Citrus reticulata*). On the elevated lands with cooler continental conditions, crops such as rye, oats, potatoes, and tobacco are grown.

Seasonally, cinnamonic forest soils have poor technologic tillage properties, and are suiTable for ploughing for a limited period. The soil structure changes to massive or friable at the plough layer. The decline of organic matter (less than 2.5 %) is evidence that natural fertility has been degraded. All of this provokes compaction and crust formation. The main problem is aeration of the subsoil.

Severe erosion, particularly in the east Rhodope Mountains, means that expensive soil conservation technologies need to be applied. In some regions, *Robinia pseudoacacia* was planted as protection against erosion and for timber.

References

Antipov-Karataev IN, Gerassimov IP (1948) The soils of Bulgaria. Zemsnab Press, Sofia (In Bulgarian)

Arinushkina E (1970) Manual on chemical soil analysis. Moscow State University Press, Moscow (In Russian)

Bascomb CL (1968) Distribution of pyrophosphate extractable iron and organic carbon in soils of various groups. J Soil Sci 19(1):251–268

Boneva K (1998) Clay minerals in Bulgarian Chromic Luvisols and their grouping according to the mineralogical composition. Soil Sci Agrochem Ecol J XXXIII(3):20–22 (In Bulgarian)

Bondev I (1991) Vegetation in Bulgaria, map at scale 1:600 000 with explanatory text. Kliment Oxridski University Press, Sofia (In Bulgarian)

Dilkova R (1985) Structure and aeration in main soil varieties in Bulgaria. Thesis, Institute of Soil Science Nikola Poushkarov, Sofia (In Bulgarian)

Dilkova R, Kerchev Zh, Kercheva M (1998) Evaluating and grouping of soils according to their susceptibility to anthropogenic degradation, advances in geoecology, vol 31. Catena Verlag, Reiskirchen, pp 125–131

FAO-UNESCO-ISRIC (1990) Revised Legend of the soil map of the world. FAO World Soil Resources Report No 60, Rome

Filcheva E, Rousseva S, Kulikov A, Nedyalkov S, Chrnogorova T (2002) Organic carbon stocks in soils of Bulgaria. In: Kimble JM, Lal R, Follett RF (eds) Agricultural practices and policies for carbon sequestration in soils, Lewis Publishers, CRC Press LLC, New York, pp 471–476

Ganev S, Arsova A (1980) Methods of determining the strongly acid and the slightly acid cation exchange in soil. Soil sci Agrochem J XV(3):22–33 (In Bulgarian)

Gyurov G, Ninov N (1964) Cinnamonic soils in Bulgaria. In: Soils of south-east Europe, Bulgarian Academy of Sciences Press, Sofia (In Bulgarian)

Gyurov G et al (1978) Some problems of genesis and classification of gray forest soils, brown forest and pseudopodzolic soils in Bulgaria. Second scientific conference on soil science, Sofia (In Bulgarian)

Hadzhiyanakiev A (1989) Iron in Bulgarian soils. Zemizdat Press, Sofia (In Bulgarian)

Keys to Soil Taxonomy (2010) USDA United States department of agriculture, NRCS Natural Recourses Conservation Service, 11th edn. Soil Survey Staff, Washington, DC, p 338

Koynov V (1964) Peculiarities of the main Bulgarian soil types and their basic soil-genetic processes. In: Gerasimov IP et al. (eds) Soils of south–eastern Europe. Proceedings of the international symposium of soil science, June 1963, Sofia. Bulgarian Academy of Sciences Press, Sofia, p 15–62 (In Russian)

Koynov V et al (1964) Classification and systematic of soils in Bulgaria. In: Gerasimov IP et al. (eds) Soils of south–eastern Europe. Proceedings of the international symposium of soil science, June 1963, Sofia. Bulgarian Academy of Sciences Press, Sofia, p 231–246 (In Russian)

Koynov V, Trashliev Ch, Yolevski M et al (1968) Soil map of Bulgaria at scale 1:400 000 with description, Institute of soil science Nikola Poushkarov, Bulgarian Academy of Sciences Press, General Office of Geography and Mapping, Sofia (In Bulgarian and English)

Mehra OP, Jackson ML (1960) Iron oxide removal from soils and clays by dithionite—citrate: system buffered with sodium bicarbonate. Clays Clay Mineral 7(317–327):2345

Penkov M et al (1992) Soil classification and diagnostic in Bulgaria in relation to the land reform fragmentation. Zemizat Press, Sofia (In Bulgarian)

Petrova L (1980) State of organic nitrogen in some soils in Bulgaria. Thesis, Institute of Soil Science N Poushkarov, Sofia, 138 (In Bulgarian)

Poushkarov N (1931) Soil map of Bulgaria at scale 1:500,000, Sofia

Tamm O (1922) Eine methode zur bestimung der anorganischen komponente ds gelcomplexes im boden. Statens Skogsforsoksanstalt, Medd.19:387–404 (by Arinoushkina E (1972) Manuel of chemical analysis of soils, Moscow)

World Reference Base for Soil Resources (2006) A framework for international classification, correlation and communication. FAO-ISRIC, Rome

Yolevski M, Hadzhiyanakiev A (1976) Agrogrouping of soils in Bulgaria. Extended systematic list. Code of groups and soil varieties, Institute of Soil Science Nikola Poushkarov, Sofia (In Bulgarian)

Yolevski M, Ninov N, Krastanov S et al (1983) Brief diagnostic and criterion of the main taxonomic soil divisions in Bulgaria. Precise Classification Scheme, Institute of Soil Science Nikola Poushkarov, Sofia (In Bulgarian)

8 Brown Forest Soils

Derived from the Italian word *cambiare,* meaning to change and the Latin word *inceptum,* meaning beginning.

Brown forest soil is a mountainous, undifferentiated, brownish colored soil, often found with rock fragments.

8.1 Concept of the Type

Main concepts

1. Brown forest soil is a mineral soil characterized by uniform texture;
2. Minimum translocation of silicate clays, iron, aluminum, and hydroxides takes place, despite the significant amount of precipitation;
3. Freely drained.

There are many processes that may be responsible for brown forest soil formation. Leaching has taken place in most years. These soils typically show little evidence of weathering in situ and oxidation of iron oxides.

8.2 Ecology of Soil Formation

8.2.1 Distribution

Brown forest soils encompass a broad variety of soils which covers a large surfaces across hilly and mountainous regions. The mountainous zone includes all mountains in the country with an altitude over 600 (800) m above sea level. Five provinces are singled out in which these soils show significant provincial differences: Rila (2,300 km^2) and Pirin (1,210 km^2) province, Balkan Mountains (25,986 km^2) province, Vitosha and Sredna Gora Mountain province, Rhodope Mountains (14,700 km^2) province, and Ossogovo-Belassitsa province.

The Balkan Mountains spread from east to west and split the country, determining a climate specific north/south division. From a morphographic point of view the Balkan Mountains system comprises the main Balkan massif and Fore Balkan in which area the hilly belt is greatly developed (with altitude of 200–600 m). The Rhodope Mountains spread from east to west and are 240 km long and 100 km wide. Rila Mountain (2,925 m) is a part of the western Rhodope anticline structure. Pirin Mountain (2,914 m) has typical alpine mountain relief.

Change in the bioclimatic conditions and in the character of land cover is linked with the altitude above sea level. The belt of brown forest soils is vast. It covers all mountains in the country, often reaching to their ridges. Brown forest soils are widely distributed in different parts of the country (Figs. 8.1, 8.2).

Bulgarian brown forest soil occupies about 163,100.7 ha or 14.7 % of the entire territory. In southern and western Bulgaria the soil is bordered by cinnamonic forest soil; on the northern slopes of the Balkan Mountains it is bordered by light gray pseudopodzolic soil. Dark colored mountain forest soils occupy lands at an altitude of 1,800–2,100 m. At altitudes over 2,100 m dark colored mountain forest soils are bordered by mountainous meadow soils. Dark colored mountain forest soils occupy about 100,000 ha or 0.94 % of the entire country's territory.

Climatic conditions

Brown forest soil is found in climatic conditions where the rate of precipitation ensures good drainage in most years. The air humidity is high (over 75 %); snow cover is typical in most years. Freezing occurs in the upper soil layer.

The temperature regime and total precipitation vary from one region to another. In Bulgaria, brown forest soils may be found in temperate as well as in humid continental climatic conditions. In different regions the altitude is significant.

The mean annual temperature is 3.7–10.8 °C. In January the mean annual temperature varies from −5.7 to 0.8 °C. In July the mean annual temperature varies from 12.7 to 20.6 °C. The mean amount of precipitation varies from 613 to 1,163 mm.

T. Shishkov and N. Kolev, *The Soils of Bulgaria*, World Soils Book Series, DOI: 10.1007/978-94-007-7784-2_8,

Fig. 8.1 Typical landscape of dark brown forest soils

Fig. 8.2 Light brown forest soil, Bulgarian classification Yolevski and Hadzhiyanakiev (1976), *Haplic Cambisol* (Eutric, Skeletic), WRB (2006), *Typic Dystrudepts,* Keys to Soil Taxonomy (2010), *Eutric Cambisol*, Revised Legend FAO-UNESCO-ISRIC (1990)

Topography

Brown forest soil is found in mountainous areas with altitudes of 700–1,800 m above sea level in any topographic position, commonly on lower or steep slopes. Dark colored mountain forest soils occur at altitudes of 1,800–2,100 m. Usually, they are developed on stable and moderate to steep sloping sites.

This is a typical forest soil, soil depends on slope exposure and elevation of landscape. Soils exposed to the north are somewhat deeper, darker and wetter; and on the southern slopes soils are shallow, lighter, and drier. A common trend is that the degree of humus content increases with altitude.

Parent materials

Brown forest soils have formed on a wide range of parent materials. Most soils have developed on the eluvium through chemical weathering of the underlying rock. Some provincial differences in soils are determined by the character of these materials. It should be mentioned that in mountain regions parent rocks can vary over short distances and are in relation with the topography of the region.

Thus in the Balkan Mountains parent rocks are of the Precambrian Age (units of pillow-lavas, of sheeted dykes, of banded cumulates, metamorphosed shales, lidites, sandstones, marbles, spmicas, keratophyres, granitized biotite and two-mica gneisses, migmatites, granite-gneisses, amphibolites, schists); of the Ordovician Age (gneiss-granites and migmatites, metamorphosed conglomerates, sandstones, siltstones, shales, quartzites); of the Carboniferous (Namurian-Westfalian) Age (granites, granodiorites, granodiorite-granite complex, breccia-conglomerates, sandstones, siltstones, shales); of the Triassic Age (limestones, dolomitic limestones, dolomites, silty shales, sandstones, siltstones, conglomerates); and of the Paleogene (Paleocene-Middle Eocene) Age (flysch alternation of sandstones, siltstones, and clays).

In the Rhodope Mountains parent rocks are of the Precambrian Age (granitized biotite and two-mica gneisses, migmatites, granite-gneisses, amphibolites, schists, quartzites, leptinites, marbles, amphibolites); of the Carboniferous (Namurian-Westfalian) Age (granites, granodiorites); of the Cretaceous Age (granites); of the Paleogene (Eocene) Age (continental molasses, breccia-conglomerates, grabens, sandstone, limestone, marl formation); and of the Paleogene (Oligocene) Age (rhyolites, rhyodacites, trachyrhyolites, delenites).

In the Sredna Gora Mountain soils are developed on parent rocks of the Precambrian Age (migmatites, granite-gneisses, amphibolites, schists, quartzites, leptinites, marbles, amphibolites, gneiss-shists); of the Carboniferous (Namurian-Westfalian) Age (granites, granodiorites); and of the Cretaceous Age (gabbro, syenomonzonites, granodiorites, quartzmonzonites).

In the Rila mountain soils are developed on parent rocks of the Precambrian Age (migmatites, granite-gneisses, amphibolites, schists, quartzites, leptinites, marbles, amphibolites, gneiss-shists); of the Carboniferous (Namurian-Westfalian) Age (granites, granodiorites); and of the Cretaceous Age (granites).

In the Osogovska, Vlahina, Malashevska, Ograzhden, Belassitsa mountains soils are developed on parent rocks of the Precambrian Age (granitized biotite and two-mica gneisses, migmatites, granite-gneisses, amphibolites, schists, quartzites, leptinites, marbles, amphibolites, gneiss-shists, metaconglomerates, porphyroblastic gneisses); of the Ordovician Age (diorites, gabrodiorites, granodiorites); of the Carboniferous (Namurian-Westfalian) Age (granites, granodiorites); of the Paleogene (Eocene) Age (continental molasses, breccia-conglomerates, grabens, sandstone, limestone, marl formation); and of the Paleogene (Oligocene) Age (rhyolites, rhyodacites, trachyrhyolites, delenites).

Vegetation

Precise species vary from place to place. In the mountains the vertical environmental zones are linked to the elevation above sea level, where the type of vegetation is in compliance with the moisture and temperature regime. The zone of deciduous forest is at 400–800 m altitude; the zone of mixed forest dominated by beech woods is at 800–1,200 m altitude; the zone of coniferous forest is at altitudes of 1,200–1,500–1,700 m (may reach to 1,800 m); and the zone of subalpine meadows is over 2,100 m altitude.

Low elevated areas are dominated by oak forest (*Querques* sp.), mixed with beech (*Fagus* sp.), birch (*Betula* sp.), maple (*Acer* sp.), chestnut (*Castanea* sp.) and hazel (*Corylus avellana*) trees, and blackberry bushes (*Rubus* sp.).

Deciduous forest dominated by beech (*Fagus* sp.) reaches up to 1,200–1,700 m above sea level.

Coniferous forest, consisting of species of spruce (*Picea* sp.), Scotch pine (*Pinus silvestris, Pinus montana*), and pine spruce (*Abies alba*) reaches to 1,800–2,100 m above sea level. A typical indicator of brown forest soil distribution is the presence of bracken (*Pteridium aquilinum*) (Bondev 1991). In the past, the greater part of these forests were destroyed and replaced with the secondary formations of *Populeta tremulae,* and *Betuleta pendulae,* and on wet soils *Alneta incanae*, with some bush formations of *Junipereta sibiricae, Chamaecytiseta absinthioides,* and *Vaccinieta myrtilli,* and grass formations of *Nardeta strictae, Agrostideta capillaris, Pteridieta aquillini,* and others. On calcareous soils on the Pirin and Slavyanka mountains xeromesophytic forest of *Pineta heldreichii* has developed.

Brown forest soils are characterized by abundance of meso and micro fauna.

The beech woods belt at 1,200–1,700 m altitude is characterized by basophytic vegetation and mixed forest of *Fageta sylvaticae,* and *Abieta albae*. In some places forests were replaced by *Populeta tremulae,* and *Betuleta pendulae* and grass formations of *Agrostideta capillaris, Nardeta strictae, Bellardiochloeta violaceae, Pteridieta aquilini,* and others.

A belt of hornbeam-oak woods reaches to an altitude of 600–1,200 m. Mesophytic and xeromesophytic vegetation is formed of *Fegeta moesiacae, Castaneeta sativae, Aesculeta hippocastani, Carpineta betuli, Querceta daleschampii, Ostryeta carpinifoliae, Pineta nigrae,* and *Tilieta tomentosae*.

In southeastern Bulgaria relict vegetation comprises *Fageta orientalis* and *Querceta polycarpae*. In places where natural vegetation has been destroyed grow formations of *Coryleta avellanae, Junipereta communis*, and grass formations of *Agrostideta capillaris, Festuceta rubrae, Cynosureta cristati*, and *Pteridieta aquilini*.

Xeromesophytic and xerothermic formations of vegetation also exist, comprising *Festuceta valesiacae, Festuceta dalmaticae, Festuceta pancicianae,* and *Chrysopogoneta grylli*.

Time

Brown forest soils are of Holocene Age. Presumably, their development started at the end of the glacial period or at the end of the Pleistocene period. Being comparatively young, these soils have a tendency to evolve.

Human influence

Most of the brown forest soils are located in hilly or mountainous regions with steep slopes and under dense deciduous and/or coniferous forest cover. Historically, lumber has been the principal industry. Afforestation is very productive. Dairy farming is quite often practiced. In the open spaces land is used for grassing.

The limitations for the agricultural use are the acid soil reaction, the exchange ions of Al^{3+} and H^{+}, and stoniness. The short growing season and low temperatures present an obstacle to growing most crops, although conditions are reasonably conducive for obtaining a high potato yield. Cultivation has caused fast deterioration of soil structure and a rapid decrease in humus content.

8.2.2 Genesis

Brown forest soil has altered horizons, but has not undergone much weathering. They show at least some development in a subsurface horizon as formation of soil structure and some oxides or clay minerals. Brown forest soil is a phase of development from the Lithosol, Ranker, or Rendzina soils. Many of the soils are shallow to bedrock and stony, and usually occur on steeply sloping terrain.

The most common diagnostic feature is a *cambic* horizon. This soil horizon is characterized by a change in structure development or absence of rock structure in at least half of the volume, compared to the lower horizon. The alteration is recognized by clay increase compared to the parent materials, change of color (darker or redder), and evidence of redoximorphic features (gleyic or mottles).

Soil processes may vary from place to place (Kolchakov 1989), but leaching remains principal. Soils are well drained. Due to the high rates of precipitation, cations are leached out of the soil by drainage. The biological circulation of the cations by means of plant roots uptake is open to discussion

Biological activity is high at the surface (Nedialkova and Shishkov 1999, 2003). Humus accumulation occurs in the horizon beneath the litter. The dark brown color is due to the presence of fulvic acids and/or brown colored humic acids. The content of organic matter sharply decreases with increasing soil depth.

Typically soils are acid where fulvic acids prevail over humic acids. The humic substances composition shows a change with altitude. The decomposition of the leaf litter under deciduous forest produces humic and fulvic acids bounded with cations, sesquioxides, and clay minerals. The decomposition of pine needles produces aggressive fulvic acids mainly bounded with sesquioxides. The maximum of iron complexes occurs in the humus horizon. The free iron and aluminum oxides released in the weathering are bounded with humus acids, neutralizing their aggression.

Principal variations within the type are in thickness, humus content, base saturation, and degree of erosion.

8.3 Soil Diagnostic and Classification

8.3.1 Morphology

It is widely accepted that the reference profile of brown forest soil shows morphology as it is presented under a beech and/or coniferous forest cover, despite the fact of great soil variability. The typical feature is the evidence of alteration from the parent material (Fig. 8.3).

Fig. 8.3 Light brown forest soil, Bulgarian classification Yolevski and Hadzhiyanakiev (1976), *Haplic Cambisol* (Eutric, Skeletic), WRB (2006), *Typic Dystrudepts,* Keys to Soil Taxonomy (2010), *Eutric Cambisol*, Revised Legend FAO-UNESCO-ISRIC (1990)

O—a thin (<5 cm) loose leafy litter at the soil surface, consisting of partly decomposed leaves (or pine needles) and possibly grass remnants, naturally occurs in woodlands; below is clear transition into

Ah(Ap)—an upper mineral humus accumulative horizon that is 5–10 cm thick, strong brown colored, homogeneous, structure is granular (spheroids or polyhedrons) with size 1–2 mm diameter, moderate pedality grade, consistence is very friable (moist) and soft (dry), or it is plough. Porosity is high. Throughout there is evidence of high biological activity; beneath is gradual transition into

A(Ap)—a mineral humus horizon that is 10–20 cm thick, dark brown or brownish colored, homogeneous, structure is granular (spheroids or polyhedrons) with size 2–5 mm diameter, moderate pedality grade, consistence is friable (moist) and soft (dry); beneath is gradual change into

Bw—a metamorphic horizon that is 50–60 cm thick, with hue 10YR brown or dark yellowish brown color, a few rock fragments of different size are in the soil mass. Structure is subangular blocky (blocks having mixed rounded and plane faces of aggregates), with fine size

(5–10 mm diameter), moderate pedality grade, consistence is friable (moist) and slightly hard (weakly resistant to pressure) dry. Porosity is high. There is gradual transition into

C—parent materials that are half weathered, yellowish brown or pale brown undifferentiated, showing a little change. Structure is massive (no structure), where large rock fragments occur. Grades into

R—continuous hard parent rock.

Horizon differentiation is only on the basis of the structural aggregates formation and color of soil mass. Weathering in situ is apparent throughout with no movement of the products.

The brown color is due to the presence of light fulvic, brownish humic acids, and iron oxides. The brownish color may also be due to the segregation of fine goethite at the surface of clay minerals forming iron-illites structures (Stefanovits 1964).

Soil profile is of uniform texture. The humid soil variety has almost uniform dark brown color to a considerable depth. Stratification of parent material in the soil profile is possible.

8.3.2 Classification

At the Second World Soil Congress in 1930, brown forest soil was defined as a single type. In the central Europe, the term/terminology of *Braunerden* after E. Ramann or *Waldboden* after G. Murgochi was applied. On the FAO-UNESCO Soil Map of the World (FAO 1971–1981) soils with a metamorphic Bw horizon were referred to as Cambisols. In central Europe these soils are commonly found in areas with flat or undulated topography (Tavernier and Smith 1957). In Bulgaria brown forest soil is naturally found on sloppy well drained terrains and is rarely found on a flat landscape.

In the systematic of brown forest soils there are no principles. The Bulgarian soil classifications of 1947/1964 included eutric and dystric soil subtypes. The classification of Yolevski and Hadzhiyanakiev (1976) recognized brown forest soil and secondary grassed soil. The classification of Yolevsky et al. (1983) renamed the type as brown mountainous forest soil, where the change was linked to the degree of development of the subtypes (haplic, dark, and light). Dark mountainous forest soils were classified by Georgiev (1972). There was classification in Penkov et al. (1992) which at the theoretic level renamed the type as brown forest metamorphic soil (eutric, dystric).

The distinguishing feature is the base of saturation, where the eutric varieties are with base saturation >50 % and dystric are <50 %. This does not influence the use of these soils.

Within the subtype soil differs in the thickness of humus horizon and in the rates of erosion.

Classification after WRB (2006) required *Cambic* diagnostic horizon for inclusion inthe reference group of Cambisols.

Keys to Soil Taxonomy (2010) required a *Histic*, *Ochric*, Umbric, *Plaggen* (a man-made layer 50 cm thick), *Mollic* epipedon, where the base saturation of the entire soil does not meet the depth criteria for Inceptisol order classification. In Soil Taxonomy there are demands for subsurface horizons: allowed is a *cambic, albic, agric, calcic, gipsic, salic, placic or duripan*.

In this case *Mollic* epipedon may occur in other orders without Mollisol classification. Thus subsurface horizon *cambic* in Inceptisols takes precedence over the Mollic horizon.

8.3.3 Brown Forest Soil Subtypes in Bulgaria

8.3.3.1 Brown Forest Soil

Dystric or **Eutric** (*subtype*) **brown forest** (*type*) soil named after the Bulgarian classification of Koynov et al. (1964), and Yolevski and Hadzhiyanakiev (1976).

Haplic Cambisol (*Dystric*) *or* (*Eutric*) (*Skeletic*) is named according to WRB (2006) and *Typic Dystrudepts* according to Keys to Soil Taxonomy (2010).

Dystric or Humic Cambisol: Revised Legend of the World Soils (FAO-UNESCO-ISRIC 1990).

The profile is not differentiated in clay and sesquioxides distribution. The clay content is low and uniformly distributed. Particle size distribution of the fine earth shows that coarse and fine sand fractions (1.0–0.25 and 0.25–0.05 mm) dominate (Tables 8.1, 8.2, 8.3 and Figs. 8.4, 8.5, 8.6). Texture is homogeneous at the depth of profile (Tables 8.4, 8.5, 8.6 and Figs. 8.7, 8.8, 8.9, 8.10, 8.11, 8.12).

Under natural conditions most of these soils have O–A–Bw–C–R horizon sequence.

The thickness of humus horizon (A) varies; it is 10–25 cm in the light varieties and 40–60 cm in the dark varieties. The stock of organic matter is 80–100 tons per hectare to a depth of 100 cm in the light varieties and 140–210 tons per hectare in dark varieties. In humus composition fulvic acids (C_f) prevail there may be fulvic acids of the aggressive type. Fulvic acids are bound with sesquioxides (Artinova and Groseva 1989). In brown forest soil (eutric) there may be a considerable amount of humic acids.

The dark forest soil is 60–80 cm thick; humus is of moder type acid where there is low biological activity.

Soil reaction is strongly to slightly acid in the upper part pH_{KCl} values 4.0–5.5 (Ganev and Arsova 1980; Ganev et al. 1990).

Table 8.1 Variables of the particle-size distribution in 15 referenced profiles of brown forest soil (dystric) in Bulgaria

Profile count 15	Horizon	Upper depth	Lower depth	Loss HCl	Particle-size distribution (mm), % (according method of Kachinskiy revised)							
					>1 mm	1–0.25	0.25–0.05	0.05–0.01	0.01–0.005	0.005–0.001	<0.001	<0.01[a]
Mean	Ap	0	22	2.2	6.4	36.0	21.8	13.7	6.2	5.5	8.4	20.2
	AB	22	49	1.4	6.7	36.3	21.9	14.2	6.1	6.1	7.4	19.9
	Bw	49	71	1.5	6.8	36.3	23.1	13.9	6.1	5.1	7.2	18.5
	BC	71	100	2.0	10.4	33.5	24.9	14.1	4.8	4.2	6.1	15.1
Stdev	Ap	0	5	3.0	8.5	17.9	10.5	6.7	2.5	3.2	5.6	9.9
	AB	5	8	1.7	7.2	16.1	8.7	7.2	2.9	3.8	5.1	9.6
	Bw	8	10	2.2	10.1	15.8	8.8	8.3	3.6	3.1	6.0	9.7
	BC	10	15	3.1	14.4	11.9	11.8	4.5	2.1	1.7	4.9	6.4

[a] Physical clay, the parameter characterized sum of the particle-size distribution, according to Kachinskiy method (mm) $\sum < 0.01 = <0.001 + (0.005–0.001) + (0.01–0.005)$

Table 8.2 Variables of the particle-size distribution in 7 referenced profiles of brown forest soil (humic, dystric) in Bulgaria

Profile count 7	Horizon	Upper depth	Lower depth	Loss HCl	Particle-size distribution (mm), % (according method of Kachinskiy revised)							
					>1 mm	1–0.25	0.25–0.05	0.05–0.01	0.01–0.005	0.005–0.001	<0.001	<0.01[a]
Mean	Ah	0	27	2.3	2.5	13.7	28.2	26.1	8.3	6.3	12.5	27.1
	AB	27	54	1.5	3.9	18.0	26.9	23.5	8.9	6.4	11.0	26.3
	Bw	54	75	1.6	6.9	19.4	30.4	20.8	6.5	5.4	8.9	20.9
	BC	75	106	0.8	0.0	14.0	51.2	19.3	5.5	4.7	4.7	14.8
Stdev	Ah	0	9	0.9	3.9	5.5	11.7	11.1	1.9	2.2	4.4	6.1
	AB	9	11	0.9	3.6	7.4	13.7	8.7	2.6	2.6	3.7	3.0
	Bw	11	11	1.0	11.0	8.7	16.0	4.1	1.3	1.4	3.3	4.1
	BC	11	35	0.5	0.0	1.5	3.6	0.9	0.6	1.6	0.3	2.5

[a] Physical clay, the parameter characterized sum of the particle-size distribution, according to Kachinskiy method (mm) $\sum < 0.01 = <0.001 + (0.005–0.001) + (0.01–0.005)$

Table 8.3 Variables of the particle-size distribution in 7 referenced profiles of brown forest soil (eutric) in Bulgaria

Profile count 7	Horizon	Upper depth	Lower depth	Loss HCl	Particle-size distribution (mm), % (according method of Kachinskiy revised)							
					>1 mm	1–0.25	0.25–0.05	0.05–0.01	0.01–0.005	0.005–0.001	<0.001	<0.01[a]
Mean	A	0	27	1.1	11.3	24.8	24.9	15.9	5.6	5.7	10.7	21.7
	Bw	27	54	1.6	11.7	24.7	27.6	13.3	4.6	4.9	11.5	21.0
	C	61	75	0.4	21.7	30.7	20.0	12.4	3.5	1.9	9.5	14.8
Stdev	A	0	3	0.7	5.3	6.4	2.6	6.0	2.2	2.0	6.7	6.1
	Bw	3	12	1.2	8.7	6.4	5.8	1.9	1.4	1.0	8.0	7.3
	C	12	9	0.1	10.6	7.9	3.7	5.3	1.4	0.0	7.6	6.1

[a] Physical clay, the parameter characterized sum of the particle-size distribution, according to Kachinskiy method (mm) $\sum < 0.01 = <0.001 + (0.005–0.001) + (0.01–0.005)$

Cation exchange capacity is 10–20 meq/100 g soil. Exchangeable cations are generally Al^{3+} and H^{+} with less Ca^{2+} and Mg^{2+}.

The qualitative composition of clay minerals, determined by X-ray diffraction analysis and transitional electronic microscopy, has shown a prevalence of hydromicas, about 57 % in the clay (<0.001 mm) at the surface and gradually increasing at the depth to 67 % in the clay, followed by the montmorillonite at rates of 22 % in the clay at the surface and gradually decreasing at the depth to 17 % in the clay.

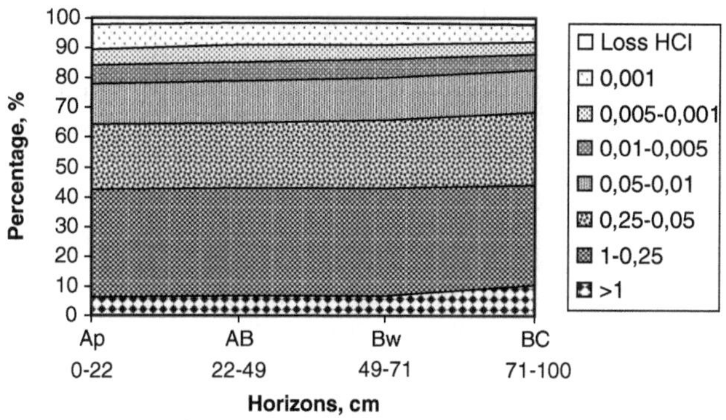

Fig. 8.4 Pattern of the particle-size distribution (mm) in brown forest soil (dystric) (according to revised Kachinskiy method)

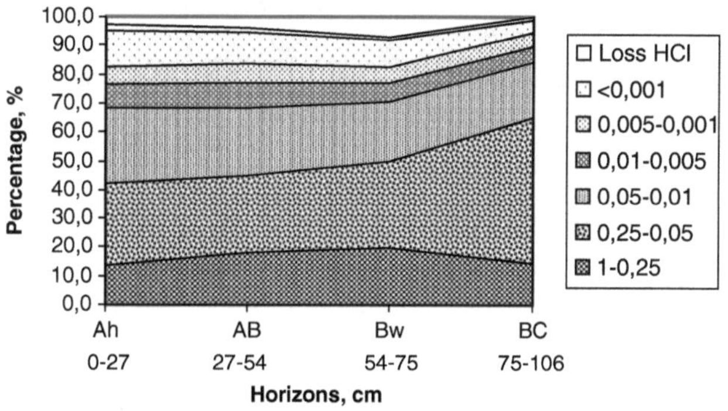

Fig. 8.5 Pattern of the particle-size distribution (mm) in brown forest soil (humic, dystric) (according to revised Kachinskiy method)

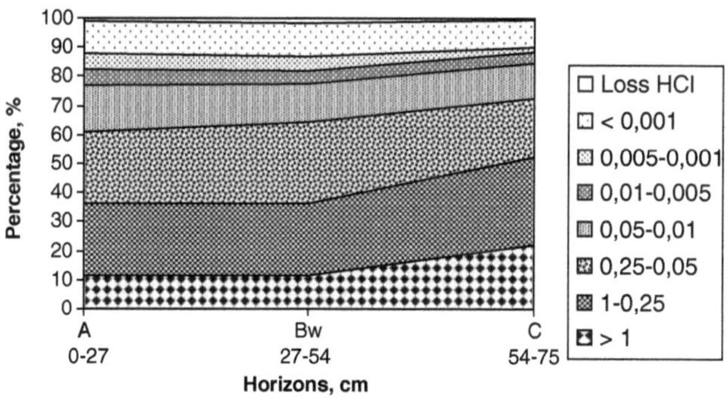

Fig. 8.6 Pattern of the particle-size distribution (mm) in brown forest soil (eutric) (according to revised Kachinskiy method)

Kaolinite content is 13 % in the clay throughout the soil, and quartz is about 10 % in the clay (<0.001 mm) at the surface and sharply decreases down the profile to 4 % in the clay (Boneva et al. 1986).

The amount of total iron (Fe_2O_3) (Arinushkina 1970) decreases with the depth. The silicate form of iron (60.0–82.6 % of the total iron Fe_2O_3) dominates in all horizons, where the distribution gradually increases and the highest values are in the parent materials. The silicate minerals are weathered near the surface (Hadzhiyanakiev 1989).

The non-silicate iron (Mehra and Jackson 1960) is about 40 % (of the total iron Fe_2O_3) in the humus horizon, and 17.5 % in the parent materials.

Brown forest soil has a specific distribution of the different forms of iron compounds. In the humus horizon the rates of crystallized and amorphous forms of iron are equally 24.0–27.0 % (of non-silicate iron). The conditions for non-silicate iron crystallization, and for the complex formation with organic substances, are favorable. This is due to the fact that fulvic acids prevail in the humus composition and are characterized by high ability of complex formation with iron.

In the cambic Bw horizon there is a significant increase in the strongly crystallized and free amorphous form of iron, but iron bounded with organic substances sharply decreases.

The hydroxide amorphous form of iron (Tamm 1922) is only 2.5–9.5 % (of the total iron) and without differentiation in the soil.

The free amorphous form of iron significantly increases in the cambic Bw horizon to about 36.3 % (of the total non-silicate amorphous iron).

The amorphous form of iron bounded with organic substances (Bascomb 1968) shows high values in the humus horizon; but in the cambic Bw horizon it is very low at only 7.6–9.6 % (of the total non-silicate amorphous iron). This is due to the fact that fulvic acids prevail in the humus composition and are characterized by a low ability to make complex formations with iron (Hadzhiyanakiev 1989).

Water stable aggregates with size >1 mm are over 85 % at the surface (0–15 cm) and decrease downward to about 20–30 %. Prevailing water stable aggregates with size over 0.25 mm are nearly 85 % at the surface and decrease to 50 % at the depth.

The bulk density of soil at the state of field water capacity ranges from 1.0 to 1.5 g/cm^3. The bulk density gradually rises with the depth (Dilkova 1985; Dilkova et al. 1998). The aeration of soil when dry affects the deeper layers, but the internal aeration of aggregates is low. Soils are characterized by high water infiltration rates when wet.

Brown forest soil is characterized by high biological activity.

The natural spatial relationship is with with regosols, lithosols, rendzinas, cinnamonic forest soils, pseudopodzolic soils, deluvial (Colluvial) soils, and dark mountainous soils.

Table 8.4 Variables of humus, pH, carbonates, and particle-size distribution (mm) according to USDA in 15 referenced profiles of brown forest soil (dystric) in Bulgaria

Profile count 15	Horizon	Upper depth	Lower depth	Humus (%)	pH_{KCl}	Bulk density g.cm^3 calc	Clay <0.002	Silt 0.002–0.05	Fine sand 0.05–0.2	Total sand 0.05–2.0
Mean	Ap	0	22	2.2	3.9	1.3	12.7	24.3	22.9	63.3
	AB	22	49	1.2	4.0	1.4	11.5	24.6	22.7	63.9
	Bw	49	71	0.7	4.0	1.5	10.9	23.7	23.9	65.4
	BC	71	100	0.4	3.8	–	8.6	20.0	22.7	58.9
Stdev	Ap	0	5	1.2	0.3	0.2	7.2	9.6	10.9	14.8
	AB	5	8	0.9	0.2	0.1	6.1	10.5	8.5	14.4
	Bw	8	10	0.5	0.4	0.1	6.7	10.4	8.2	14.0
	BC	10	15	0.0	0.4	–	6.1	10.4	13.6	25.6

Table 8.5 Variables of humus, pH, carbonates, and particle-size distribution (mm) according to USDA in 7 referenced profiles of brown forest soil (humic, dystric) in Bulgaria

Profile count 7	Horizon	Upper depth	Lower depth	Humus (%)	pH_{KCl}	Bulk density g.cm^3	Clay <0.002	Silt 0.002–0.05	Fine sand 0.05–0.2	Total sand 0.05–2.0
Mean	Ah	0	27	9.4	3.8	–	17.1	39.2	28.7	43.8
	AB	27	54	5.5	3.9	–	15.3	37.1	27.4	47.6
	Bw	54	75	2.2	4.1	–	12.9	32.1	31.4	55.0
	BC	75	106	0.2	4.3	–	7.5	27.4	51.2	65.2
Stdev	Ah	0	9	2.4	0.2	–	5.0	12.4	11.4	9.1
	AB	9	11	3.3	0.1	–	4.0	11.5	13.5	8.5
	Bw	11	11	0.9	0.1	–	3.3	6.1	15.3	5.1
	BC	11	35	–		–	1.4	0.7	3.6	2.1

Table 8.6 Variables of humus, pH, carbonates, and particle-size distribution (mm) according to USDA in 7 referenced profiles of brown forest soil (eutric) in Bulgaria

Profile count 15	Horizon	Upper depth	Lower depth	Humus (%)	pH_{KCl}	Bulk density g.cm^3 calc	Clay <0.002	Silt 0.002–0.05	Fine sand 0.05–0.2	Total sand 0.05–2.0
Mean	A	0	27	1.8	5.0	1.4	15.2	26.4	26.8	58.4
	Bw	27	54	1.0	5.0	1.4	15.7	22.6	29.6	61.7
	C	61	75	0.2	4.6	1.7	12.7	19.2	22.9	68.1
Stdev	A	0	3	0.4	0.5	0.1	7.2	7.5	1.9	7.3
	Bw	3	12	0.5	0.6	0.1	7.7	3.1	5.0	7.1
	C	12	9	–	0.2	–	10.1	6.5	2.7	3.5

8.4 Soil Properties and Soil Management

8.4.1 Soil Properties

Typically, brown forest soil is brown colored and gets lighter with the depth. The gradual transition of horizons produces a homogeneous profile. The content of humic substances can be 10–12 % at the surface, but is usually 3–4 %. The process of transformation of organic residues is accompanied by nitrogen enrichment of humic substances shown by the ratio C/N 10–20 which means medium to very low rates. In the case of cultivated land use, the amount of humus content drops to 2–3 % being of the same composition. The content of humus compounds distribution is typical for the forest soils in Bulgaria, in that it sharply decreases from the surface to the depth. The type of the humus composition is fulvic or

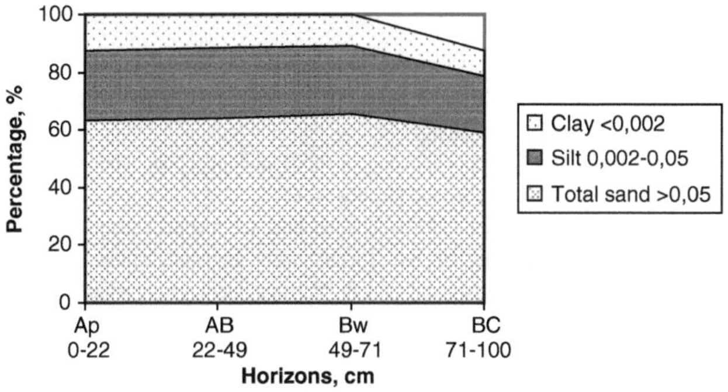

Fig. 8.7 Pattern of the particle-size distribution (mm) in brown forest soil (dystric) (USDA)

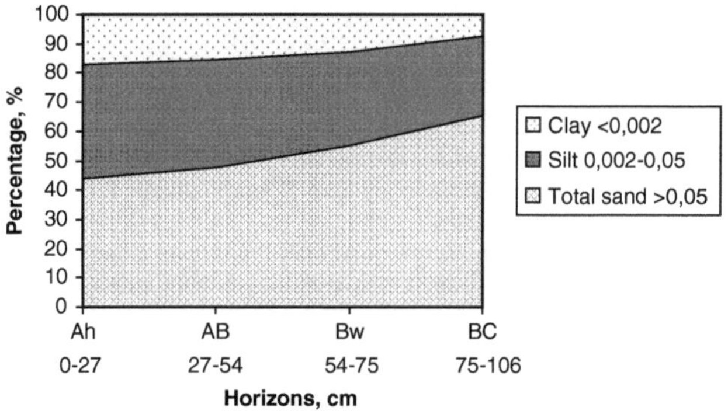

Fig. 8.8 Pattern of the particle-size distribution (mm) in brown forest soil (dystric, humic) (USDA)

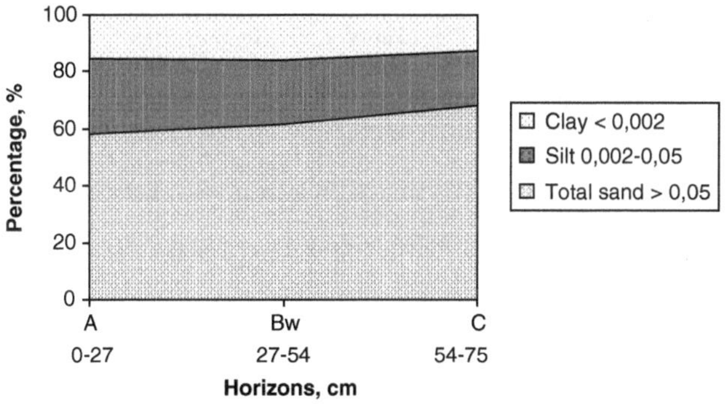

Fig. 8.9 Pattern of the particle-size distribution (mm) in brown forest soil (eutric) (USDA)

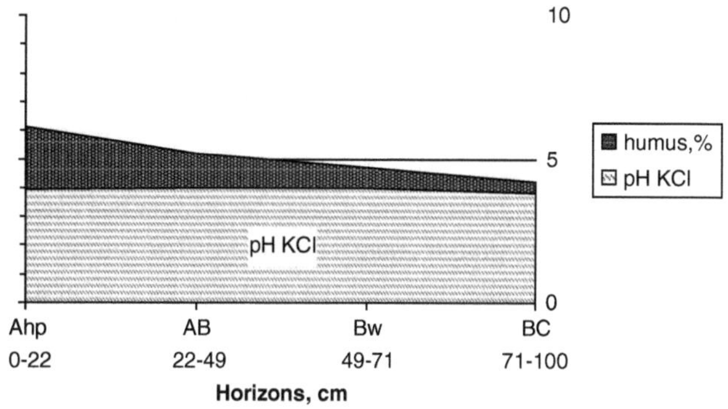

Fig. 8.10 Pattern of the pH and humus content distribution in brown forest soil (dystric)

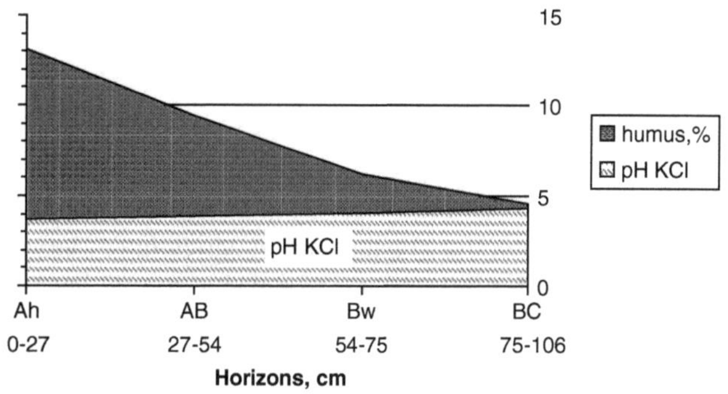

Fig. 8.11 Pattern of the pH and humus content distribution in brown forest soil (dystric, humic)

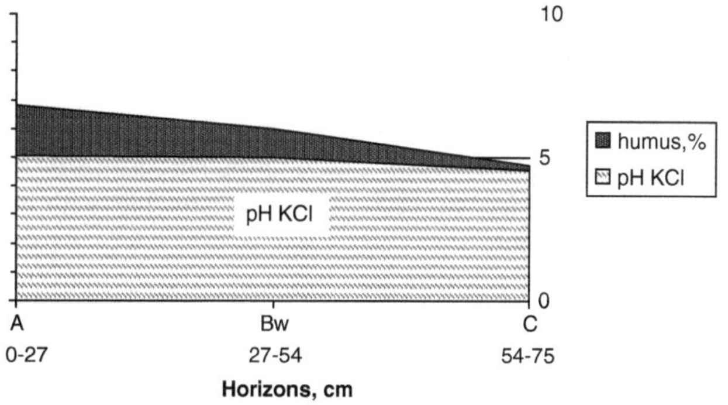

Fig. 8.12 Pattern of the pH and humus content distribution in brown forest soil (eutric)

humic-fulvic (Shishkov and Petrova 1997). The brownish colored fraction is dominant in humic acids. A large amount of humic substances are bounded with clay minerals.

The free Fe_2O_3 is uniformly distributed throughout; somewhere it can prevail at the surface layer. Oxides, hydrous oxides, hydroxides SiO_2, and R_2O_3 distribution is uniform in the profile.

Hydrolysis is active and most of the released iron and aluminum remains in the soil while some is removed in the drainage or by leaching. Mineral colloids composition is uniform, consisting of mixed type and amorphous substances. This determines comparatively low sorption capacity.

The main process is leaching of the basic cations. Sodium, potassium, and magnesium are washed out of the soil into the drainage water. Lateral movement carries dissolved cations which contribute to the base saturation. Commonly easily soluble salts are washed out of the profile.

In the profile the upper horizon usually has the maximum content of clay which decreases gradually with the depth, or has uniform clay distribution. This is evidence of the rate of hydrolysis related to the high content of organic matter. The amount of clay content in the Bw horizon is larger than that in the parent materials.

The dystric soil variety is characterized by strong acid soil reaction in the humus horizon. There is a permanent leaching of the cations. Naturally, this soil variety is typical for coniferous and mixed forest areas formed on hard acidic rocks quartzite, granite, and sandstone. The eutric soil variety, with slightly acid soil reaction and base saturation over 50 %, is typical for broadleaf forests.

Brown forest soil is characterized by favorable physical properties. Soil aeration and water movement is high throughout the soil. In cases where these soils are cultivated, the exposed surface horizon is prone to be loosely structured when dry. This is due to the very fine granular structure.

8.4.2 Soil Management

Brown forest soil is commonly used for forestry, and the open areas are used for grazing and for agriculture. The vast areas occupied by this soil are woodland. Forests have an important role in preventing soil erosion by means of flood regulation. Natural environment conditions are conducive to forestry and afforestation is effective. In some areas endemic trees have been replaced by other species.

Arable land now occupies large areas that were woodland comprising hornbeam-oak such as *Fagus sylvatica, Carpinus betulus, Quercus dalechampii*, and *Pinus nigra*. In southeastern Bulgaria agricultural lands have replaced forests of *Fagus orientalis*, and *Quercus polycarpa* and mixed forests of *Quercus polycarpa,* and *Quercus frainetto*.

When agriculture is practiced artificial fertilizers are applied on an annual basis, which is possible with the addition of manure.

Brown forest soil is characterized by favorable physical properties. Soils are not prone to compaction. Aeration and water movement is high throughout the soil. Dryness is common for a short period during the summer.

Brown forest soil is highly fertile, and suitable for orchards and berries. Potatoes, beans, alfalfa, and flax are successfully grown on soils with acid soil reaction.

In cases of tree felling or forest fires, these soils are susceptible to erosion and rapid deterioration. Systematic preventative measures need to be taken, and good practices must be applied in forest management. Tree felling must be organized in such a way as to provide protection for soil and water resources.

References

Antipov-Karataev IN, Gerassimov IP (1948) The soils of Bulgaria. Zemsnab Press, Bulgarian

Arinushkina E (1970) Manual on chemical soil analysis. Moscow State University Press, Moscow (In Russian)

Artinova N, Grozeva M (1989) Group and fractional composition of humus in mountain brown forest soils formed under different conditions of soil formation. IVth National soil science conference, problems of soil science under intensive agriculture, 28–30 May 1986, pp 101–108

Bascomb CL (1968) Distribution of pyrophosphate extractable iron and organic carbon in soils of various groups. J Soil Sci 19(1):251–268

Boneva K, Houbenov G, Behar A (1986) Chemical and mineralogical composition of soils of the Central Balkan Mountains and the Eastern Rhodopes. Soil Sci Agrochem Plant Prot J XXI(6):3–11, Sofia (In Bulgarian)

Bondev I (1991) Vegetation in Bulgaria, map at scale 1:600 000 with explanatory text. St. Kliment Oxridski University Press, Sofia (In Bulgarian)

Dilkova R (1985) Structure and aeration in main soil varieties in Bulgaria. Theses, Institute of Soil Science Nikola Poushkarov, Sofia (In Bulgarian)

Dilkova R, Kerchev Zh, Kercheva M (1998) Evaluating and grouping of soils according to their susceptibility to anthropogenic degradation, Advances in GeoEcology, 31. CATENA VERLAG, Reiskirchen, pp 125–131

FAO-UNESCO-ISRIC (1990) Revised Legend of the Soil Map of the World. FAO World Soil Resources Report No 60, Rome

Ganev S, Arsova A (1980) Methods of determining the strongly acid and the slightly acid cation exchange in soil. Soil Sci Agrochemistry J XV(3):22–33, Sofia (In Bulgarian)

Ganev S et al (1990) Physic-chemical properties of soils in Bulgaria and other world's regions. Department of chemistry, Institute of Soil Science Nikola Poushkarov, Sofia (In Bulgarian)

Georgiev AG (1972) Classification of the mountain-forest dark-colored soils in Bulgaria. First national soil science congress, Bulgarian Academy of Sciences Press, Sofia (In Bulgarian), 23–25 Sept 1969

Hadzhiyanakiev As (1989) Iron in Bulgarian soils. Zemizdat Press, Sofia (In Bulgarian)

Keys to Soil Taxonomy (2010) USDA United States department of Agriculture, NRCS natural recourses conservation service, 11th edn. Soil Survey Staff, Washington, DC, p 338

Kolchakov I (1989) Soil mapping under mountain conditions. Lectures on Soil Science. In: FAO project TCP/BUL/4502 (T), Institute of Soil Science Nikola Poushkarov, Sofia, pp 157–165 (In English)

Koynov V et al (1964) Classification and systematic of soils in Bulgaria. In: Gerasimov IP et al. (eds) Soils of south–eastern Europe. Proceedings of the international symposium of soil science, June 1963, Sofia. Bulgarian Academy of Sciences Press, Sofia: 231–246 (In Russian)

Mehra OP, Jackson ML (1960) Iron oxide removal from soils and clays by dithionite—citrate—system buffered with sodium bicarbonate. Clays Clay Miner 7:317–327

Nedialkova K, Shishkov T (1999) Arbuscular micorrhizal fungi and phosphate solubilizing microorganisms in natural ecosystem. J Balkan Ecol 2(2):66–71

Nedialkova K, Shishkov T (2003) Phosphate-solubilizing microorganisms population in mountain soils. J Balkan Ecol 6(3):266–273
Penkov M et al (1992) Soil classification and diagnostic in Bulgaria in relation to the land reform fragmentation. Zemizat Press, Sofia (In Bulgarian)
Shishkov T, Petrova L (1997) Organic matter particularities of south-eastern Rhodope Mountain soils. In: Genesis, classification, quality and sustainable use of soils in regions with Mediterranean Type Climate, 4th International meeting on red Mediterranean soils, Plovdiv, Bulgaria, 27 May–2 June 1997 (Bulgarian J Agric Sci 5(1):93–96
Stefanovits P (1964) The processes in brown forest soils as a basic for their classification. In: Soils of south-east Europe, Bulgarian Academy of Sciences Press, Sofia
Tamm O (1922) Eine methode zur bestimung der anorganischen komponente ds gelcomplexes im boden. Statens Skogsforsoksanstalt, Medd 19:387–404 (by Arinoushkina E (1972) Manuel of chemical analysis of soils, Moscow)
Tavernier R, Smith GD (1957) The concept of braunerde (brown forest soil) in Europe and United States. Adv Agron IX:217–289 (New York)
World Reference Base for Soil Resources (2006) A framework for international classification, correlation and communication. FAO-ISRIC, Rome
Yolevski M, Hadzhiyanakiev A (1976) Agro-grouping of soils in Bulgaria. Extended systematic list. Code of groups and soil varieties, Institute of Soil Science Nikola Poushkarov, Sofia (In Bulgarian)
Yolevski M, Ninov N, Krastanov S et al (1983) Brief diagnostic and criterion of the main taxonomic soil divisions in Bulgaria. Precise Classification Scheme, Institute of Soil Science Nikola Poushkarov, Sofia (In Bulgarian)

9 Dark Mountainous Forest and Mountainous Meadow Soils

Derived from the Latin word *umbra,* meaning dark.

Dark mountainous forest and mountainous meadow soils are undifferentiated black or dark colored soils found in the high mountain landscape.

9.1 Concept of the Type

Main concepts

1. Dark mountainous forest and mountainous meadow soils are mineral soils distinguished by characteristics inherited from the parent rock materials;
2. Characterized by high values of organic substance content, and low base saturation;
3. Developed in a wet and cool upland environment.

Dark mountainous forest and mountainous meadow soils have formed in humid subalpine and alpine areas, where soils show evidence of high rates of organic substance accumulation.

9.2 Ecology of Soil Formation

9.2.1 Distribution

Bulgarian subalpine mountainous soils are distributed at the upper boundary of the coniferous forest or about 1,800 (1,900) m above sea level; and mountainous meadow soils are found above 2,100 m. The belt of mountainous meadow soil is situated at the highest altitude (Fig. 9.1). These areas have cold climatic conditions and short vegetation period. Two provinces are singled out: the Balkan Mountains province and Vitosha-Rila-Pirin mountains province.

Soils occupy the subalpine and alpine plateaus and ridges of the Balkan Mountains (peak Botev 2,376 m), Vitosha mountain (peak Cherni vrah 2,290 m), Rila mountain (peak Musala 2,925 m), Pirin mountain (peak Vihren 2,914 m), Belassitsa mountain (peak Radomir 2,029 m), and the Rhodope mountains (peak Golyam Perelic 2,190 m).

Dark mountainous forest soils occupy about 11,011 ha or 0.99 % of the entire country's territory; and mountainous meadow soils occupy about 16,820 ha or 1.52 %.

Climatic conditions

Natural conditions vary in the mountains. Commonly, the climate in the upland mountains is characterized by contrasts in daily and seasonal cycles such as lower temperature during the whole year, high amount of precipitation, and an almost equal seasonal amount of surface moisture, high air moisture, and high solar radiation.

Altitude is the most important factor in the formation of the subalpine and alpine mountain environment. In the highest mountain lands the average air temperature and amount of precipitation changes with every 100 m of rising altitude. In the subalpine and alpine zones, solar radiation increases and air humidity decreases. Soils remain snow covered for several months of most years. Seasonally, soils are deeply frozen or are in the ambient conditions of low temperature. The average vegetation period is usually 90–120 days in the alpine zone and 130–160 days in the subalpine zone. There is often a strong wind.

The mean annual temperature is –0.7 (4.9) °C. In January the mean annual temperature is –4.1 °C (–9.3 °C). In July the mean annual temperature is 7.4–13.6 °C. Average annual precipitation varies from 830 to 1,249 mm with maximum falls during winter.

Topography

Relief in the high mountain regions is alpine dissected. It is the most significant factor in soil formation, determining the translocation of the weathering materials of rocks and erosion. The type of relief is determined by a specific geology. The dominant surfaces are slopes of different exposure and gradient. Dark mountainous forest and mountainous meadow soils are found on the mountain ridges and plateaus of the subalpine and alpine zone. This relief determines conditions for the intensive lateral geochemical translocation. Soils are naturally affected by slope exposure and elevation. Commonly, soil cover is ruptured by rocks and boulders.

T. Shishkov and N. Kolev, *The Soils of Bulgaria*, World Soils Book Series, DOI: 10.1007/978-94-007-7784-2_9,

Fig. 9.1 Highland landscape

Mountain exposure also has a very strong influence on water and moisture regimes. Differences in soils and soil formation are determined by solar and wind exposure, and the effect that this has on natural vegetation growth.

Parent materials
Continuous hard rock has some provincial differences related to the soils, and determines their specific character. Acid igneous rock contains more than 60 % silica and free quartz.

Thus in the Balkan Mountains parent rocks are of the Carboniferous (Namurian-Westfalian) Age granites and granodiorites.

In the Rila-Pirin-Rhodopi mountains soils are on the Carboniferous (Namurian-Westfalian) Age granites and granodiorites; on the Cretaceous Age Mediterranean type intrusive granites; and on the Paleogene Age intrusive granites.

In the Rhodope Mountains they are on the Paleogene (Oligocene) Age ignimbrites, rhyolites, trachyrhyolites, and delenites.

In the Vitosha mountain parent rocks are of the Cretaceous Age granites, syenomonzonites, granodiorites, quatzmonzonites, and monzonites.

Vegetation
In the mountains where coniferous forest has been turned into upland meadows, grass formations dominate together with juniper and dwarf mountain pine.

In the highest parts of the Rila, Pirin, and Balkan Mountains with altitudes of 2,300–2,500 m, psychrophytic and cryophytic hecistothermic vegetation is distributed, forming an alpine belt where grass with some dwarf bush formations predominating. On the silicate soils are acidophylic formations such as *Cariceta curvulae, Festuceta riloensis, Seslerieta comosae, Junceta trifidi, Festuceta airoides, Agrostideta rupestris*, and much more rarely *Saliceta retusae, Saliseta herbaceae, Empetreta nigrae,* and *Vaccinieta uliginosi. Nardus stricta* communities have invaded on thicker soils (Bondev 1991).

Above 2,000 m altitude in the Rila, Pirin, Balkans, Vitosha, Rhodopes, Osogovo, Belasitsa, and Slavyanka Mountains is a subalpine belt with mesophytic and hygrophytic vegetation. Indigenous vegetation is shrubby, including *Pineta mugi* and on wet soils *Alneta viridis.*

Peaty vegetation is formed from *Cariceta acutae, Cariceta stellulatae, Cariceta rostratae, Sphagneta* sp., *Saliceta lapponi*, and *Hygronardeta* sp.

The subalpine shrubs are *Junipereta sibiricae, Junipereta pygmeae, Chamaecytiseta absinthioides*, dwarf shrubs *Vaccinieta myrtilli, Vaccinieta vitis-ideae, Bruckenthalieta spiculifoliae,* and *Dryeta octo petalae*, and grass formations are *Nardeta strictae, Agrostideta capillaris, Festuceta validae, Bellardiochloeta violaceae, Festuceta penzesii,*, and *Seslerieta korabensis.*

Time
Being young soils, their development presumably started at the end of glaciations.

Human influence
The mountain uplands are suitable for grazing. Overgrazing can provoke erosion. Shallow soils should be left to develop a natural vegetation and wild life.

9.2.2 Genesis

In the high mountain lands soil formation is commonly on hard rock. The natural upland environment does not have an analogy in the plains. This fact determines comparatively shallow soil development and high rates of skeletal performance, with poor assortment of soil mass. In alpine areas the so-called dry peat horizon overlaps the mineral part. The soil has inherited many of the characteristics of the parent rock. Both physical and chemical weathering is important in the transformation of parent materials. Primary minerals prevail in the soil mass; the amount of clay minerals is low. The soil profile is not differentiated. Soils with southern exposure are shallow and with a lower amount of humus; soils with northern exposure are deeper and darker. Duchaufour (1968) described the soil as alpine ranker.

In the mountain regions with significant amounts of precipitation, the ridges are usually free of forest. Strong winds and fogs are common. Soils are seasonally overwetted in most years and accumulate a considerable amount of organic substances. The period with temperatures above freezing is short. In these conditions biological activity is mainly in the top soil. Humification is of sufficient intensity to produce a considerable amount of organic substances, and grass roots form well-developed sod. Humus acids penetrate into the mineral part, where there is some evidence that transformation takes place. This fact indicates that the soil is the early stage of soil formation.

In the subalpine zone, the decomposition of plant remains is more complete and the synthesis of new compounds is more advanced where dark mountainous forest soil and mountainous meadow soils have developed. Duchaufour (1968) described dark mountainous forest soil as pseudo-alpine ranker or humus-silicate soil. Soils have formed under the acidophilic tall-grass vegetation of the upland meadows.

Many of these soils are shallow to bedrock and usually occur on steeply sloping terrains. Due to this fact, the soils have been considered as a phase of development from the Lithosol. The principal variation within the type is the soil thickness; within the limits of a single slope, thickness can vary greatly from between 10 and 80 cm.

Weathering is a result of primary minerals transformation, and clay synthesis occurs through the ions components. Duchaufour defined this soil as of quite long evolution and the product of an intensive weathering process. Migration is in the upper part of profile, which has been diminished by organic matter. Mineralization of organic substances is low. Organic matter is linked with the mineral part. Biological activity is strongly pronounced. Dominant fulvic acids and so-called brownish humic acids are complexed with iron. Organic matter penetration reaches to a considerable depth.

9.3 Soil Diagnostic and Classification

9.3.1 Morphology

The most common diagnostic feature is *umbric* horizon. Most soils have an A/C horizon pattern profile (Fig. 9.2). Dark mountainous forest and mountainous meadow soils show the following morphological features:

Fig. 9.2 **a** Dark mountainous forest soil, Bulgarian classification Yolevski and Hadzhiyanakiev (1976), *Cambic Umbrisol* or *Haplic Umbrisol (Skeletic)*, WRB (2006), *Humic Dystrudepts* or *Humic Pachic Dystrudepts*, Keys to Soil Taxonomy (2010), *Dystric* or *Humic Cambisol*, Revised Legend FAO-UNESCO-ISRIC (1990). **b** Mountainous meadow soil, Bulgarian classification Yolevski and Hadzhiyanakiev (1976), *Leptic Umbrisol*, WRB (2006), *Humaqueptic Epiaquents*, Keys to Soil Taxonomy (2010)

Ah1—a tightly bound root mat or peated, commonly 5–10 cm thick. The fine earth is very dark brown colored, between the roots there are friable structural aggregates of fine granular structure (spheroids or polyhedrons) with size 1–2 mm diameter, loose (non-coherent) or single grains with size <1 mm diameter; beneath there is clear transition to the lower

Ah2—a humus accumulative horizon where the organic part dominates over the mineral, homogeneous, color is very dark brown to dark brown. Structure is granular (spheroids or polyhedrons) with medium size (2–5 mm) diameter, weak pedality grade, consistence is friable (moist) and soft (very weakly cohered and fragile) dry, many quartzite grains are in the soil mass, abundance of fine (0.5–2.0 mm) and medium size (2–5 mm) grass roots; beneath it gradual changes to

AC—a mineral transition to the upper part of the parent materials horizon, 15–25 cm thick, yellowish brown, light colored compared to the upper horizon. The structural aggregates are granular coarse size (5–10 mm diameter), weak pedality grade, consistence is very friable (moist) and soft (dry), many rock fragments of different size are in the soil mass; there is gradual transition into

CR—half-weathered parent materials consisting of different size pebbles and a very small amount of structureless fine earth, not differentiated. Rock fragments show little change.

R—hard continuous rock.

Soil profile is homogeneous and uniform in texture.

The humus horizon is with low bulk density, and characterized by high water saturation that decreases with depth.

9.3.2 Classification

The single type of mountainous soils is distinguished/defined as specific to natural upland environments which do not have an analogy in other geographical regions of the country.

Dark mountainous forest and mountainous meadow soils have not been well studied. Is not clear, whether an initial development of primitive (B) horizon takes place. The diagnostic is based on the pronouncement of humus horizon of moder type.

The Bulgarian soil classifications of 1947/1964 included chernozem-like (subalpine), sod (subalpine), and peat-like (alpine) subtypes. The classification of Yolevski and Hadzhiyanakiev (1976) gave the same order. Dark mountainous forest soil was classified by Georgiev (1972). There was classification in 1992 (Penkov et al.) which at the theoretic level renamed the sod subtype as haplic.

Within the subtype, soil differs in the thickness of A + AC horizon (low thick 30 cm, thick 40–60 cm and very thick over 70 cm) and in the rates of erosion (non-eroded, slightly to moderately eroded).

Classification after World Reference Base for Soil Resources (2006) required *Umbric* diagnostic horizon in the reference group of Umbrisols. Umbric is similar to the Mollic horizon except that it has base saturation <50 %.

Keys to Soil Taxonomy (2010) required *Umbric* epipedon for Inceptisol order classification. In Soil Taxonomy there are demands for subsurface horizons allowing a *cambic, albic, agric, calcic, gipsic, salic, placic, or duripan.*

9.3.3 Dark Mountainous Forest and Mountainous Meadow Soils Types in Bulgaria

9.3.3.1 Dark Mountainous Forest Soil

Dark (*subtype*) **mountainous forest** (*type*) soil is named after the Bulgarian classification of Koynov et al. (1964), and Yolevski and Hadzhiyanakiev (1976).

Cambic Umbrisol or Haplic Umbrisol (*Dystic*) (*Skeletic*) is named according to WRB (2006) and *Humic Dystrudepts, Humic Pachic Dystrudepts* according to Keys to Soil Taxonomy (2010).

Dystric or Humic Cambisol: Revised Legend of the World Soils (FAO-UNESCO-ISRIC 1990).

9.3.3.2 Mountainous Meadow Soil

Meadow (*subtype*) **mountainous** (*type*) soil is named after the Bulgarian classification of Koynov et al. (1964), and Yolevski and Hadzhiyanakiev (1976).

Umbric Leptosol or Leptic Umbrisol is named according to WRB (2006) and *Humic Lithic Dystrudepts or Typic Humicryepts* according to Keys to Soil Taxonomy (2010).

Dystric or Umbric Leptosol: Revised Legend of the World Soils (FAO-UNESCO-ISRIC 1990).

Soils are highly skeletal. The clay content is low and uniformly distributed. The profile is not differentiated in clay and sesquioxides distribution. Particle size distribution of the fine earth shows that the coarse and fine sand fractions (1.0–0.25 and 0.25–0.05 mm) dominate (Tables 9.1, 9.3, and Fig. 9.3). The texture is homogeneous at the depth (Tables 9.2, 9.4, and Fig. 9.4).

The thickness of the humus horizon and the content of humus greatly vary. Commonly, the organic matter content is high; 8–20 % in the upper part and 2–7 % at the depth beneath 30 cm (Fig. 9.5). Fulvic acids prevail in the humus composition with considerable enrichment at the depth; there may be some of the aggressive type of fulvic acids (Achkov et al. 1998). Fulvic acids are characterized by low molecule condensation. Some amounts of humic acid may occur at the top of the soil.

Soils are strongly acid with pH_{KCl} values of about 3.3–4.5 (Ganev and Arsova 1980; Ganev et al. 1990). H^+

Table 9.1 Variables of the particle-size distribution in 15 referenced profiles of dark mountainous forest soil in Bulgaria

Profile count 15	Horizon	Upper depth	Lower depth	Loss HCl	Particle-size distribution (mm), % (according method of Kachinskiy revised)							
					>1 mm	1–0.25	0.25–0.05	0.05–0.01	0.01–0.005	0.005–0.001	<0.001	<0.01[a]
Mean	Ah1	0	25	2.3	1.4	22.6	23.6	18.6	8.9	8.6	13.8	31.3
	Ah2	25	53	2.2	4.5	28.5	23.8	11.4	7.5	9.5	12.6	29.6
	AC	53	97	1.6	5.9	31.4	27.9	10.1	7.6	8.2	7.3	23.1
Stdev	Ah1	0	5	11.1	2.3	13.2	7.6	6.5	3.6	4.0	12.0	14.2
	Ah2	5	15	0.7	6.8	15.0	9.1	2.2	2.7	2.6	21.6	25.5
	AC	15	14	0.7	10.3	4.8	3.0	4.4	1.8	2.7	2.9	6.3

[a] Physical clay, the parameter characterized sum of the particle-size distribution, according to Kachinskiy method (mm) $\sum < 0.01 = <0.001 + (0.005 - 0.001) + (0.01 - 0.005)$

Table 9.2 Variables of humus, pH, carbonates, and particle size distribution (mm) according to USDA in 15 referenced profiles of dark mountainous forest soil in Bulgaria

Profile count	Horizon	Upper depth	Lower depth	Humus (%)	pH_{KCl}	Bulk density $g.cm^3$	Clay <0.002	Silt 0.002–0.05	Fine sand 0.05–0.2	Total sand 0.05–2.0
15										
Mean	Ah1	0	25	11.10	4.3	–	19.6	33.1	24.0	47.3
	Ah2	25	53	3.60	4.5	–	19.0	25.2	24.6	55.8
	AC	53	97	1.36	4.5	–	12.7	23.0	29.0	64.3
Stdev	Ah1	0	5	3.80	0.5	–	17.0	9.2	7.0	18.8
	Ah2	5	15	1.48	0.9	–	22.3	4.7	9.4	24.2
	AC	15	14	0.66	0.3	–	4.1	5.1	2.0	6.5

Table 9.3 Variables of the particle-size distribution in 14 referenced profiles of mountainous meadow soil in Bulgaria

Profile count 14	Horizon	Upper depth	Lower depth	Loss HCl	Particle size distribution (mm), % (according method of Kachinskiy revised)							
					>1 mm	1–0.25	0.25–0.05	0.05–0.01	0.01–0.005	0.005–0.001	<0.001	<0.01[a]
Mean	A	0	19	1.1	20.3	20.5	22.1	15.7	6.8	5.3	8.2	20.4
	R	–	–	–	–	–	–	–	–	–	–	–
Stdev	A	0	4	0.8	19.1	7.7	12.9	5.5	2.1	2.1	4.0	6.5
	R	–	–	–	–	–	–	–	–	–	–	–

[a] Physical clay, the parameter characterized sum of the particle-size distribution, according to Kachinsky method (mm) $\sum < 0.01 = <0.001 + (0.005 - 0.001) + (0.01 - 0.005)$

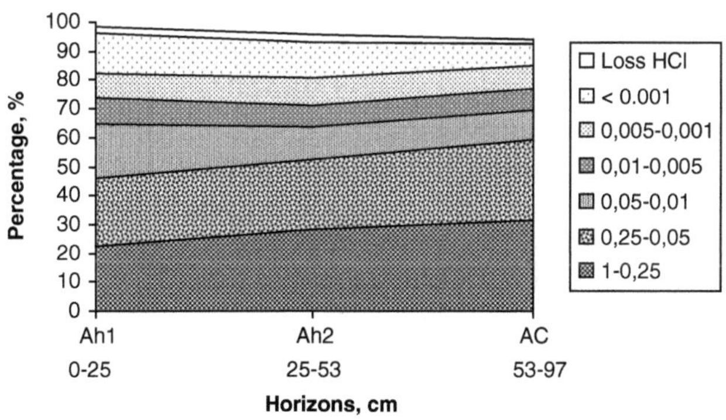

Fig. 9.3 Pattern of the particle-size distribution (mm) in dark mountainous forest soil (according to revised Kachinskiy method)

and Al^{3+} dominate among the exchangeable cations which determines the base saturation as low. Cation exchange capacity is 8–25 meq/100 g soil.

The qualitative composition of clay minerals, determined by X-ray diffraction analysis and transitional electronic microscopy, has shown a prevalence of hydromicas, about 40 % in the clay (<0.001 mm) throughout the soil. Montmorillonite and kaolinite content is 30 and 17–21 % in the clay throughout the soil, and quartz is about 9–15 % in the clay (<0.001 mm), (Boneva et al. 1986).

The amount of mobile phosphorous and potassium is low.

Table 9.4 Variables of humus, pH, carbonates, and particle size distribution (mm) according to USDA in 14 referenced profiles of mountainous meadow soil in Bulgaria

Profile count 14	Horizon	Upper depth	Lower depth	Humus (%)	pH_{KCl}	Bulk density $g.cm^3$	Clay <0.002	Silt 0.002–0.05	Fine sand 0.05–0.2	Total sand 0.05–2.0
Mean	A	0	19	5.7	3.9	–	13.0	29.2	24.2	57.8
	R	–	–	–	–	–	–	–	–	–
Stdev	A	0	4	2.7	0.2	–	5.6	7.1	11.6	10.4
	R	–	–	–	–	–	–	–	–	–

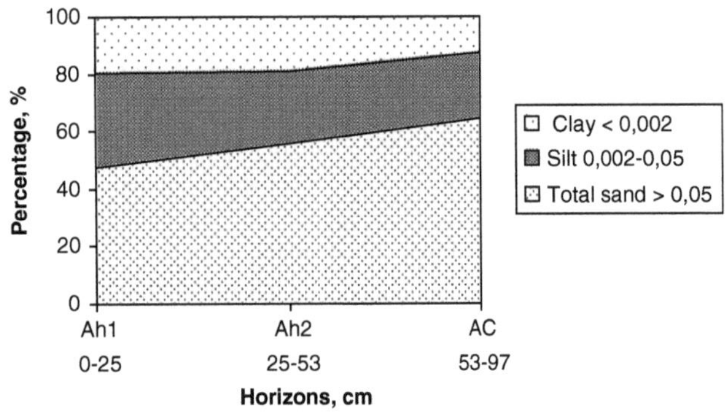

Fig. 9.4 Pattern of the particle-size distribution (mm) in dark mountainous forest soil (USDA)

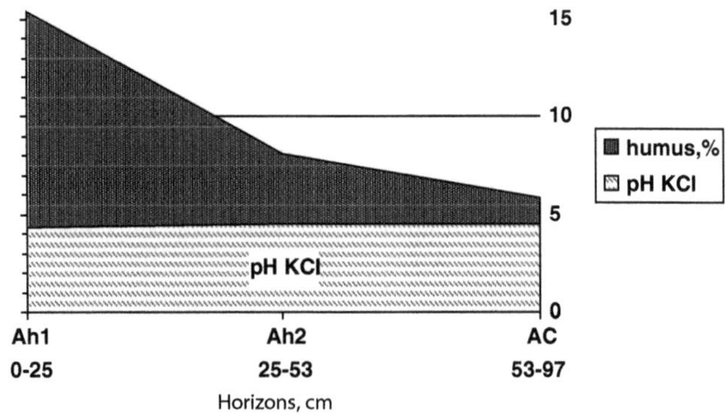

Fig. 9.5 Pattern of the pH and humus content distribution in dark mountainous forest soil

The average particle density is 2.0–2.1 g/cm^3 in the upper part and 2.4 g/cm^3 at the depth. Bulk density is about 0.5 g/cm^3 in the middle part.

Water stable aggregates with size >1 mm are about 50–55 %. Prevailing water stable aggregates with size over 0.25 mm are 80–85 %.

In dark mountainous forest and mountainous meadow soils nearly 20 % of pore space is represented by macropores or the pores of aeration (with diameter >300 μm); and 20 % is mesopores or pores with rapid moisture drainage (>30 μm diameter). Micropore space or water-filled is about 30 %, represented by slow drainage pores (diameter 0.2–30.0 μm) and 20 % of micropores is with none remaining available for roots moisture (diameter <0.2 μm), (Dilkova 1985; Dilkova et al. 1998).

Total porosity is high at the surface (70 % at field capacity) and decreases with depth. Soils are characterized by extremely high water permeability.

The amount of total iron (Fe_2O_3) (Arinushkina 1970) is very low and decreases with the depth. The non-silicate form of iron (Mehra and Jackson 1960) is nearly 60.0 % (of the total iron Fe_2O_3) and dominates in the humus horizon. Its distribution sharply decreases in the parent materials (Hadzhiyanakiev 1989).

The silicate iron in the humus horizon is about 40 % (of the total iron Fe_2O_3), and is nearly 66.5 % in the parent materials.

Dark mountainous forest and mountainous meadow soils have a specific distribution of different forms of iron compounds. In the humus horizon the rates of crystallized and amorphous forms of iron are differentiated; crystallized iron is 60.0 % and amorphous is 27–40.0 % (of non-silicate iron). The conditions for strong crystallization are favorable in the humus horizon. The slightly crystallized (non-silicate iron) is much lower in the humus horizon, and sharply increases in the parent materials by nearly 50.0 %.

There is a significant increase in the slightly crystallized and free amorphous forms of iron in the parent materials.

The hydroxide amorphous form of iron (Tamm 1922) is differentiated in the soil. It is about 35.0 % (of the total iron) in the humus horizon and nearly 13 % in the parent materials.

The free amorphous form of iron significantly increases in the humus horizon, about 47.0–52.0 % (from the total non-silicate amorphous iron).

The amorphous form of iron bounded with organic matter (Bascomb 1968) is very low, only 12.0 % (of the total non-silicate amorphous iron); and it is regularly distributed in both the humus and parent materials.

The aeration of soil when dry affects the deeper layers. Soils are characterized by high water infiltration rates when wet and have very high water-holding capacity. Soils have seasonal biological activity.

The natural spatial relationship is with Lithosols and brown forest soils.

9.4 Soil Properties and Soil Management

9.4.1 Soil Properties

Typically, the dark mountainous forest and mountainous meadow soils are dark brown colored corresponding to the high amount of organic substances of humic-fulvic type, and the considerable amount of lipids and non-extractable organic carbon. The profile is highly skeletal.

Most of the released iron and aluminum, as well as sodium, potassium, and magnesium have been washed out of the soil into the drainage water.

Soils are characterized by strong acid soil reaction in the humus (A) horizon. Sorption capacity is mostly due to the organic colloids. Despite the low base saturation, the cation exchange capacity is about 10–20 meq/100 g soil. The upper humus horizon is characterized by high content of free iron compounds in the form of concretions, segregations, and coatings.

Soils are characterized by favorable physical properties. Soil aeration (when dry) and water movement is high throughout the soil.

9.4.2 Soil Management

Many mountain meadow soils are seasonal grasslands. Overgrazing can provoke erosion. Due to the limitations of topography, shallow profile composition, considerable skeletal part, and rock protrusion these soils are not suitable for agriculture. The dense root mat horizon protects the top soil and minimizes sheet erosion. The environmentally proper management of these valuable grasslands demands a scheme for grass species selection and the application of fertilizers.

In the past, a great deal of the alpine pine shrub *Pinus mugo* in the Rila and Pirin mountains and the Balkans was destroyed in order to enlarge the alpine pastures and to produce charcoal. Nowadays, the upland environment is protected and some of sub-alpine endemic species have restored.

References

Achkov N, Nedyalkov S, Filcheva E (1998) To the diagnostic of some mountainous meadow soils. Soil Sci Agrochem 24(1):38–43 (In Bulgarian)

Antipov-Karataev IN, Gerassimov IP (1948) The soils of Bulgaria. Zemsnab Press, Sofia (In Bulgarian)

Arinushkina E (1970) Manual on chemical soil analysis. Moscow State University Press, Moscow (In Russian)

Bascomb CL (1968) Distribution of pyrophosphate extractable iron and organic carbon in soils of various groups. J Soil Sci 19(1):251–268

Boneva K, Houbenov G, Behar A (1986) Chemical and mineralogical composition of soils of the Central Balkan Mountains and the Eastern Rhodopes. Soil Sci Agrochem Plant Prot J (Sofia) 21(6):3–11 (In Bulgarian)

Bondev I (1991) Vegetation in Bulgaria, map at scale 1:600 000 with explanatory text. St. Kliment Oxridski University Press, Sofia (In Bulgarian)

Dilkova R (1985) Structure and aeration in main soil varieties in Bulgaria. Theses, Institute of Soil Science Nikola Poushkarov, Sofia (In Bulgarian)

Dilkova R, Kerchev Zh, Kercheva M (1998) Evaluating and grouping of soils according to their susceptibility to anthropogenic degradation. Advances in GeoEcology 31. Catena Verlag, Reiskirchen, pp 125–131

Duchaufour Ph (1968) L'evolution des soils: Essair sur la dynamique des profils. Masson et cie, Paris

FAO-UNESCO-ISRIC (1990) Revised legend of the soil map of the world. FAO World Soil Resources Report No 60, Rome

Hadzhiyanakiev As (1989) Iron in Bulgarian soils. Zemizdat Press, Sofia (In Bulgarian)

Ganev S, Arsova A (1980) Methods of determining the strongly acid and the slightly acid cation exchange in soil. Soil Sci Agrochem J (Sofia) 15(3):22–33 (In Bulgarian)

Ganev S et al (1990) Physic-chemical properties of soils in Bulgaria and other world's regions. Department of chemistry, Institute of Soil Science Nikola Poushkarov, Sofia (In Bulgarian)

Georgiev AG (1972) Classification of the mountain-forest dark-colored soils in Bulgaria. In: First national soil science congress, 23–25 Sept 1969, Bulgarian Academy of Sciences Press, Sofia (In Bulgarian)

Keys to Soil Taxonomy (2010) USDA United States Department of Agriculture, NRCS Natural Recourses Conservation Service, 11th edn. Soil Survey Staff, Washington, DC, p 338

Koynov V et al (1964) Classification and systematic of soils in Bulgaria. In: Gerasimov IP et al. (eds) Soils of south–eastern Europe. Proceedings of the international symposium of soil science, June 1963, Sofia. Bulgarian Academy of Sciences Press, Sofia: 231–246 (In Russian)

Mehra OP, Jackson ML (1960) Iron oxide removal from soils and clays by dithionite—citrate—system buffered with sodium bicarbonate. Clays Clay Miner 7:317–327

Penkov M et al (1992) Soil classification and diagnostic in Bulgaria in relation to the land reform fragmentation. Zemizat Press, Sofia (In Bulgarian)

Tamm O (1922) Eine methode zur bestimung der anorganischen komponente ds gelcomplexes im boden. Statens Skogsforsoksanstalt, Medd.19:387–404 (by Arinoushkina E (1972) Manuel of chemical analysis of soils, Moscow)

World Reference Base for Soil Resources (2006) A framework for international classification, correlation and communication. FAO-ISRIC, Rome

Yolevski M, Hadzhiyanakiev A (1976) Agro-grouping of soils in Bulgaria extended systematic list. Code of groups and soil varieties. Institute of Soil Science Nikola Poushkarov, Sofia (In Bulgarian)

Alluvial and Deluvial Soils

10

Derived from the Latin word *fluvius,* meaning river or stream;

And from the Latin word *colluvies,* meaning to wash.

Soils are without diagnostic horizons. They are characterized by the influence of local factors such as parent material and/or topography.

10.1 Concept of the Type

Main Concepts

1. Alluvial and deluvial (colluvial) soils are mineral soils characterized by properties inherited from the parent materials;
2. These soils occur in many environments;
3. Sedimentary layers can lack evidence of alteration.

The main processes that have taken place are stratification of materials and organic matter distribution.

10.2 Ecology of Soil Formation

10.2.1 Distribution

Alluvial soils are formed in accumulative landscapes, where the geochemical migration of different elements takes place (Fig. 10.1). In Bulgaria, alluvial soils occupy 585,037.3 ha, or about 5.29 % of the entire territory. The riparian areas traditionally include a much larger range of soils rather than with contemporary renewal of materials.

In northern Bulgaria, the main areas of distribution are in the alluvium valleys of the following rivers Lom (total length 92.5 km and catchment area—1,140 km^2), Beli Lom (total length 140.7 km and catchment area—1,279 km^2), Vitt (total length 188.6 km and catchment area—3,220 km^2), Ogosta (total length 144.1 km and catchment area—3,157 km^2), Iskar (total length 368 km and catchment area—8,646 km^2), Osam (total length 314 km and total area—2,824 km^2), Yantra (total length 285 km and catchment area—7,862 km^2), Rossitsa (total length 164 km and catchment area—2,265 km^2), Kamchia (total length 244.5 km and catchment area—5,358 km^2), and their tributaries.

In the southern part of the country, soils are distributed in the alluvium valleys of the following rivers; Maritsa (length 321 km and catchment area—21,084 km^2 on Bulgarian territory), Struma (length 290 km and catchment area—10,797 km^2 on Bulgarian territory), Mesta (length 126 km and catchment area—2,767 km^2 on Bulgarian territory), Tundzha (length 349.5 km and catchment area—7,884 km^2 on Bulgarian territory), Sazliyka (total length 145.4 km and catchment area—3,293 km^2 on Bulgarian territory), Topolnitsa (total length 154.8 km and catchment area—1,789 km^2 on Bulgarian territory), Arda (length 241.3 km and catchment area—5,201 km^2 on Bulgarian territory, Veleka (total length 147 km and total catchment area—994.8 km^2 on Bulgarian territory), and their tributaries.

Usually the size of living flood plain is 20–50 m, but it could be a few kilometers in the case of big rivers. During soil evolution alluvium has changed the location. Thus the Thracian Lowland has much larger alluvium cover than the contemporary flooded area of Maritsa River.

Deluvial (colluvial) soils are located on talus drift and creep fans in the hilly regions. Deluvial soils occupy 140, 308.9 ha, or 1.27 % of the entire territory. This single group of soils is absent from both WRB (2006) and FAO-UNESCO-ISRIC (1990).

The Rose Valley represents the deluvial non-calcareous and calcareous sediments.

Climatic Conditions

Alluvial and deluvial (colluvial) soils are formed in many environments under various climate conditions. The temperature regime and total precipitation vary from one region to another.

The flooding of the living flood plain is seasonally dependent on the amount of thaw water collected in the river basin and on rainfall intensity.

T. Shishkov and N. Kolev, *The Soils of Bulgaria*, World Soils Book Series, DOI: 10.1007/978-94-007-7784-2_10,

Fig. 10.1 Landscape of alluvial and deluvial (colluvial) soils

The level of the groundwater table is seasonally dependent and influences hydromorphic conditions.

Topography

Commonly alluvial soils are located in riparian valleys with topographical depressions (basins). The relief is plain or graded at the flood plain (alluvium deposited along the river due to repeated flooding), upper and low flood plain (with higher or lower elevation), upper and lower alluvial plain (level plain due to the deposit of alluvium by major rivers), deltaic plain (alluvial plain formed by the tributaries of the rivers), alluvial fans (fan-shaped deposit of sediments along the river), river terraces (flat uplands at different levels along the river), cut-off meandering valleys (meanders filled with river sediments), tributary valleys, graded valleys, straths, old stream beds, and stream or river valleys.

Usually the level of groundwater is 1.0–3.5 m.

For example, the terrace system of the Danube River near to the town of Ruse comprises two flooded terraces of 4–7 m altitude; and three terraces above the flooded one which are at the altitudes of 15–22, 30–66, and 54–65 m.

The discovery of fossil soils indicates that in northern Bulgaria, loess has covered Pleistocene Age river terraces (Minkov 1968).

Deluvial (colluvial) soils are confined to the creep fans on the foot slopes which are handled by gravity. Soils are formed on talus cones (cone-shaped deposits of coarse debris at foot hills), drift beds, slide rocks, hillside waste, and downhill creep. They are commonly located at the bottom of low-grade slopes and/or at the foot of steep slope formed fan-shaped deposits.

Parent Materials

Alluvial and deluvial (colluvial) soils are confined to sedimentary calcareous, basic, or acid parent materials. Sometimes it is not possible to be certain about the origin of all these materials. Usually, lowlands are filled with mixed deposits.

Whereas alluvium is sediment sorted by running water, colluvium is not.

Most materials are deposits of various textures. Some provincial differences of soils are determined by the character of the parent materials.

Alluvial materials often have thick layers (strata) of different size materials. Because the river bed is permanently meandered parts of the floodplain can change over space and time. This is the reason for the great variability and layer formation of various deposits.

Fast-moving water deposits coarse material such as rocks and gravel. The finer particles such as sand and silt are deposited by water of lower velocity, whereas fine silt and clay particles are deposited by slow-moving water.

Colluvium is an assorted mixture of material with or without rock fragments consisting of pebbles, sand, silt, and clay that accumulates at the base of steep slopes.

Thus, in the central part of the Thracian Lowland the deposits are of the Quaternary Age alluvial-drift and talus-drift sediments (boulders, pebbles, sands); the Quaternary alluvial sediments (pebbles and sands of floodplain and higher terraces), and swamp deposits.

In the western part of the country, parent materials are of the Quaternary Age drift sediments—larger drift fans; and of the Quaternary alluvial sediments (pebbles and sands of flood-plain and higher terraces).

In the stream and river valleys, the Quaternary Age alluvial sediments are pebbles and sands of floodplain and higher terraces.

Vegetation

The vegetation at the foot of the mountains, and in the river valleys is of mesophytic and hydrophytic formations. Common original vegetation was forest substituted with meadows and dry grasslands. Alluvial soils are formed under the influence of diverse herbaceous and tree hydrophilic vegetation. This is the most favorable environment for high biological activity and where the permanent accumulation of nutrition elements takes place. On river banks grow *Alneta glutinosae, Saliceta albae, Saliceta fragilis, Populeta nigrae,* and *Populeta albae*, and in the southern regions near Struma, Mesta, Tundzha, and Arda rivers *Plataneta orientalis* is found.

The formation of deluvial (colluvial) soils has been influenced by various grass associations and wood vegetation. Native vegetation is *Querceta pedunculiflorae*, and in some areas *Querceta roboris, Guercus cerris, Quercus virgiliana, Ulmeta minori,* and *Fraxineta oxycarpae*.

Common are associations of sow thistle (*Sonchus* asper), timothy (*Phleum* sp.), foxtail (*Alopecurus* sp.), fescue

(*Festuca* sp.), wheat grass (*Agropyron Gaerth*), meadow grass (*Poa* sp.), slough grass (*Beckmannia* sp.), canary grass (*Phalaris* sp.), peavine (*Lathyrus* sp.), clover (*Trifolium*), vetch (*Vicia sativa*), march cane's bill (*Geranium palusrtre*), water dock (*Rumax hydrolapathum*), butter cup (*Ranunculus* sp.), deer vetch (*Lotus* sp.), hairgrass (*Deschampsia caespitosa*), sedge (*Carex* sp.), willow (*Salix* sp.), and poplar (*Popupus* sp.), (Bondev 1991).

In some lowlands, mainly on alluvial soils and partly on some deluvial soils in humid environments, a meadow mesophytic vegetation comprising *Festuceta pratensis, Poaeta sylvicolae, Alopecureta pratensis, Lolieta perennis, Elymeta repentis, Agrostideta stoloniferae,* and others formed after the destruction of the forest.

Time

The chronological age of alluvial and deluvial (colluvial) soils is young, but this is not always the case. In the areas where soils have formed in early mature valleys, they may be young. The low floodplain is the youngest in the river terrace system. In the areas where soils have formed in late-mature valleys, where old products or other sediments have been transported, they are of later age.

Alluvial and deluvial (colluvial) soils date from the Quaternary Age, as indicated by the age of the material in which they were formed.

According to the Bulgarian systematic, some varieties are comparatively young, but still probably thousands of years old. Soils developed in alluvium of recent origin are young.

Human Influence

As regards agricultural practice, alluvial soils have the drawback of periodic flooding. The stock of river water can be managed, collected, and maintained by a system of hydrologic constructions.

In the case of seasonal drop in the level of the groundwater table, the main source for moisture is rainfall. Water for irrigation is pumped from a shallow water table when needed. Some alluvial soils need to be drained for cropping.

Irrigation practices provoke negative soil processes and secondary salinity.

In this case, alluvial soils implies that are as a repository for the salt in the drainage water from the irrigated fields. High rates of fertilizer application can cause contamination of groundwaters, for example with nitrates.

Repeated cultivation and irrigation cause deterioration of the soil structure.

Historically, deluvial (colluvial) soils have been widely cultivated in the country. At the foot slopes sheet erosion is naturally a threat.

10.2.2 Genesis

Alluvial soil is a general term for those soils developed on recent alluvium and derived from geologic materials such as alluvium. Soils have formed on the alluvium in hydromorphic conditions (presence of excess water) and can have periodic sedimentation in accordance with seasonal fluctuation of the groundwater table. Thus, soil formation takes place simultaneously with sedimentation.

Periodical sedimentation caused by water is typical, and the deposition of new materials can be an ongoing process. Yearly, the flooded soils accumulate new sediments load with water. The balance of soil formation is accumulative.

The formation of alluvial deposits is a result of products (chemical and mechanical) migration throughout the territory of the river basin as follows: unweathered primary materials, clay minerals, organic substances compounds, carbonates and soluble salts, compounds of P, K, N, Fe, Mn, microelements.

If continued flooding and deposition of alluvium is absent, a normal leaching and gradual decrease in organic matter content occurs with increasing soil depth. This is the main difference of the Bulgarian concept with WRB (2006) and FAO-UNESCO-ISRIC (1990), where only the recent deposition of geologic materials is admitted.

The formation of alluvial soil tends to be complex rather than simple. The fact is that all great civilizations in antiquity have grown up on fertile soils in areas of the rivers Tigris and Euphrates in Mesopotamia, the Nile in Egypt, the Ganges in India, and the Yellow River in China.

Traditionally, in Bulgaria the alluvial soils encompass a large group of soils that have developed in alluvial terraces and valleys, produced by pedogenic processes that have been active over time in river areas, and with varying degrees of spatial pronouncement. This was confirmed over the years during the large soil survey of the country. If the requirement for periodic recent deposition of material on the soil surface be respected under conditions of irregular river flow in Bulgaria, then this area would be negligible.

The Bulgarian systematic takes into consideration that the level of the groundwater table does not influence soil development on the highest terraces. However, on the lower terraces, where the level of groundwater is below 2 m, typically the lower part of the soil is seasonally overwetted and develops either gleyic or mottled color pattern. During the dry period oxidation takes place.

On the low flood terrace, where the level of the groundwater table is below 2 m, alluvial soils have a simple stratified profile, are saturated during the year and have strong gley color.

Gleyic properties are the result of reduction caused by anaerobic microbiological activity. The irregular distribution of the microorganisms correspond to the soil porosity. Thus the green–blue and rusty motley color pattern is developed.

In humid environments soils, accumulate organic matter, and compounds of iron, silica, manganese, and phosphorous; in more or less arid environments soils accumulate lime, gypsum, and soluble salts.

Deluvial (colluvial) soils are formed by downhill creep, where the sorting of materials comes about through gravity. Creep is the slow movement of soil masses down slopes that are usually steep. The process takes place in response to gravity where there is pronounced water saturation. In the deluvial soils, organic matter content naturally decreases over time.

10.3 Soil Diagnostic and Classification

10.3.1 Morphology

The depth of grass roots penetration in alluvial and alluvial meadow soils influences productive aggregates formation.

In profile composition deluvial (colluvial) soils show many graded layers.

Both alluvial and deluvial soils have ochric and non-diagnostic features (Figs. 10.2, 10.3).

***A*(*k*), (*Ap*)**—a thin upper mineral humus horizon, 5–10 cm thick with evidence of humification, more or less dark colored. Structure is granular (spheroids or polyhedrons) of medium size (2–5 mm diameter), moderate pedality grade, consistence is friable (moist) and hard (dry) or is single grain; commonly loose when ploughed. Abundance of roots with fine size (0.5–2.0 mm); gradually grade down to

***1AC*(*k*)**—a mineral sediment layer that can be subdivided into prime, second, etc. Common thickness is 50–100 cm. Upper part can be with granular coarse size (5–10 mm diameter) structure, medium pedality grade, consistence is friable (moist), and hard (dry). A few iron-manganese concretions can be observed. Coarse material of different origin and size (grains, pebbles, and stones) is in the soil mass. Beneath is gradual change with mixed pattern to the parent materials.

***2C*(*k*)**—a mineral layer, there is a change of sediments. Structure is massive (no structure) or structureless (no aggregation or orderly arrangement), consistence is friable (moist) and hard (dry). Soft spots and concretions of calcium carbonate are often of the calcareous variety; the carbonate-free varieties can have iron-manganese concretions. Rusty and grayish spots due to gleyic conditions are observed. Many stones and pebbles of different size are in the soil mass. There is clear transition to the second C horizon beneath.

***3C*(*k*)**—a mineral layer, there is a change of sediments. The structure is massive or structureless, consistence is

Fig. 10.2 Deluvial (colluvial) soil, Bulgarian classification Yolevski and Hadzhiyanakiev (1976), *Haplic Cambisol (Colluvic)*, WRB (2006), *Typic Orthents*, Keys to Soil Taxonomy (2010), *Cambisol*, Revised Legend FAO-UNESCO-ISRIC (1990)

Fig. 10.3 Alluvial soil, Bulgarian classification Yolevski and Hadzhiyanakiev (1976), *Haplic Fluvisol*, WRB (2006), *Typic Fluvents*, Keys to Soil Taxonomy (2010), *Fluvisol*, Revised Legend FAO-UNESCO-ISRIC (1990)

friable (moist) and hard (dry). Usually there is a gley color pattern and some iron manganese concretions can be observed. An abundance of stones and pebbles is typical. Here, there can be the most prominent accumulation of secondary carbonates, mostly soft spots and concretions.

10.3.2 Classification

The Fluvisols order comprises a large group of soils; after FAO-UNESCO-ISRIC (1990), classification requires the recent deposition of materials. Colluvial soils are not recognized in the FAO classification. In the Revised Legend of the World Soils (FAO-UNESCO-ISRIC 1990), the Fluvisols group is subdivided into dark colored, eutric, dystric, and salic.

Traditionally, the Bulgarian systematic recognizes a much larger group of soils formed in alluvium than only those containing recently deposited materials. The common criterion is evidence of sedimentation throughout the alluvial plain and stratification in the soil profile.

The Bulgarian systematic approach is that alluvial (or alluvial meadow) soils may develop in the absence of continued flooding and deposition of recent alluvium. For this reason, many soils located in large river valleys in Bulgaria are not Fluvisols, according to WRB and FAO-UNESCO-ISRIC (1990). This is the main difference between the national Bulgarian systematic and other international systems. Bulgarian soil classification encompass a broad variety of soils formed on alluvial sediments that display a layered sequence in soil.

The Bulgarian concept of the process of meadow soil formation includes humification in a favorable wet soil regime and a sufficient period of aeration.

Bulgarian classification of Koynov et al. (1964) subdivided the type of alluvial soils into periodically flooded and non-flooded.

The diagnostic of deluvial (colluvial) soils is not clearly defined. Within the subtype soil differs in the thickness of humus horizon (low thick <10 cm, medium 10–40 cm, and thick >40 cm).

The classification of Yolevski and Hadzhiyanakiev (1976) identified subtypes of alluvial and alluvial meadow, deluvial and deluvial meadow, and alluvial-deluvial soil, and also included cultivated sands and natural sand subtypes. Soils characterized as medium swamped are not included.

The soil classifications of Yolevski et al. (1983) defined typical and meadow subtypes within the type of alluvial and deluvial soil.

There was classification in Penkov et al. (1992) which renamed the subtypes (underdeveloped, anthropogenic, calcaric, eutric, dystric, mollic) but this has not been applied so far.

Within the subtype soil differs in the thickness (low thick <40 cm, medium 40–80 cm, thick >80 cm, and very thick >120 cm) and in the appearance of swamp (slight).

Classification after WRB (2006) required *Fluvic* diagnostic material of recent deposition in the upper 50 cm for inclusion in the reference group of Fluvisols.

Keys to Soil Taxonomy (2010) required only *Ochric and Histic* epipedon for Entisols order classification. In Soil Taxonomy subsurface horizons allowed are other horizons if buried but most have none. *Fluvents*, alluvial soils affected by wetness, are of fluvial origin and contain sediments recently deposited by water.

10.3.3 Alluvial and Deluvial Soil Subtypes in Bulgaria

10.3.3.1 Alluvial and Alluvial Meadow Soils

Alluvial, **alluvial meadow,** and **alluvial (deluvial)** (*subtype, type*) soils are named after the Bulgarian classification of Koynov et al. (1964), and Yolevski and Hadzhiyanakiev (1976).

Haplic Fluvisol is named according to WRB (2006) and *Typic Fluvents* according to Keys to Soil Taxonomy (2010).

Fluvisol is after Revised Legend of the World Soils (FAO-UNESCO-ISRIC 1990).

The basic feature is the appearance of layer sedimentation in the alluvium, where organic matter is irregularly distributed. Content of organic matter can range from 1 to 6 %.

The calcareous variety has secondary carbonates. Carbonate-free varieties are eutric (with base saturation >50 %), and dystric (with base saturation <50 %).

Particle size distribution shows an irregular pattern where the fine sand fraction (0.25–0.05 mm) dominates followed by the coarse silt fraction (0.05–0.01 mm), (Tables 10.1, 10.2 and Figs. 10.4, 10.5). The texture reflects the sedimentation distribution (Tables 10.3, 10.4 and Figs. 10.6, 10.7).

10.3.3.2 Deluvial and Deluvial Meadow Soils

Deluvial and **Deluvial meadow** (*subtype, type*) soil is named after the Bulgarian classification of Koynov et al. (1964), and Yolevski and Hadzhiyanakiev (1976).

Haplic Cambisol (Colluvic) is named according to WRB (2006) and *Typic Orthents* according to Keys to Soil Taxonomy (2010).

Cambisol: Revised Legend of the World Soils (FAO-UNESCO-ISRIC 1990).

Particle size distribution shows a more or less regular pattern, where the fine sand fraction (0.25–0.05 mm)

Table 10.1 Variables of the particle-size distribution in 19 referenced profiles of calcareous alluvial soils

Profile count 19	Horizon	Upper depth	Lower depth	Loss HCl	Particle-size distribution (mm), % (according to method of Kachinskiy revised)							
					>1 mm	1–0.25	0.25–0.05	0.05–0.01	0.01–0.005	0.005–0.001	<0.001	<0.01[a]
Mean	Apk	0	29	8.8	0.6	8.4	28.6	22.9	8.2	8.5	13.9	30.7
	1ACk	29	56	10.2	0.7	7.1	26.7	24.9	8.4	8.5	13.4	30.3
	2ACk	56	84	10.3	0.8	7.7	28.1	21.8	7.2	7.6	16.4	31.3
	1Ck	84	112	10.2	1.1	6.0	25.4	23.0	8.0	8.3	18.0	34.3
	2Ck	112	145	11.0	0.7	8.0	28.7	21.7	7.0	7.3	15.5	29.9
Stdev	Apk	0	3	8.4	1.0	6.4	16.7	6.2	3.6	6.1	8.4	14.6
	1ACk	3	8	9.3	3.0	7.8	15.0	9.1	4.0	6.0	8.5	14.9
	2ACk	8	11	9.6	2.7	7.2	14.8	8.7	3.2	4.2	12.4	18.4
	1Ck	11	13	10.9	4.2	6.9	12.3	9.2	3.2	4.1	12.7	16.7
	2Ck	13	11	10.7	1.2	10.5	15.9	8.1	3.0	3.8	12.3	17.0

[a] Physical clay, the parameter-characterized sum of the particle-size distribution, according to Kachinskiy method (mm) $\sum < 0.01 = <0.001 + (0.005 - 0.001) + (0.01 - 0.005)$

Table 10.2 Variables of the particle-size distribution in 34 referenced profiles of alluvial soils (eutric)

Profile count 34	Horizon	Upper depth	Lower depth	Loss HCl	Particle-size distribution (mm), % (according to method of Kachinskiy revised)							
					>1 mm	1–0.25	0.25–0.05	0.05–0.01	0.01–0.005	0.005–0.001	<0.001	<0.01[a]
Mean	Ap	0	28	2.0	4.2	19.2	28.7	19.1	6.5	6.7	13.6	26.8
	1AC	28	55	1.9	6.5	18.1	24.3	18.3	7.2	6.8	16.9	30.9
	2AC	55	83	1.9	5.0	20.4	24.5	17.2	6.5	6.3	18.1	30.9
	1C	83	112	2.1	8.2	20.7	25.7	15.9	5.9	5.5	16.0	27.4
	2C	112	143	3.2	9.8	19.5	24.0	16.3	6.1	5.9	15.2	27.2
Stdev	Ap	0	4	1.6	6.0	12.5	9.4	6.1	2.9	2.8	5.9	10.2
	1AC	4	8	1.7	11.8	15.7	9.2	7.0	3.5	3.4	9.1	14.1
	2AC	8	10	1.8	11.1	18.3	10.1	7.4	3.0	3.4	11.5	16.0
	1C	10	13	2.0	13.4	14.1	10.8	7.7	3.4	3.2	11.1	15.7
	2C	13	17	5.7	17.0	14.9	12.2	7.9	4.4	3.4	11.2	16.9

[a] Physical clay, the parameter characterized sum of the particle-size distribution, according to Kachinskiy method (mm) $\sum < 0.01 = <0.001 + (0.005 - 0.001) + (0.01 - 0.005)$

dominates followed by the coarse sand fraction (1.0–0.25 mm), (Table 10.5 and Fig. 10.8). The texture reflects the sedimentation distribution (Table 10.6).

10.4 Soil Properties and Soil Management

10.4.1 Soil Properties

The typical feature of alluvial and alluvial meadow soils is variability (Shishkov and Filcheva 2009, 2010). Alluvial soils have properties inherited from local factors. Their spatial distribution has resulted in the formation of areas with various textures.

These soils typically show little evidence of weathering in situ and oxidation of iron oxides.

Alluvial and alluvial meadow soils can be full-wetted for most of the year, or wet in the lower part of soil, or the level of underground water does not affect soil development.

Periodically anaerobic conditions provoke chemical or microbiological processes of reduction resulting in the development of gley features. Oxidation takes place during the dry period.

Alluvial and alluvial meadow soils are characterized by a low store of nutrient elements. Soils are poor in organic matter content and nitrogen. Low content of organic colloids and coarse texture characterizes soils as with low CEC 10–15 meq/100 g and low buffer properties.

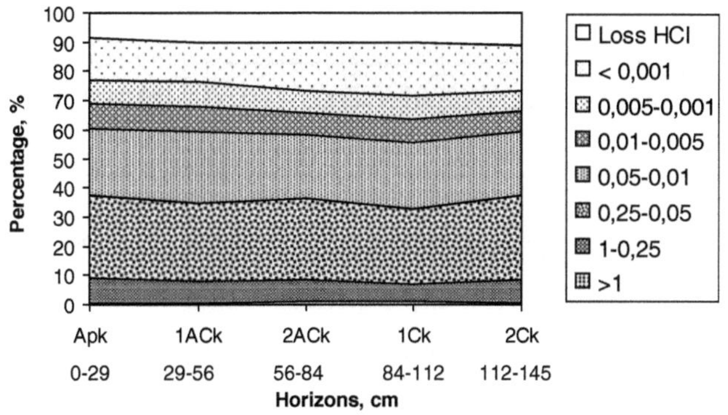

Fig. 10.4 Pattern of the particle-size distribution (mm) in calcareous alluvial soil (according to revised Kachinskiy method)

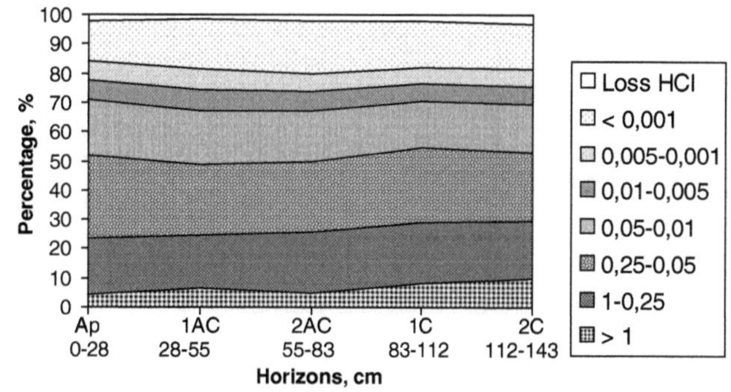

Fig. 10.5 Pattern of the particle-size distribution (mm) in alluvial soils (eutric), (according to revised Kachinskiy method)

Table 10.3 Variables of humus, pH, carbonates, and particle-size distribution (mm) according to USDA in 19 referenced profiles of calcareous alluvial soils

Profile count	Horizon	Upper depth	Lower depth	Humus (%)	pH_{KCl}	Bulk density g.cm^3 calc	Clay <0.002	Silt 0.002–0.05	Fine sand 0.05–0.2	Total sand 0.05–2.0
19										
Mean	Apk	0	29	2.0	7.0	1.4	23.0	38.8	29.4	38.1
	1ACk	29	56	1.5	7.0	1.4	23.2	41.7	27.6	35.2
	2ACk	56	84	1.2	7.0	1.5	25.8	37.0	29.1	37.2
	1Ck	84	112	1.2	7.0	1.5	27.7	39.1	26.5	33.1
	2Ck	112	145	1.0	7.0	1.5	25.4	36.8	29.8	37.8
Stdev	Apk	0	3	0.8	0.3	0.1	10.7	9.9	16.6	18.1
	1ACk	3	8	0.6	0.3	0.1	11.1	13.2	14.8	20.0
	2ACk	8	11	0.7	0.3	0.1	14.0	10.4	14.7	19.0
	1Ck	11	13	0.5	0.4	0.1	13.5	12.4	12.6	19.4
	2Ck	13	11	0.6	0.4	0.1	14.4	11.2	16.1	20.2

Table 10.4 Variables of humus, pH, carbonates, and particle-size distribution (mm) according to USDA in 34 referenced profiles of alluvial soils (eutric)

Profile count	Horizon	Upper depth	Lower depth	Humus (%)	pH_{KCl}	Bulk density g.cm^3 calc	Clay <0.002	Silt 0.002–0.05	Fine sand 0.05–0.2	Total sand 0.05–2.0
34										
Mean	Ap	0	28	1.7	5.7	1.4	18.5	30.4	29.5	51.2
	1AC	28	55	1.3	5.7	1.4	21.8	30.4	25.3	47.8
	2AC	55	83	1.0	5.8	1.5	22.5	28.1	25.2	49.3
	1C	83	112	0.7	5.9	1.5	20.3	26.1	26.9	53.6
	2C	112	143	0.6	6.0	–	20.4	27.5	25.4	52.1
Stdev	Ap	0	4	0.9	0.8	0.2	6.9	9.1	9.3	14.2
	1AC	4	8	0.7	0.7	0.1	10.2	11.0	9.0	18.6
	2AC	8	10	0.6	0.7	0.1	12.6	11.2	9.9	20.8
	1C	10	13	0.5	0.7	0.1	11.9	11.5	10.5	21.2
	2C	13	17	0.5	0.7	–	12.5	12.6	11.8	23.3

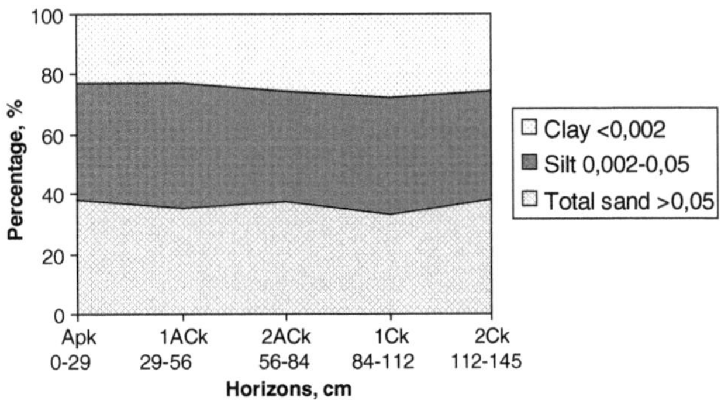

Fig. 10.6 Pattern of the particle-size distribution (mm) in calcareous alluvial and alluvial meadow soils (USDA)

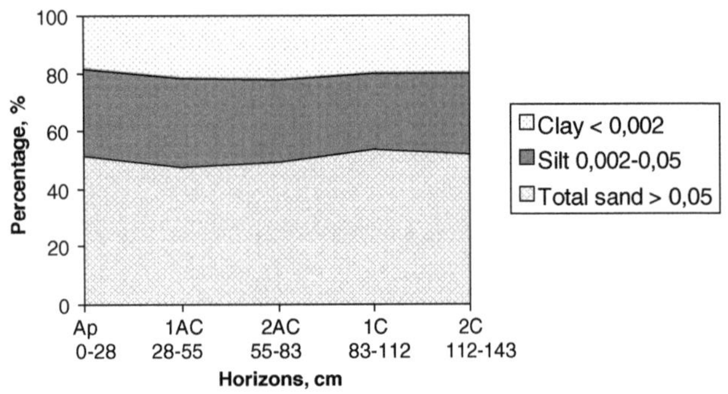

Fig. 10.7 Pattern of the particle-size distribution (mm) in alluvial soils (eutric) (USDA)

Usually, easily soluble salts are washed out of the profile, but sometimes they can accumulate in the lower part. In this case the soil cover is composite mosaic.

Deluvial (colluvial) soils have a better balance of nutrients due to the lateral stock. Soils respond well to the application of manure and compost.

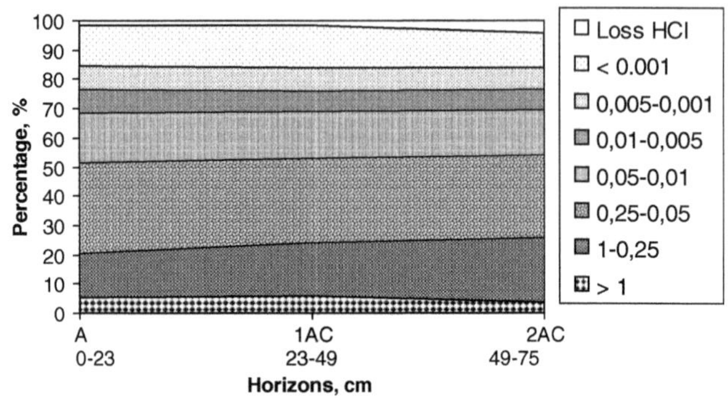

Fig. 10.8 Pattern of the particle-size distribution (mm) in deluvial soils (eutric), (according to revised Kachinskiy method)

10.4.2 Soil Management

Favorable water, air, and soil thermal properties mean that these soils have been one of the most popular for agriculture since ancient times. Environments with an available supply of water and nutrient elements are characterized by natural fertile soils.

The amount of available nitrogen is not sufficient for producing high yields. Conditions for nitrogen mineralization are not favorable because of poor aeration. Surface stoniness can be a drawback for agriculture.

These soils are suitable for growing vegetables and fruit trees, including apple, plum, pear, peach, and quince. Popular woods quite often occur where the level of the groundwater table is high, but for most crops this is a limitation.

The water stable aggregates decrease dramatically after a few years of cultivation.

The oil-bearing rose and many other oil-bearing plants are grown in deluvial (colluvial) soils (notably in the Rose Valley). Commonly tobacco, cereals, maize, and sunflower are grown, along with orchards of plum, cherry, walnut, apricot, peach, and almond trees; viticulture can also be established.

Table 10.5 Variables of the particle-size distribution in 9 referenced profiles of deluvial soils (eutric)

Profile count 9	Horizon	Upper depth	Lower depth	Loss HCl	Particle-size distribution (mm), % (according to method of Kachinskiy revised)							
					>1 mm	1–0.25	0.25–0.05	0.05–0.01	0.01–0.005	0.005–0.001	<0.001	<0.01[a]
Mean	A	0	23	1.3	5.3	15.0	31.0	17.1	8.2	7.9	14.1	30.2
	1AC	23	49	1.4	5.6	18.4	28.8	16.3	6.6	8.4	14.5	29.0
	2AC	49	75	4.3	4.0	21.8	28.2	15.5	6.7	7.5	12.0	26.9
Stdev	A	0	4	0.5	10.6	10.0	18.3	7.5	3.2	4.6	9.0	11.6
	1AC	4	8	0.6	11.2	11.0	17.4	6.0	2.5	3.3	6.8	9.5
	2AC	8	12	9.2	7.6	19.0	18.5	7.2	3.4	3.4	5.0	7.9

[a] Physical clay, the parameter characterized sum of the particle-size distribution, according to Kachinskiy method (mm) $\sum$ <0.01 = <0.001 + (0.005 − 0.001) + (0.01 − 0.005)

Table 10.6 Variables of humus, pH, carbonates, and particle-size distribution (mm) according to USDA in 9 referenced profiles of deluvial soils (eutric)

Profile count	Horizon	Upper depth	Lower depth	Humus (%)	pH_{KCl}	Bulk density $g.cm^3$ calc	Clay <0.002	Silt 0.002–0.05	Fine sand 0.05–0.2	Total sand 0.05–2.0
9										
Mean	A	0	23	2.24	4.2	1.41	19.2	30.8	31.7	50.0
	1AC	23	49	1.39	4.3	1.50	20.1	28.6	29.3	51.4
	2AC	49	75	1.37	4.5	–	18.4	28.4	28.9	53.2
Stdev	A	0	4	1.37	0.7	0.12	9.1	11.7	17.8	18.1
	1AC	4	8	1.14	0.7	0.07	6.7	9.1	17.0	13.8
	2AC	8	12	1.25	1.1	–	7.8	11.1	18.1	17.3

References

Bondev I (1991) Vegetation in Bulgaria, map at scale 1:600,000 with explanatory text. St. Kliment Oxridski University Press, Sofia (In Bulgarian)

FAO-UNESCO-ISRIC (1990) Revised legend of the soil map of the world. FAO World Soil Resources Report No 60, Rome

Keys to Soil Taxonomy (2010) USDA United States Department of Agriculture, NRCS Natural Recourses Conservation Service, 11th edn. Soil Survey Staff, Washington, DC, p 338

Koynov V et al (1964) Classification and systematic of soils in Bulgaria. In: Gerasimov IP et al (eds) soils of south–eastern Europe. Proceedings of the international symposium of soil science, June 1963, Sofia. Bulgarian Academy of Sciences Press, Sofia: 231–246 (In Russian)

Minkov M (1968) Loess in the Northern Bulgaria. Sofia (In Bulgarian)

Penkov M et al (1992) Soil classification and diagnostic in Bulgaria in relation to the land reform fragmentation. Zemizat Press, Sofia (In Bulgarian)

Shishkov T, Filcheva E (2009) Soil organic matter: a key aspect of quality in agri-environment system. 1. Case of soils from Bulgaria and Italy. Soil tillage and ecology. In: International conference, 1–5 Sept 2009, Albena, Bulgaria, pp 423–431

Shishkov T, Filcheva E (2010) Content and composition of organic matter of soils with different land use, developed on the old and contemporary alluvial Maritsa river terraces in Southern Bulgaria. In: The world conference on humic substances and natural organic matter, 15th international meeting of the humic substances society, Tenerife, Canary Islands, 27 June, 2 July 2010, pp 378–383

World Reference Base for Soil Resources (2006) A framework for international classification, correlation and communication. FAO-ISRIC, Rome

Yolevski M, Hadzhiyanakiev A (1976) Agro-grouping of soils in Bulgaria. Extended systematic list. Code of groups and soil varieties. Institute of Soil Science Nikola Poushkarov, Sofia (In Bulgarian)

Yolevski M, Ninov N, Krastanov S et al (1983) Brief diagnostic and criterion of the main taxonomic soil divisions in Bulgaria. Precise classification scheme, Institute of Soil Science Nikola Poushkarov, Sofia (In Bulgarian)

11 Hydromorphic Soils with Gleyic Properties

Derived from the Russian word *gley*, meaning grayish-blue and rusty mottled mass, introduced in 1905 by G. N. Visotskiy.

Many soils are characterized by temporal or permanent wetland conditions.

11.1 Concept of the Type

Main concepts

1. Mineral soils developed under natural or maintained conditions of excess moisture;
2. Soils occur in many environments with various moisture regimes;
3. May have an organic layer under constant wetland conditions;
4. Wetlands are areas of high biodiversity and are under the regulation of the Ramsar Convention and Corine Biotop.

The main processes that have taken place are linked to prolonged exposure to anaerobic conditions in water saturated soils, and seasonal alteration of reduction and oxidation conditions.

11.2 Ecology of Soil Formation

11.2.1 Distribution

Hydromorphic soils with gley properties comprise a large group of soils with different genesis and soil types distribution throughout the country. Soils are confined to over-wetted terrains in many environments (Figs. 11.1, 11.2). The distribution of soils is related to different conditions, as follows:

- permanent high level of groundwater, which is typical for all swamped, boggy, and peat-boggy soils, naturally occurring in closed topographic areas. Soils are often linked to the local lakes, deltas, and river valleys;
- periodic high level of groundwater, which is typical for all meadow soil varieties in Bulgaria;
- internal slope flows periodically existent in well-drained soil substrates, overlying consolidated rock or an impermeable layer, occurring in many environments;

Climatic conditions
Soils occur in many environments with various climate conditions; most are seasonally dependant on precipitation.

Topography
Topography is various and can be alluvial terraces in the case of meadow alluvial soils, fluvial plain in the case of alluvial soils, closed natural depression in the case of swamped, boggy and peat-boggy soils, gently sloping terrain in the accumulative landscape where water-logging occurs, lower foot slopes, or dissected mountain relief.

Parent materials
Parent materials can be various sediments or any calcareous, basic, or acid unconsolidated materials. Most materials are deposits of various textures of the Quaternary Age sediments. Consolidated hard rock is typical in mountainous regions.

Vegetation
Vegetation is confined to the waterside environment. Wetland soils are formed under the influence of diverse hydrophilic vegetation. This is the most favorable environment for high biological activity, where the permanent accumulation of nutrition elements takes place. Meadow soils commonly have favorable conditions of moisture regime and high aerobic status, which promotes humus formation under tall meadow grasses. Various grass associations have influenced the formation of these soils. Common are associations of timothy (*Phleum* sp.), foxtail (*Alopecurus* sp.), fescue (*Festuca* sp.), wheat grass (*Agropyron Gaerth*), meadow grass (*Poa* sp.), clover (*Trifolium*), and vetch (*Vicia sativa*).

In lowlands at the seaside the original hydrophytic herbaceous vegetation is of the following formations:

T. Shishkov and N. Kolev, *The Soils of Bulgaria*, World Soils Book Series, DOI: 10.1007/978-94-007-7784-2_11,

Fig. 11.1 Permanent wetland nearby water bodies

Fig. 11.2 Swamped land with permanent high level of groundwater and soil gleyic properties

Fig. 11.3 Sandy gleyic soil, Bulgarian classification Yolevski and Hadzhiyanakiev (1976), *Haplic Gleysol Arenic*, WRB (2006), *Aquic Quartzipsamments*, Keys to Soil Taxonomy (2010) *Gleyic Arenosol*, Revised Legend FAO-UNESCO-ISRIC (1990)

Phragmiteta australis, Typheta angustifoliae, Typheta latifoliae, Schoenoplecteta lacustris, Schoenoplecteta triquetre, and *Schoenoplecteta tabermontanae*.

Time
Soils developed in alluvium of recent origin are young. Meadow alluvial and deluvial (colluvial) soils are of the Quaternary Age, as indicated by the age of the material in which they were formed. The swamped and boggy soils date from the Holocene Age.

Human influence
The Ramsar Convention (1971) regulated for the protection of wetland zones. Some of the dikes that had been constructed along the Danube River were removed in order to restore the natural flooding of lands, thus promoting the water balance in wetland zones and restoring ecosystems (Usunov 2012).

In cases of drought, when the main source of moisture (rainfall) is scarce, the meadow lands, usually with a higher level of groundwater, are preferable for the growing of most crops. In other cases, soils need to be drained for cropping. An efficient drainage system can prevent the rise of groundwater.

11.2.2 Genesis

Soils with gleyic color pattern commonly have a water impermeable horizon deep inside the soil. In clayey soils, the so-called water-resistant horizon is in the top soil.

Under conditions of poor drainage, where there is a pronounced permanent stock of water for most of the year, Gleysoils develop (Fig. 11.3). The water impermeable layer is quite deep and is not involved in soil evolution. Gleysoils are characterized by quite prolonged exposure to anaerobic conditions. Naturally, gley soils have an irregular blue-green mottle pattern with blue-olive colored horizon, which is structureless and of low porosity.

The nature of gley is complex due to the biochemical and microbiological nature of its origin. Bivalent forms of iron, manganese, and aluminum are mobile. If conditions change to aerobic, then the oxidation of iron and manganese takes place.

Swamped, boggy, and peat soils (Histosols) are characterized by a permanently high level of groundwater (Fig. 11.2). Organic remnants are in different stages of decomposition. The decomposition of organic tissues takes place only in dry conditions. Anaerobic conditions are typical in these soils, where the content of oxygen is very poor. Swamped soils can have an organic (H) horizon overlying the mineral gleyic subsoil, while the boggy and peat soils are entirely organic.

11.3 Soil Diagnostic and Classification

11.3.1 Morphology

Typical gley is formed at the depth of the groundwater table.

Fig. 11.4 Meadow cinnamonic gleyic soil, Bulgarian classification Yolevski and Hadzhiyanakiev (1976), *Gleyic Luvisol,* WRB (2006), *Vertic Endoaqualfs,* Keys to Soil Taxonomy (2010), *Gleyic Luvisol,* Revised Legend FAO-UNESCO-ISRIC (1990)

Gleysoils do not have any other diagnostic horizons apart from humus (A) and histic (H) and gleyic horizons. The gleyic horizon is of homogeneous blue-grayish color with tiny black concretions.

Meadow soils encompass a broad variety of soils denoted with presence of reduction and oxidation processes. This includes the following soils: alluvial meadow, deluvial meadow, meadow cinnamonic (Fig. 11.4), meadow calcareous or typical smolnitsa, meadow calcareous or typical chermozem-like, meadow chernozem—calcareous, typical and leached, meadow solonetz, and slonetz-solonchak. The depth of grass roots penetration influences valuable aggregates formation.

The profile composition of alluvial meadow and deluvial (colluvial) meadow soils shows many graded layers.

In all cases the texture is important.

11.3.2 Classification

Soil with excess moisture content may have different systematic solutions. For example, in French classification these soils are confined to a single class, and in Soil Taxonomy a Histosols suborder is included. The Bulgarian soil classification is based on the occurrence of soil forming processes, so different soil types exhibit so-called gley,

Table 11.1 Variables of particle-size distribution in calcareous gleysoil, village Sindel, district city of Varna

Profile count	Horizon	Upper depth	Lower depth	Loss HCl	Particle-size distribution (mm), % (according method of Kachinsky revised)							
3					>1 mm	1–0.25	0.25–0.05	0.05–0.01	0.01–0.005	0.005–0.001	<0.001	<0.01[a]
Mean	Ak	0	30	20.1	0.0	1.1	9.1	17.7	7.3	10.8	33.9	52.1
	Agk	30	57	20.1	0.0	0.8	10.7	19.4	7.6	8.9	32.6	49.1
	AGk1	57	80	20.8	0.0	0.7	11.6	19.4	6.7	8.7	32.1	47.5
	AGk2	80	108	21.2	0.0	0.6	11.4	18.9	7.5	9.7	30.7	47.9
	AGk3	108	135	20.8	0.0	0.5	20.4	12.6	7.4	8.6	29.7	45.7
	AGk4	135	166	22.2	0.0	0.7	13.0	20.1	7.2	9.3	27.6	44.0
Stdev	Ak	0	2	6.1	0.0	1.8	2.0	8.3	1.5	0.6	6.6	6.7
	Agk	2	5	4.5	0.0	1.3	4.1	5.9	0.1	1.9	6.4	8.1
	AGk1	5	7	3.5	0.0	1.2	2.8	6.0	1.6	2.1	5.8	6.9
	AGk2	7	7	3.6	0.0	1.0	5.7	10.3	1.4	1.3	11.8	13.5
	AGk3	7	8	3.0	0.0	0.9	19.4	11.0	1.7	4.3	17.5	23.3
	AGk4	8	7	3.2	0.0	1.2	6.5	10.2	1.4	3.1	15.6	17.2

[a] Physical clay, the parameter characterized sum of the particle-size distribution, according to Kachinskiy method (mm) $\sum$ <0.01 = <0.001 + (0.005 - 0.001) + (0.01 - 0.005)

gleyic, and gleyic-like characteristics depending on soil evolution.

1. Permanently hydromorphic conditions with stagnic gleyic pattern.
   Surface stagnic conditions are linked to the occurrence of gleyic color pattern in the top soil due to precipitation. Soil Taxonomy defined the suborder *Umbraquepts*, and while there is no analogy in Bulgarian classification, some soils can resemble the so-called mountainous meadow soils. Clay is always stable in the profile.
2. Permanent hydromorphic conditions with gleyic horizon in boggy soils. Soil Taxonomy and also WRB defined the suborder *Histosols*, and analogous in Bulgarian classification are the so-called swamped, boggy, and peat soil varieties.
3. Permanent hydromorphic conditions at the depth due to the high level of the groundwater table where soils may be:
   - Alluvial soils with gleyic color pattern at the depth and seasonal fluctuation of the level of groundwater. Soil Taxonomy defined the suborder *Fluvaquents*, *Aquic Fluvents* which can be analogous to the Bulgarian classification of alluvial soils and some so-called alluvial meadow soils.
   - Soils with advanced evolution and gleyic color pattern at the depth, which is due to the high level of the groundwater table. Soil Taxonomy defined all *Aqualfs, Aquents, Aquepts, Aquolls* and particularly the suborder *Calciaquerts, Endoaquerts, Endoaquepts, Endoaqualfs, Endoaquolls* (in the case of base saturated soils), and *Umbraquepts* (in the case of acid soils), and analogous in Bulgarian classification are the so-called meadow varieties of the base saturated soils: alluvial, deluvial, cinnamonic, smolnitsa, chernozem, chernozem-like, some swamp and peat subtypes, solonetz-like, and solonchak-like soils. Not all of these soils are of typically hydromorphic origin

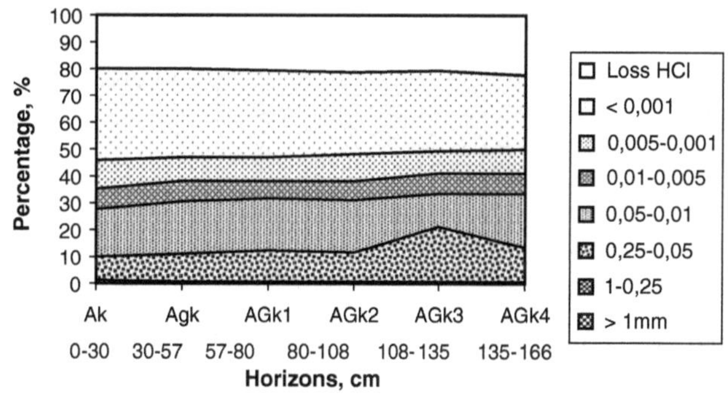

Fig. 11.5 Pattern of the particle size distribution (mm) in calcareous gleysoil, village Sindel, district city of Varna (according to revised Kachinskiy method)

11.3.3 Hydromorphic Soils with Gley in Bulgaria

Traditionally, in Bulgarian systematic the gley color pattern is linked with its pronouncement in the profile and the depth of location is so-called typical gley, gley-like, and gleic.

Table 11.2 Variables of humus, pH, carbonates, and particle-size distribution (mm) according to USDA in 3 referenced profiles of calcareous gleysoil, village Sindel, district city of Varna

Profile count	Horizon	Upper depth	Lower depth	Humus (%)	pH H_2O	Clay <0.002	Silt 0.002–0.05	Fine sand 0.05–0.2	Total sand 0.05–2.0
3									
Mean	Ak	0	30	2.94	7.2	50.1	38.2	10.7	11.7
	Agk	30	57	2.35	7.2	47.9	39.2	12.2	13.0
	AGk1	57	80	1.98	7.2	47.6	38.5	13.2	13.9
	AGk2	80	108	1.90	7.3	47.0	39.4	13.0	13.6
	AGk3	108	135	1.39	7.3	45.2	32.3	22.0	22.5
	AGk4	135	166	1.32	7.4	44.2	40.4	14.7	15.4
Stdev	Ak	0	2	1.08	0.1	8.7	5.7	1.5	3.4
	Agk	2	5	0.86	0.1	8.8	3.7	3.9	5.2
	AGk1	5	7	0.74	0.1	8.3	5.0	2.6	3.8
	AGk2	7	7	1.04	0.1	13.3	7.8	5.5	5.9
	AGk3	7	8	0.78	0.2	19.2	10.5	19.3	19.1
	AGk4	8	7	0.91	0.1	15.6	8.8	6.4	6.9

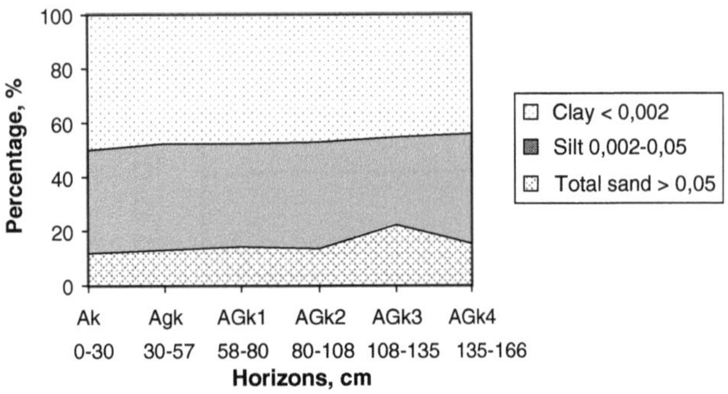

Fig. 11.6 Pattern of the particle-size distribution (mm) in calcareous gleysoil, village Sindel, district city of Varna (USDA)

11.3.3.1 Meadow Soil Varieties

All varieties of **meadow** soils (*subtype, type*) are named after the Bulgarian classification of Koynov et al. (1964), and Yolevski and Hadzhiyanakiev (1976) and are so-called alluvial meadow, deluvial meadow, meadow cinnamonic, meadow smolnitsa carbonate or typical, meadow chermozem-like carbonate or typical, meadow chernozem (may be carbonate, typical, leached), meadow solonetz, and meadow slonetz-solonchak soil varieties.

11.3.3.2 Meadow Soils: Swamped, Boggy, and Peat-Boggy

Varieties of **meadow swamped, boggy, and peat-boggy** soils (*subtype, type*) are named after the Bulgarian classification of Koynov et al. (1964), and Yolevski and Hadzhiyanakiev (1976).

Gleysol and *Rheic* (*Fibric, Hemic, Sapric*) *Histosol* is named according to WRB (2006) and *Typic Haplohemists, Typic Haplowassists, Typic Haplosaprists, Sapric Haplowassists* according to Keys to Soil Taxonomy (2010).

Histosols: Revised Legend of the World Soils (FAO-UNESCO-ISRIC 1990).

Gleysoils do not have any other diagnostic horizons apart from humus (A) and histic (H) horizons. The process of transformation of organic residues is accompanied by nitrogen enrichment of humic substances shown by the ratio C/N 20 which means very low rates. The horizon sequence in the profile is H–(A)–G. The content of humus gradually decreases with depth. In humus composition humic acids (C_h) prevail over fulvic acids (C_f) but this is not always the case.

Particle-size distribution shows that the clay fraction (<0.001 mm) dominates followed by the coarse silt fraction (0.05–0.01 mm) (Table 11.1 and Fig. 11.5). Data according to the USDA are shown in Table 11.2 and Fig. 11.6.

11.4 Soil Properties and Soil Management

The main drawback in hydromorphic soils is poor oxygen supply. There is also a shortage of available nitrogen. Hydrophilic species are part of the wetland environment.

Environments with an available supply of water and nutrition elements are characterized by natural fertile soils.

Soils with gleyic color pattern are usually with anaerobic conditions and have never been dry. The limitation is the level of the groundwater table. If the groundwater table is quite deep, soils may be aerated and thus constitute valuable pastures with meadow tall-grasses.

Drainage can significantly improve the soils with gley because of better aeration conditions. Biological activity increases and roots penetration is deeper. Drainage is not efficient if the compacted and clayey horizons are on the surface.

References

FAO-UNESCO-ISRIC (1990) Revised Legend of the soil map of the world. FAO World Soil Resources Report No 60, Rome

Keys to Soil Taxonomy (2010) USDA United States Department of Agriculture, NRCS Natural Recourses Conservation Service, 11th edn. Soil Survey Staff, Washington, DC, p 338

Koynov V et al (1964) Classification and systematic of soils in Bulgaria. In: Gerasimov IP et al. (eds) soils of south–eastern Europe. Proceedings of the international symposium of soil science, June 1963, Sofia. Bulgarian Academy of Sciences Press, Sofia: 231–246 (In Russian)

Uzunov Yo et al (2012) Ecosystems of the biosphere reserve Srebarna lake. Bulgarian academy of sciences, Institute of biodiversity and ecosystem research. Prof. Marin Drinov Academic Publishing House, Sofia, p 217 (In English)

World Reference Base for Soil Resources (2006) A framework for international classification, correlation and communication. FAO-ISRIC, Rome

Yolevski M, Hadzhiyanakiev A (1976) Agro-grouping of soils in Bulgaria. Extended systematic list. Code of groups and soil varieties. Institute of Soil Science Nikola Poushkarov, Sofia (In Bulgarian)

Saline Soils

12

Derived from the Russian word *sol,* meaning salt.

The saline group of soils comprises solonchak and solonetz soils that have been influenced by occasional local factors responsible for the level of the groundwater table.

12.1 Concept of the Type

Main Concepts

1. Salinization is the natural or manufactured cause for the formation of salt bearing soils;
2. Solonchak refers to any mineral soil with evidence of soluble salt loaded in the profile that inhibits plant growth;
3. Solonetz is a mineral soil of individual evolution, in which exchangeable Na^+ and Mg^{2+} ions are dominant and the value of pH_{H2O} is 8–10;
4. The complexity and spotting of soil cover indicates the demand for a detailed soil survey at scale 1:200–1:5,000.

The main process that takes place in the formation of solonchaks is circulation of solutions with dissolved salts.

The main processes that take place in the formation of solonetz are leaching out of soluble salts, soda appearance, dispersion of colloids, and their translocation to the lower horizon.

12.2 Ecology of Soil Formation

12.2.1 Distribution

Salinity in the soil cover appears to be due to the local factor responsible for the height of groundwater. Usually, solonetz is a component of the complex of other saline soils. The nature of the soils is of complex soil cover, where there is spotted soil distribution. This indicates the demand for a detailed soil survey at a scale of at least 1:5,000.

Saline soils occupy approximately 33,310.03 ha of the territory, including 14,186.05 ha occupied by solonetz, solonchak, and solonetz-solonchak, and 19,123.98 ha occupied by soils recognized as saline-like (Table 12.1, Figs. 12.1 and 12.2).

Saline soils are distributed in the lowland of the city of Plovdiv (near the villages of Belozem, Radinovo, Benkovski, Tsaratsovo, Saedinenie, Stryama, and Rakovski); in the lowland of the city of Stara Zagora, municipality of Radnevo; in the Burgas lowland, near the city of Nesebar; in the Black Sea region (near the villages of Chernomorets and Atia, and the towns of Sozopol, Primorsko, and Kiten); in the Tundzha hilly region (near the city of Straldzha and the villages of Zimnitsa, Lozen, and Atolovo); in the region of the town of Nova Zagora; in the Devnya region; in the regions of Danube River on the Chernopolska lowland (near the villages of Gulyantsi, Dabovan, Zagrazhden, Gigen, Iskar, and Brest).

Saline soils are localized on the terraces of the rivers Mochuritsa, Ovchritsa, Sredetska, Rusokastrenska, Chadzhiyska, Skat, Yantra, Provadiyska, Dyavolska, Dvoynitsa, and Lesnovska.

Saline soils have developed in accumulative landscapes, where salts are in the profile due to the enhanced evaporation of shallow water bodies like saline lakes, swamps, and old river beds. Often salinity occurs in the periphery of swamped areas.

In northern Bulgaria, these areas are located in the Chernopolska low valley, near the villages of Gigen, Iskar, Brest, and Dabovan (where soil pH value is 9.4), and near the village of Zagrazhden (where soil pH value is 9.5).

Areas of saline soils are near the village of Kamenar, in the district of the city of Varna (where soil pH value is 9.8);

T. Shishkov and N. Kolev, *The Soils of Bulgaria*, World Soils Book Series, DOI: 10.1007/978-94-007-7784-2_12,

Table 12.1 Data of the inventory areas are in compliance with the GIS of soil cover based on the large soil survey of the country's territory at scale 1:10,000, 1:25,000

Saline soil varieties occured on the territory of Bulgaria	Area, ha
Alluvial, solonetz-like	104.07
Alluvial, solonetz-like, slightly swamped	30.02
Alluvial, solonchak-like	23.29
Alluvial, solonchak-like, slightly swamped	54.66
Alluvial (deluvial) meadow, solonetz-like	1,185.81
Alluvial (deluvial) meadow, solonetz-like, slightly swamped	1,178.78
Alluvial (deluvial) meadow, solonetz-like, medium swamped	69.23
Alluvial (deluvial) meadow, solonchak-like	533.42
Alluvial (deluvial) meadow, solonchak-like, medium swamped	207.29
Alluvial meadow, solonetz-like	1,069.89
Alluvial meadow, solonetz-like, slightly swamped	1,285.10
Alluvial (deluvial) meadow, solonetz-like, medium swamped	453.95
Alluvial meadow, solonchak-like	139.88
Alluvial meadow, solonchak-like, slightly swamped	97.09
Alluvial meadow, solonchak-like, medium swamped	454.24
Deluvial meadow, solonetz-like	214.40
Deluvial meadow, solonchak-like, slightly swamped	4.81
Meadow solonetz, thick, soda	156.46
Meadow solonetz, medium thick, soda	1,945.69
Meadow solonetz, shallow, soda	843.60
Meadow solonetz, shallow, sulphate-chloride	34.48
Meadow solonetz, medium thick, sulphate-chloride	83.50
Meadow solonetz, thick, chloride-sulphate	180.94
Meadow solonetz, medium thick, chloride-sulphate	900.49
Meadow solonetz, shallow, chloride-sulphate	253.85
Meadow solonetz-solonchak, thick, soda	244.41
Meadow solonetz-solonchak, medium thick, soda	1,320.11
Meadow solonetz-solonchak, shallow, soda	625.22
Meadow solonetz-solonchak, shallow, sulphate-chloride	1,566.34
Meadow solonetz-solonchak, medium thick, chloride-sulphate	918.15
Meadow solonetz-solonchak, shallow, chloride-sulphate	322.07
Meadow solonetz-solonchak, medium thick, chloride-sulphate	1,394.15
Meadow solonetz-solonchak, thick, chloride-sulphate	31.47
Meadow solonchak, soda	1,539.20
Meadow solonchaks, sulphate-chloride	25.72
Meadow solonchaks, chloride-sulphate	1,800.11
Meadow smolnitsa, thick, solonetz-like	295.10
Meadow smolnitsa, solonetz-like	376.72
Meadow smolnitsa, solonetz-like, slightly swamped	544.69
Meadow smolnitsa, solonetz-like, medium swamped	194.02
Meadow smolnitsa, solonchak-like	713.49
Meadow smolnitsa, solonchak-like, slightly swamped	1,074.82
Meadow smolnitsa, solonchak-like, medium swamped	1,136.34

(continued)

Table 12.1 (continued)

Saline soil varieties occured on the territory of Bulgaria	Area, ha
Meadow chernozem, solonetz-like	997.44
Meadow chernozem, solonchak-like, medium swamped	32.75
Meadow boggy, solonetz-like	475.13
Meadow boggy, solonchak-like	1,205.37
Meadow cinnamonic, solonetz-like	154.92
Meadow cinnamonic, solonchak-like	688.93
Meadow cinnamonic, solonchak-like, slightly swamped	573.39
Medow chernozem-like smolnitsa, solonetz-like	13.76
Medow chernozem-like smolnitsa, solonetz-like, medium swamped	28.38
Medow chernozem-like, solonetz-like	667.54
Medow chernozem-like, solonetz-like, slightly swamped	124.02
Medow chernozem-like, solonchak-like	765.32
Peat-boggy, solonchak-like	977.54
Peat-boggy, solonchak-like	978.22
Total	33,310.03

Author Nedyalkov Simeon from the Agency of Soil Resources (former) at the Ministry of Agriculture and Food

Fig. 12.1 Map-scheme of saline soils distribution in Bulgaria modeled at scale 1:1,500,000 based on the large soil surveys 1:10,000 and 1:25,000, author Nedyalkov Simeon from the Agency of Soil Resources (former) at the Ministry of Agriculture and Food

near the village of Zhitnitsa, in the district of the city of Dobrich (where soil pH value is 7.9–9.5); near the village of Novgrad, in the district of the city of Ruse (where soil pH value is 8.0); and near the town of Elin Pelin, in the district of the city of Sofia (where soil pH value is 7.0).

In southern Bulgaria, the saline soils are confined to the Thracian Lowland, in the district of the city of Plovdiv. Soils are located on the alluvial terraces of Maritsa River (where soil pH value is 8.6). The height of the groundwater table is 1–3 m, drained by the Maritsa River.

In the surrounding areas of the villages of Mladovo, Lubenova mahala, and Mlekarovo and the town of Kermen, in the district of the town of Sliven (where soil pH value is 8.5–9.5).

In the surrounding areas of of the villages of Lozenets, Zimnitsa, and Atolovo and the town of Straldzha, in the district of town of Yambol, where soils are in the Pliocene saline bearing sediments of Tundzha River.

In eastern Bulgaria, the shelf sea solonchaks are the youngest among the saline soils. Commonly, the wet fluffy

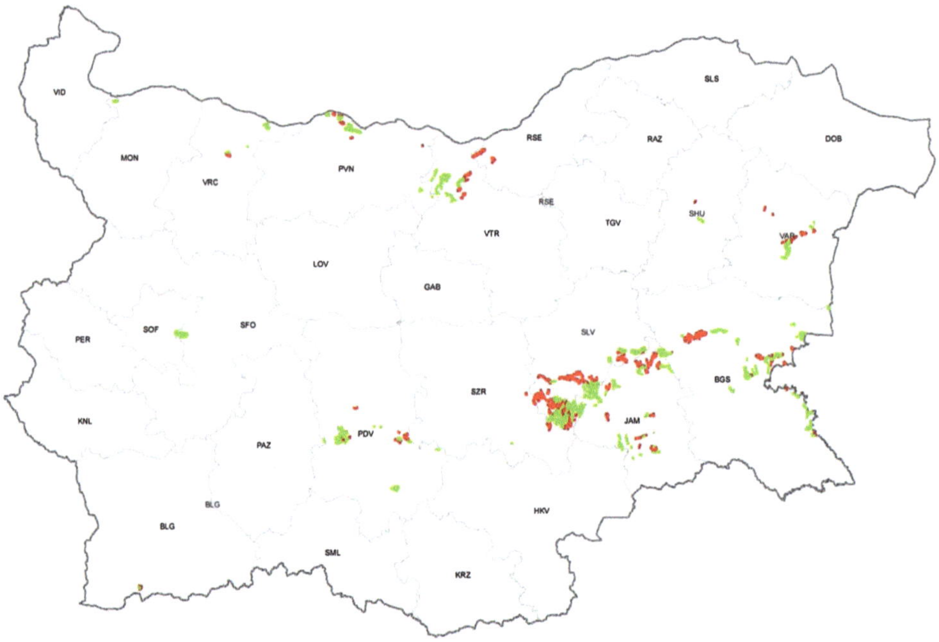

Fig. 12.2 Saline soils in Bulgaria at scale 1:1,500,000 based on the large survey 1:10,000 and 1:25,000, author Nedyalkov Simeon from the Agency of Soil Resources (former) at the Ministry of Agriculture and Food. Typical saline soils are shown as *red-colored* contours and saline-like soils are shown as *green-colored* contours

crust of salts is at the soil surface and beneath is a sandy gleyic layer of chloride salinity (Cl^-: $SO_4^{-2} > 1$).

Boggy solonchaks are of an entirely gleyic color pattern near the Nesebar swamp. The salt accumulation in the region can be designated as sulfate–chloride (Cl^-: $SO_4^{-2} > 1$).

Some solonchaks of lacustrine origin are located near to the lake of Burgas, and in the surrounding areas of the villages of Yasna Polyana and Rudnic (where soil pH value is 8.7).

Saline soils have formed in the old sediments, characterized by the hydromorphic phase of evolution. In the area of the village of Staro Oryahovo, there is a closed depression collecting water, salts, and fine materials from the catchments and plains of Kamchia River. This is a basin, being a transition area for geochemical products, and the height of the groundwater table is 1.4 meters. The salt accumulation in the region can be designated as chloride–sulfate (Cl^-: $SO_4^{-2} > 0.2–1.0$).

Saline soils occur in the most southeastern part of the country, in the valley of Struna River, and near the village of Marikostenovo, in the district of the town of Blagoevgrad.

Secondary solonchaks have developed in the irrigated lands as a result of poor irrigation maintenance (Popandova 1978). Secondary industrial salinization (NaCl mainly), which affected chernozems and alluvial soils, was reported along the Provadiya-Devnya salt pipeline and around salterns and salt mines, where soil pH value significantly increased at the surface due to the appearance of Na^+ at the CEC (Raykov et al. 1989).

Climatic conditions
Saline soils are formed in many environments with hot and dry periods.

Hydrology takes precedence over the climatic factor.

The level of the groundwater table is seasonally dependent and salts migration has a local circulation trend.

Topography
Naturally saline soils are confined to plain relief and are usually located in sea shelf regions, some old lake beds, alluvial terraces, plain depressions, and slopes, where the height of the groundwater table is <3.0 m. Commonly, the topography of saline soil is a flat or low situated valley, where the local flow of discharging drainage waters has a link with the surrounding hills.

Usually, solonetz is formed on flat terrains within the complex of other saline soils; its position in microrelief is somewhat elevated.

Parent materials
Parent materials of saline soils can be various unconsolidated deposits or may be old soils in the case of old landforms.

Vegetation
Halophytic vegetation is a primary indicator that an area is affected by salts. In lands that contain little salt, the natural vegetation remains the same as that in the surrounding areas. Under the influence of salinity, vegetation is of the herbaceous species to the halophytic species. The degree of salinity has a direct link with the suppressed plant

community. Solonetz is characterized by the suppressed xerophytes plant species.

Common are associations of weeping alkali-grass (*Puccinellia distance*), barley (*Hordeum maritimum*), plantain (*Plantago salsa*), fourwing saltbush (*Atriplex cana*), echinopsylon (*Echinopsilon sedoides*), colonial bent grass (*Agrostis verticillata*), beard-grass (*Polypogon monspeliensis*), slender foxtail (*Alopecurus agrestis*), salt grass (*Puccinellia distance*), glasswort (*Salicornia* L.), Russian thistle (*Salsola L*), aster (*Aster tripolium*), sand spurry (*Spergularia Presl.*), lyme grass (*Elymus europaeus*), rush (*Juncos maritimus*), fescue (*Festuca* L.), barley (*Hordeum* L.), bulbiferous meadowgrass (*Poa bulbosa*), Bermuda grass (*Cynodon dactylon*), spear-leaved orache (*Atriplex hastate*), scorzonera (*Scorzonera* L.), ribwort (*Plantago lanceolata*), cushion gypsophila subspecies (*Gypsophila muralis*), thoroughwax (*Bupleurum* L.), and narrow-leaf cress (*Lepidium ruderale*), (*Limonium vulgare*) (Bondev 1991).

In southeastern Bulgaria formations are of *Puccinellieta convoletae, Limonieta gmelini, Aeluropeta littoralis, Saliorneta europaeae,* and others.

Time

The chronological age of saline soils is young. The accumulation of gypsum and soluble salts may occur within a very short period.

Saline soils are of the Holocene Age. Some varieties are comparatively young. The origin of the soils is indicated by the age of the material in which they are formed. Some soils have formed on the transported old products or old-weathered crusts of a later age.

Solonetz is dated to the Holocene Age and developments of the original soil have occurred very rapidly.

Human influence

Saline soils can develop quickly under human intervention. Irrigation practices provoke negative soil processes and secondary salinity.

Soils that are irrigated with water containing salt are most prone to secondary salinity. Often the water for irrigation is pumped from a shallow water table when needed.

A good drainage system should maintain the height of the groundwater table. Some saline soils need to be drained for cropping. Amelioration of the saline soils can be energy-consuming resources. The practice is that the secondary saline soils have expanded. Periodic infiltration of the irrigation water through pipes and channel walls can be a precondition for the appearance of salinity.

12.2.2 Genesis

It is not clear how the free salts in the groundwater appear. Soluble salts can derive from the weathering of rocks followed by geochemical migration. They can be carried on sea air and may occur on the mineral artesian discharge, in mud mineral lake or salt-bearing basins, as well as on the salt outcrop. In nature, sodium is derived from the weathering of rock.

Saline soils or solonchaks are formed due to the high level of the groundwater table. Salts may have accumulated through the evaporation of groundwater that contains high levels of soluble salts. The fluctuation of the soluble salts has a periodic trend, during which the circulated amount of the salts is never the same. Gypsum-bearing soils were recognized as sulphate soils by Dokuchaev in 1896.

As well as solonchak type Bulgarian classification recognized the meadow solonchak variety. Meadow solonchak has preserved the genetic horizons of previous soils. Natural vegetation species grown on these soils usually forms a good dense cover. Commonly, the level of the groundwater table is >2.5 m, but the total salt content is <1.0 %, and for this reason the soil is named meadow solonchak saline-like.

According to theory, solonetz originates from solonchak, after salts have been leached out. The balance of ions shows a dominance of sodium and magnesium and a shortage of calcium. The exchangeable cations can adsorb the negatively charged soil colloids (clay and humus). Sodium releases a small amount of free soda Na_2CO_3 forms. Because soda (Na_2CO_3) is less soluble than sodium chlorite (NaCl) and sulfate (Na_2SO_4), it accumulates in the soil. This determines alkaili soil reaction and the dispersal of colloids. The dispersal of the colloids takes place together with the degradation of soil structure. Mineral colloids dispersed as SiO_2, R_2O_3, and other oxides as well as humus colloids migrate from the upper layer downward to the middle part. This results in a massive compacted horizon, with low water infiltration rates and very low hydraulic conductivity. If salt appears the colloids coagulate.

12.3 Soil Diagnostic and Classification

12.3.1 Morphology

Solonchak soil has the morphology of any soil affected by salinity (meadow chernozems, meadow smolnitsa, meadow cinnamonic, meadow deluvial, and alluvial soils). The structure at the surface horizon can vary. In the profile salt efflorescence is typical. If salt of NaCl prevails, a crust is formed on the surface; if salt of $CaSO_4$ and $MgSO_4$ prevails, wet solonchak is formed; if salt of Na_2SO_4 prevails, the surface layer is fluffy.

Solonetz soil is saline, alkaline, clayey, viscous (when moist), and compacted with prismatic structure sodic/natric horizon (Fig. 12.3). The pattern of horizon sequence in the profile of solonetz is AE-E(g)-Btng1-Btng2-Byzg-C(zg).

Fig. 12.3 Meadow solonetz soil

AE—a mineral humus-eluvial horizon, with thickness of about 10 cm. Color is grayish brown. Structure can be platy or single grain (individual grains), of medium pedality grade, consistence is very friable (moist) to loose and slightly hard (dry). Coarse grains in the soil mass are covered with dark skins. Many fine roots. Beneath is clear change to the lower

E(g)—very thin mineral eluvial horizon, with thickness close to 3 cm or none. Color is bleached grayish compared to the upper layer; structure is loose or friable platy. Coarse grains in the soil mass are bleached. Change of oxidation and reduction forms rusty and grayish mottles in the soil mass. Beneath is sharp change to the lower

Btn(z,g)—illuvial sodic/natric horizon. Structure is strong prismatic, and consistence is very firm (moist) and very hard (dry). Very few cutans may be on the ped faces. Very dark grayish brown, gleyic color pattern, and large cracks are common in the soil mass. There is clear transition to the horizon beneath.

Byz(g)—a thin saline horizon, often with gypsum threads. The structure is angular blocky (blocks having mixed rounded and sharp-angled faces of aggregates), consistence is firm (moist) and hard (dry). Usually, there is gleyic color pattern. Pedogenic gypsum threads are common. The presence of carbonate, gypsum, and salts responds at the lesser compaction pronounce.

Cy(g,k)—parent materials are of strong subangular blocky structure (blocks having mixed rounded and plane faces of aggregates), consistence is firm (moist) and slightly hard (dry). Usually there is gleyic color pattern. This horizon can contain the most prominent accumulation of gypsum and secondary carbonates, mostly soft spots.

12.3.2 Classification

High variability over short distances is problematic for saline soils classification, where only ions elucidate the type of salinity and amount of salts. Principally, pH values and electrical conductivity lead to solonchak classification. Total salts are calculated and presented as the weight percentage of the total water soluble salts in the soil. Sodium adsorption ratio (SAR) is a measure of irrigation water quality.

The proportion of soluble salt content in saline soil is >1.0 %, which appears to be toxic for most agricultural crops; this proportion in the saline-like varieties is <1.0 %.

The electrical conductivity of the saturated paste extract (ECe) is >4 dS/m. The electrical conductivity of the saturated paste extract (ECe) is >4 dS/m (decisiemens per metre). In the international System of Units (SI) 1 dS/m is equal to 1 mmho/cm (milli-mhos per centimeter).

Soil is classified as solonchak, if the total salt content (in dry residue) is >1 % of soluble salt content in the upper layer (0–50 cm). If the salt accumulation is beneath 150 cm of the soil surface, the soil is considered non-saline.

On the lower taxonomic levels, solonchaks differ in the type of salinity: chloride-sulfate (Cl^-: $SO_4^{-2} > 0.2$–1.0);

sulfate–chloride (Cl^-: $SO_4^{-2} > 1$), and soda (HCO^{3-}: $Cl^- + SO_4^{2-} > 1$). In the case that soda Na_2CO_3 is in the solonchak profile, this transitional variety is designated as solonetz-solonchak soil.

In 1982 at the XII Soil Congress, solonetz was defined as alkali soil with columnar structure and was referred to the group of sodic soils.

Soda is the most toxic salt; in cases where Na_2CO_3 is >0.6 % soil is totally unfertile; and if the quantity is >0.1 % plant growth is restricted.

Soil is classified as solonchak, if the content of exchanged Na^+ is >20 % of the cation-exchange capacity.

In Bulgaria, natural conditions determine the meadow solonetz variety (recognized if exchangeable Na^+ is >20 % of the CEC), the solonetz-like variety (recognized if exchangeable Na^+ is <20 % of the CEC), the solonchak-solonetz variety (if exchangeable Na^+ is >20 % of the CEC and the total salt content (dry residue) >1 %), the meadow solonchak variety (recognized if the total salt content is >1 %) and the solonchak-like variety (if the total salt content is <1.0 %).

The Bulgarian classification of 1964 (Koynov et al.) defined the subtypes of meadow-steppe and meadow solonetz, solonchak and solonetz-solonchak. The classification of Yolevski and Hadzhiyanakiev (1976) recognized alluvial and meadow subtypes of solonetz and solonchak. The meadow varieties can be of the alluvial, deluvial, cinnamonic, smolnitsa, chernozem, swamp, and peat subtypes of the type of solonetz and solonchak soils.

The soil classifications of 1983 (Yolevski et al.) defined three subtypes in the type of solonchak (typical, meadow, and lacustrine), and four subtypes in the type of solonetz (meadow, chernozem-like, cinnamonic-like, and smolnitza-like).

There was classification in 1992 (Penkov et al. 1992) that proposed the solonetz subtypes of typical, solonchak-like, gleyic, pseudopodzolic, chernozemic-like; and the solonchak subtypes of typical and gleyic.

The solonchak order after FAO-UNESCO-ISRIC (1990) comprises a large soil group with the requirement of >1 % of soluble salt content in the upper layer (0–15 cm). The solonetz order requires >15 % of Na^+ of CEC in the subsurface; and in the case that Na^+ is <15 % the requirement of Mg^{2+} is >40 % of CEC.

Classification after WRB (2006) required *Salic* diagnostic horizon in the upper 50 cm in the reference group of Solonchaks; and *Natric* diagnostic horizon in the layer of 100 cm in the reference group of Solonetz. In the reference group of Solonetz *albic, gypsic, and calcic* horizons are also admitted. Solonchaks are recognized as external (if salic horizon is in the upper part) and internal (if salic horizon is in the lower part).

According to the Keys to Soil Taxonomy (2010) saline soil varieties are referred to the different orders, where they are distinguished on a subgroup hierarchy level.

12.3.3 Saline Soils in Bulgaria

12.3.3.1 Solonchak Soils

Alluvial and **meadow** *(subtype)* **solonchak** *(type)* is named after the Bulgarian classification of Yolevski and Hadzhiyanakiev (1976).

Salic (Folic or Histic) Fluvisol (Humic); Endosalic Phaeozem; Haplic Salic Vertisol;Endosalic Chernozem; Mollic *Salic Vertisol; Salic (Fibric, Hemic, Sapric) Histosol* is named according to WRB (2006); and *Typic Halaquepts, Vertic Halaquepts, Salidic Natrustalfs, Aquic Natrxerolls, Halic Endoaquerts, Sodic Endoaquerts, Aquic Salaquerts, Salidic Haplustolls, Salidic Calciustolls, Typic Salusterts, Halic Haplusterts, Sodic Salusterts, Halic Haploxererts, Halic Terric Haplosaprists, Halic Haplosaprists* is named according to Keys to Soil Taxonomy (2010).

Solonchak (Haplic, Mollic, Calcic, Gypsic, Sodic, Gleyic, Gelic) is after Revised Legend of the World Soils (FAO-UNESCO-ISRIC 1990).

Soil Taxonomy recognized saline soils on a level suborder in the different orders depending on other soil properties. Bulgarian classification recognized typical meadow solonchak (if >1 % of soluble salt) and meadow solonchak-like soils in different soil types (<1 % of soluble salt).

The solonchak type comprises a wide variety of soils with preserved horizons in the profile and salt appearance at the surface. A general property of solochak is soluble salt accumulation within 0–100 (150) cm. Usually the amount of $CaCO_3$ is significant in the soil. Various compositions of dissolved salts can be found in the soil solution (carbonates, bicarbonates, sulfates, chlorides), but never soda Na_2CO_3. The value of pH_{H2O} is 7.0–8.5. Cation exchange capacity is dominated by Ca^{2+} and Mg^{2+} ions.

There is no restriction of textural class in the soil group of solonchak. The data presented characterized natural areas in the surroundings of the Sindel swamp in the district of the

Table 12.2 Example of the variables of particle size distribution in three profiles of meadow gleyic solonchak, areas of village Sindel, district of city of Varna

Profile count 3	Horizon	Upper depth	Lower depth	Loss HCl	Particle-size distribution (mm), % (according method of Kachinskiy revised)							
					>1 mm	1–0.25	0.25–0.05	0.05–0.01	0.01–0.005	0.005–0.001	<0.001	<0.01[a]
Mean	Apz	0	30	22.7	0.0	0.0	5.4	10.8	9.5	13.8	37.7	61.0
	Azg1	30	52	23.4	0.0	0.0	6.0	11.3	7.6	12.4	39.4	59.3
	Azg2	52	76	21.7	0.0	0.0	9.2	12.6	7.1	12.1	37.3	56.6
	ACzg1	76	103	22.4	0.0	0.0	6.7	12.3	9.0	10.8	38.9	58.6
	ACzg2	103	129	21.3	0.0	0.0	6.4	11.4	8.7	11.8	40.5	60.9
	Cgz	130	162	26.7	0.0	0.0	5.0	8.0	7.0	13.1	40.5	60.5
Stdev	Apz	0	2	11.9	0.0	0.0	3.0	4.8	2.7	1.5	9.4	6.0
	Azg1	2	1	6.7	0.0	0.0	1.7	1.7	1.6	1.9	6.3	3.7
	Azg2	1	1	2.8	0.0	0.0	5.6	4.4	0.9	2.7	10.6	11.7
	ACzg1	1	2	0.8	0.0	0.0	4.0	5.6	1.0	3.3	7.5	10.2
	ACzg2	2	4	0.6	0.0	0.0	3.7	4.3	1.8	2.9	5.6	7.4
	Cgz	6	2	5.3	0.0	0.0	1.3	3.9	1.5	2.1	3.6	0.1

[a] Physical clay, the parameter characterized sum of the particle-size distribution, according to Kachinskiy method (mm) $\sum < 0.01 = < 0.001 + (0.005 - 0.001) + (0.01 - 0.005)$

Table 12.3 Example of the variables of humus, pH, carbonates, and particle-size distribution (mm) according to USDA in three profiles of meadow gleyic solonchak, areas of village Sindel, district of city of Varna

Profile count 3	Horizon	Upper depth	Lower depth	Humus (%)	pH H_2O	Clay <0.002	Silt 0.002–0.05	Fine sand 0.05–0.2	Total sand 0.05–2.0
Mean	Apz	0	30	2.4	7.5	56.9	35.9	7.2	7.2
	Azg1	30	52	2.3	6.9	58.2	34.0	7.8	7.8
	Azg2	52	76	2.8	6.8	55.0	34.1	10.9	10.9
	ACzg1	76	103	3.1	7.3	56.3	35.3	8.4	8.4
	ACzg2	103	129	3.1	7.2	57.8	34.2	8.0	8.0
	Cgz	130	162	1.6	7.2	61.3	31.7	7.0	7.0
Stdev	Apz	0	2	0.8	0.6	4.1	4.3	2.2	2.2
	Azg1	2	1	0.3	–	2.1	3.4	1.4	1.4
	Azg2	1	1	2.0	–	9.9	4.2	5.7	5.7
	ACzg1	1	2	3.2	0.7	8.7	4.8	4.0	4.0
	ACzg2	2	4	3.2	0.5	6.9	3.4	3.6	3.6
	Cgz	6	2	1.6	0.5	0.3	0.7	0.9	0.9

city of Varna. Also, in this region secondary industrial salinezation was reported by Raykov et al. (1989) along the salt-pipeline, around the salterns and salt mine. Any other solonchak in the country with its own unique set of properties may be presented as an example. The example of the particle size distribution shows an irregular pattern where the clay fraction (<0.001 mm) dominates (Tables 12.2, 12.3 and Fig. 12.4).

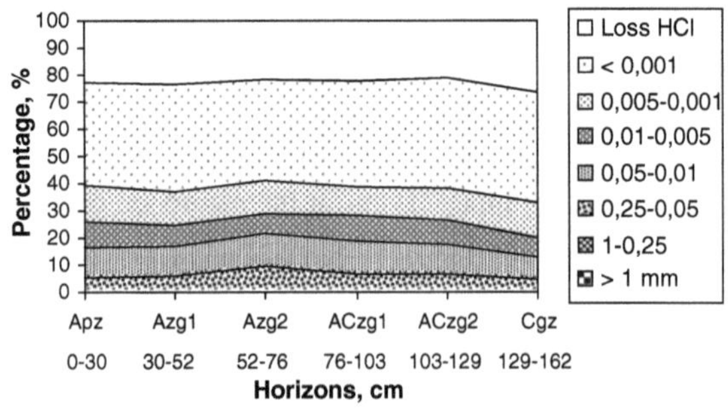

Fig. 12.4 Example of the pattern of particle-size distribution (mm) in the meadow gleyic solonchak, areas of village Sindel, district of city of Varna (according to revised Kachinskiy method)

12.3.3.2 Solonetz Soils

Alluvial and **meadow** (*subtype*) **solonetz** (*type*) is named after the Bulgarian classification of Yolevski and Hadzhiyanakiev (1976).

Table 12.4 Example of variables of particle size distribution in 9 referenced profiles of meadow gleyic solonetz-solonchak

Profile count 9	Horizon	Upper depth	Lower depth	Loss HCl	Particle-size distribution (mm), % (according method of Kachinskiy revised)							
					>1 mm	1–0.25	0.25–0.05	0.05–0.01	0.01–0.005	0.005–0.001	<0.001	<0.01[a]
Mean	Az	0	13	7.6	0.3	9.1	17.2	23.8	9.4	8.9	23.8	42.1
	Abzg	13	39	9.7	0.3	4.7	15.8	19.3	11.5	10.1	28.7	50.3
	Bzg	39	67	13.9	0.2	8.6	14.0	16.8	7.6	9.7	29.5	46.7
	Bczyg	67	87	16.9	4.1	8.2	18.9	18.7	7.7	7.4	18.2	30.7
	Czyg	87	114	10.5	5.5	10.8	16.2	19.8	10.0	10.5	16.6	32.9
Stdev	Ahz	0	11	6.6	0.7	10.1	11.1	5.0	1.9	6.6	7.6	13.1
	Abzg	9	17	6.7	0.8	4.7	6.5	6.3	3.6	7.0	9.2	11.4
	Bzg	4	7	9.1	0.3	6.2	0.4	6.2	1.1	1.1	9.1	9.1
	BCzg	17	23	10.8	8.1	12.2	11.2	9.8	3.5	5.6	11.7	16.9
	Czg	30	37	4.1	12.6	12.2	7.8	15.5	7.4	9.4	10.4	16.5

[a] Physical clay, the parameter characterized sum of the particle-size distribution, according to Kachinskiy method (mm) $\sum < 0.01 = < 0.001 + (0.005 - 0.001) + (0.01 - 0.005)$

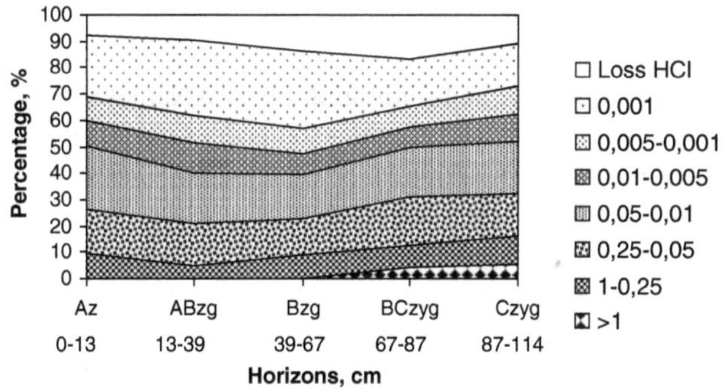

Fig. 12.5 Pattern of the particle-size distribution (mm) in meadow gleyic solonetz-solonchak (according to revised Kachinskiy method)

Salic Folic or Histic Fluvisol (Humic)(Sodic); Haplic Cambisol (Colluvic, Sodic); Haplic Luvisol (Sodic, Chromic); Haplic Salic Vertiso Sodic; Endosalic Chernozem Sodic; Endosalic Phaeozem Sodic; Gleysol (Sodic); Mollic Salic Vertisol; Salic (Fibric, Hemic or Sapric) Histosol (Sodic); Salic Solonetz is named according to WRB (2006) and *Salidic Natrustalfs, Aquic Natrxerolls, Aquic Natrustolls, Aquic Natrxerolls, Vertic Natraquolls, Typic Natraquolls, Typic Natrustolls, Typic Natralbolls, Typic Natraquerts, Sodic Haplusterts, Sodic Haploxererts, Salidic Natrustalfs, Sodic Endoaquerts, Sodic Salusterts* is according to Keys to Soil Taxonomy (2010).

Solonetz (Haplic, Mollic, Calcic, Gypsic, Sodic, Gleyic) is after Revised Legend of the World Soils (FAO-UNESCO-ISRIC 1990).

In the case that soda Na_2CO_3 is in the profile of solonchak, this transitional variety is designated as solonetz-solonchak soil. Usually dissolved NaCl occurs. Cation exchange capacity is dominated by Ca^{2+} but the appearance of Na^+ ions is significant. The structure of soil aggregates is stable, the clay dissociation is low, and the value of pH_{H2O} is below 8.5.

The example of particle size distribution in meadow gleyic solonetz-solonchak shows that the clay fraction (<0.001 mm) dominates (Table 12.4 and Fig. 12.5); and according to the USDA (Table 12.5 and Fig. 12.6).

Solonetz is a soil with dispersed colloides, where clay is accumulated at the compacted natric/sodic (Bt) horizon. The typical structure of the aggregates is prismatic with black humus-iron skins on the ped faces. The soil profile is free of soluble salts. Values of pH_{H2O} are near 7 at the surface horizon and can be over 9 at the natric/sodic (Bt) horizon.

Occurrence of soda Na_2CO_3 provokes an increase of pH_{H2O} at values 9–10. Clay dispersal and structure of soil aggregates at the surface (AE) horizon has been destroyed. Montmorillonite clay minerals prevail at the natric/sodic horizon. The ratio between the values of pH_{H2O} and the percentage of Na^+ in the CEC is: at pH_{H2O} 9 content of Na^+ is 20–40 % in the CEC, and at pH_{H2O} 10 content of Na^+ is near 50 % in the CEC.

The example of particle-size distribution in meadow solonetz shows that the clay fraction (<0.001 mm) dominates (Table 12.6 and Fig. 12.7); and according to the USDA (Table 12.7 and Fig. 12.8).

Table 12.5 Variables of humus, pH, carbonates, and particle-size distribution (mm) according to USDA in 9 referenced profiles of meadow gleyic solonetz-solonchak in Bulgaria

Profile count 9	Horizon	Upper depth	Lower depth	Humus (%)	pH H_2O	Clay <0.002	Silt 0.002–0.05	Fine sand 0.05–0.2	Total sand 0.05–2.0
Mean	Az	0	13	2.1	6.7	32.4	40.6	17.8	27.0
	ABzg	13	39	1.3	7.3	39.0	39.6	16.5	21.3
	Bzg	39	67	0.3	7.6	41.8	34.5	15.0	23.7
	BCzyg	67	87	0.6	7.5	31.5	37.1	20.7	31.4
	Czyg	87	114	0.8	7.5	–	–	–	–
Stdev	Ahz	0	11	1.0	1.2	12.5	4.6	10.8	15.2
	ABzg	9	17	0.7	0.5	10.2	5.3	6.3	9.8
	Bzg	4	7	0.0	0.3	3.8	3.3	1.0	7.1
	BCzg	17	23	0.4	0.4	11.3	14.3	11.1	21.9
	Czg	30	37	1.0	0.5	–	–	–	–

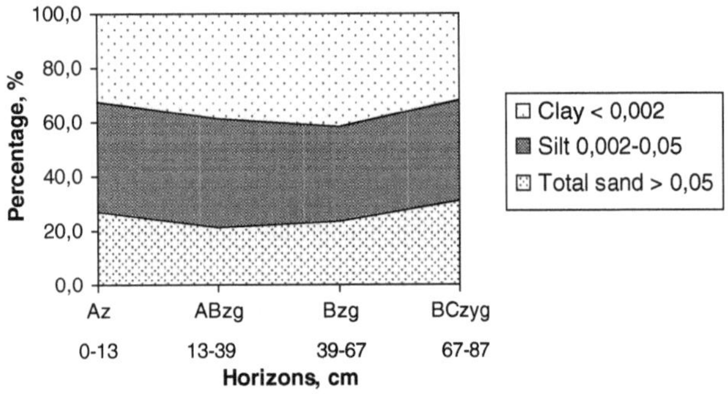

Fig. 12.6 Pattern of the particle-size distribution (mm) in meadow gleyic solonetz-solonchak (USDA)

12.4 Soil Properties and Soil Management

12.4.1 Soil Properties

The most important property of solonchak is the high content of salt in the upper layer (0–50 cm). The ions can be chloride, sulphate, carbonate, sodium, calcium, magnesium, and potassium. Type of salinity is based on the ratio of anions and cations.

Individual distribution is highly variable, but more important is the ratio of different ions in a soil. The chemical and physical soil properties have relation to the amount, proportion, and nature of salts. Together with the easily dissolved salts, gypsum and carbonates often occur.

Solonchaks are characterized by low content of humus (about 1–3 %), with low cation exchange capacity 10–20 meq/100 g. Soil reaction is neutral if chloride and sulfate salt prevail in the soil, and alkaline, if soda occurs. Soils with neutral salinity are characterized by favorable structure, porosity, and infiltration rates. Alkaline solonchaks can be compacted. When the soil is dry large cracks can open; under wet conditions the soil is muddy and sticky.

Solonetz is a deeply saline soil.

Typical features of solonetz are clay enrichment in the middle part, compaction pronounce, alkaline pH values, and presence of salts at the depth.

The solonetz profile is strongly differentiated in terms of clay, cation exchange capacity, clay mineralogy and sesquioxides, calcium, and other elements distribution. The texture shows large clay enrichment beneath the humus-eluvial (AE) horizon. Exchangeable sodium and magnesium cations dominate over calcium. Sodium can disperse soil colloids and lead to the development of poor physical properties.

Solonetz soils are non-saline and have adsorbed or exchangeable sodium (ESP) over 15. At depths the situation can change and calcium can take precedence among the exchangeable cations. Usually due to leaching, the salts accumulate at depth. The exchangeable Na^+ and Mg^{2+} cations prevail over Ca^{2+} and H^+. Soil reaction is alkali in the middle and lower part of the profile and can be neutral or slightly acid in the upper part.

The sodium adsorption ratio (SAR) of a soil saturated paste extract is pH > 8.5.

Solonetz soil is characterized by low bioactivity. Humus content is low 1.5–2.0 %, and fulvic acids, which are highly mobile, prevail in the humus composition.

Table 12.6 Variables of the particle size distribution in 5 referenced profiles of meadow solonetz

Profile count 5	Horizon	Upper depth	Lower depth	Loss HCl	Particle-size distribution (mm), % (according to method of Kachinskiy revised)							
					>1 mm	1–0.25	0.25–0.05	0.05–0.01	0.01–0.005	0.005–0.001	<0.001	<0.01[a]
Mean	Aez	0	22	4.5	0.8	5.7	14.2	25.4	9.0	11.2	29.1	49.4
	Btn	22	48	3.6	0.6	4.3	9.9	20.3	8.4	11.5	41.5	61.4
	Btzn	48	85	3.9	2.5	7.2	13.3	19.8	7.4	9.6	36.4	53.4
	Bczyg	85	118	4.7	1.8	5.1	15.3	23.3	7.5	9.7	32.8	49.9
	Czyg	118	145	9.8	6.8	6.8	9.4	17.6	5.8	8.9	34.9	49.6
Stdev	Aez	0	8	2.6	1.9	6.4	7.9	7.2	3.1	5.1	14.8	16.4
	Btn	9	7	1.7	1.2	4.0	3.0	6.9	2.2	6.0	5.4	6.9
	Btzn	11	12	1.8	4.6	6.8	6.5	6.7	2.2	5.1	16.7	18.4
	Bczyg	12	16	3.4	3.4	5.2	10.0	8.0	2.4	5.5	11.3	13.4
	Czyg	14	9	7.3	11.7	7.9	3.7	8.8	3.2	8.4	6.2	17.3

[a] Physical clay, the parameter characterized sum of the particle-size distribution, according to Kachinskiy method (mm) $\sum < 0.01 = < 0.001 + (0.005 - 0.001) + (0.01 - 0.005)$

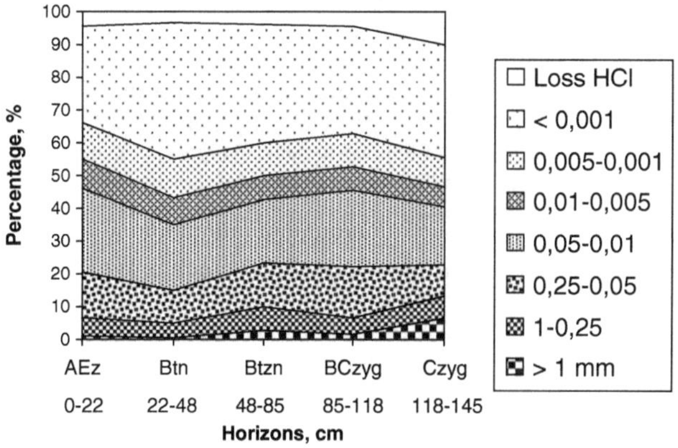

Fig. 12.7 Pattern of the particle-size distribution (mm) in meadow solonetz (according to revised Kachinskiy method)

12.4.2 Soil Management

The Thematic Strategy for Soil Protection (2002) recognized soil salinity as a threat to soil functions. Saline soils have low natural fertility. The drawback for the use of soil affected by salts is that plant roots have to compensate for the created osmotic potential. Plants have poor nutrition and moisture uptake. All saline soils have poor conditions for plant growth. Due to physiological drought only plants of the halophyte community are grown.

Drainage is the most effective procedure for reducing the level of the groundwater table. The most effective practice is the prevention of capillary movement.

Salts can vary greatly over short distances. The risk of salinity calls for the imposition of measures for salt control.

Saline soils without irrigation are used as low productive grasslands or abandoned. Where there is irrigation, saline soils should have effective drainage.

Soil management practices have to be adapted so as to guard against second salinity development, and to eliminate the infiltration of irrigation water from channel walls.

Reclamation of saline soils requires the removal of salts by means of leaching and an effective drainage system. The most important question is how to utilize the salt contained in drainage waters. Soils that have high clay content and low hydraulic conductivity are not reclaimed.

The drawback of alkaline soils is aerobic deficit during the wet period and shortage of moisture during the dry period. In soils with sodium the amount of magnesium is important. The purpose of reclamation is the permanent replacement of the exchangeable sodium and magnesium with calcium. Usually, the ground gypsum $CaSO_4 \cdot 2H_2O$ is mixed with the maintained soil layer.

Soils that have high clay content and low hydraulic conductivity are not reclaimed.

The poplar and acacia plant varieties are resistant to salt content.

Crops have varying degrees of salt tolerance; the growth of wheat, barley, oats, and millet is possible on salt-affected lands.

Table 12.7 Variables of humus, pH, carbonates, and particle-size distribution (mm) according to USDA in 5 referenced profiles of meadow solonetz in Bulgaria

Profile count 5	Horizon	Upper depth	Lower depth	Humus (%)	pH H_2O	Bulk density g.cm^3	Clay <0.002	Silt 0.002–0.05	Fine sand 0.05–0.2	Total sand 0.05–2.0
Mean	Aez	0	22	2.1	7.87	1.51	37.3	41.9	14.6	20.8
	Btn	22	48	1.3	8.41	1.90	49.3	35.9	10.3	14.8
	Btzn	48	85	0.3	9.08	1.79	43.6	33.9	14.0	22.5
	BCzyg	85	118	0.6	8.49	1.97	40.6	37.7	15.8	21.7
	Czyg	118	145	0.8	8.55	1.91	46.3	32.4	10.7	21.3
Stdev	Aez	0	8	1.0	0.78	–	15.7	10.8	8.0	14.6
	Btn	9	7	0.7	0.92	–	5.6	9.8	3.1	7.6
	Btzn	11	12	0.0	0.74	–	17.6	10.1	7.2	15.8
	BCzyg	12	16	0.4	0.62	–	13.7	10.6	9.9	12.5
	Czyg	14	9	1.0	0.59	–	4.9	12.4	4.6	17.3

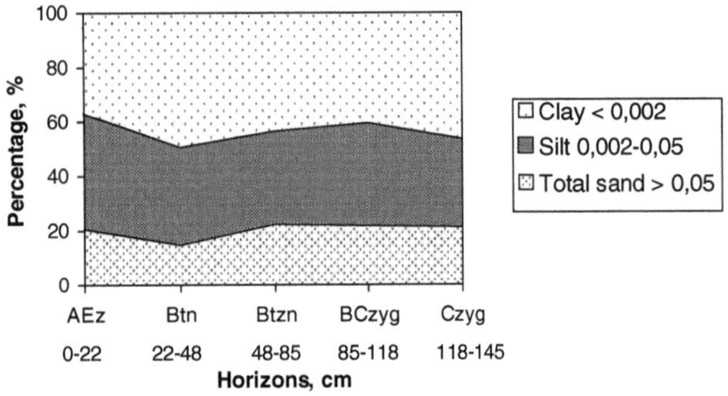

Fig. 12.8 Pattern of the particle-size distribution (mm) in meadow solonetz (USDA)

References

Bondev I (1991) Vegetation in Bulgaria, map at scale 1:600 000 with explanatory text. St. Kliment Oxridski University Press, Sofia (In Bulgarian)

European Commission (2002) Communication of 16 April 2002 from the Commission to the Council. In: The European Parliament, the Economic and Social Committee and the Committee of the Regions: towards a thematic strategy for soil protection, COM 2002, p 179

FAO-UNESCO-ISRIC (1990) Revised Legend of the soil map of the world. FAO World Soil Resources Report No 60, Rome

Keys to Soil Taxonomy (2010) USDA United States Department of Agriculture, NRCS Natural Recourses Conservation Service, 11th edn. Soil Survey Staff, Washington, DC, p 338

Koynov V et al. (1964) Classification and systematic of soils in Bulgaria. In: Gerasimov IP et al. (ed) Soils of south–eastern europe. Proceedings of the international symposium of soil science, June 1963, Sofia, Bulgarian Academy of Sciences Press, Sofia pp 231–246 (In Russian)

Penkov M et al (1992) Soil classification and diagnostic in Bulgaria in relation to the land reform fragmentation. Zemizat Press, Sofia (In Bulgarian)

Popandova S (1978) Change of salt composition in the saline soils under irrigation. Theses, Institute of Soil science Nikola Poushkarov, Sofia (In Bulgarian)

Raykov L, Kavarjiev Y, Popandova S (1989) Secondary salinezation of soils in the region of Provadiya-Devnya. Soil Sci Chem J 24(1):58–64 (In Bulgarian)

World Reference Base for Soil Resources (2006) A framework for international classification, correlation and communication. FAO-ISRIC, Rome

Yolevski M, Hadzhiyanakiev A (1976) Agro-grouping of soils in Bulgaria, extended systematic list. Code of groups and soil varieties. Institute of Soil Science Nikola Poushkarov, Sofia (In Bulgarian)

Yolevski M, Ninov N, Krastanov S et al (1983) Brief diagnostic and criterion of the main taxonomic soil divisions in Bulgaria. Precise Classification Scheme, Institute of Soil Science Nikola Poushkarov, Sofia (In Bulgarian)

13 Soils Manifested Slight Development

Derived from the Greek word *leptos,* meaning thin and *solum,* meaning soil.

Derived from the Greek word *rhegos,* meaning cover and *solum,* meaning soil.

Derived from the Latin word *arena,* meaning sand and *solum,* meaning soil.

Derived from the Italian word *terra,* meaning soil/land and *rossa,* meaning red.

Derived from the Greek technikos, meaning skill-fully made and *solum,* meaning soil.

The large group of soils manifesting morphologically slight development is of different origin and formed in natural or anthropogenic landscapes.

13.1 Concept of the Soils Manifesting Slight Development

Main concepts
1. Few changes have taken place.
2. Soils have developed under all types of conditions and with various inherited properties of the parent materials.
3. Leptosols, Regosols, Arenasols, Terra rossa, and Technosols are the popular names for these soils and indicate a particular soil formation in the soil science terminology.

13.2 Ecology of Soil Formation

13.2.1 Distribution

Soils manifesting slight development are widely distributed in many landscapes, where a fresh exposure of parent material occurs.

In Bulgaria, terra rossa can be found scarcely scattered in some places and less often in compact areas.

Climatic conditions

Soils manifesting slight development is distributed in different environments with very different climatic conditions.

Natural conditions in the mountains where they develop are various. Climate in mountainous regions is characterized by contrasts in daily and seasonal cycles.

The northern Bulgaria lowlands and plains belong to the temperate climatic region, which is a continuation partly of the Middle European temperate continental region and partly of the South Russian steppe climatic region.

The southern Bulgaria lowlands and plains belong to the xerothermal zone characterized by semi-continental climate type and transitional continental type with Mediterranean climatic influence.

Topography

The initial stage of soil evolution is determined by the different landforms and landscapes.

In the case of shallow soils on consolidated rock (Leptosols), this is usually hill slope areas and mountain ranges, but development on plain terrain is possible. The topography specific provokes geological erosion processes on sloping lands and this is a starting point of soil development. Relief is the most significant factor, determining translocation of the weathering materials of rocks and the process of erosion. The majority of shallow to bedrock soils usually occur on steeply sloping terrains (Fig. 13.1). Mountain exposure strongly affects the temperature and moisture regime.

In the case of sands and sandy soils (Arenosols), soils mainly occur on flat terrains, commonly near to water bodies (Fig. 13.2).

In the case of severely eroded soils on fine unconsolidated parent materials (Regosols), soils mainly occupy undulating and rolling terrains, where water-induced erosion is manifest.

Terra rossa soil occupies undulating terrains with limestone bedrock as well as karst terrains.

T. Shishkov and N. Kolev, *The Soils of Bulgaria*, World Soils Book Series, DOI: 10.1007/978-94-007-7784-2_13,

Fig. 13.1 Bedrock exposition with shallow mountainous soils

Reclaimed soils (Technosols) occupy man-made anthropogenic landscapes near to ores, quarries, and mining areas.

Parent materials
Provincial differences in soils manifesting slight development are determined by the character of these materials.

In mountains, soil formation is commonly on consolidated rock. Consolidation of parent materials has a marked influence on the soil development. Shallow soils on hard rock (Leptosols) are formed on bedrock or on unconsolidated skeletal materials which are naturally deep.

In the case of severely eroded soils on unconsolidated parent materials (Regosols), soils are formed of the fine earth materials of different origin and composition. Regosols may develop on base saturated materials, as well as on low base saturation materials.

Terra rossa occurs only on hard and permeable limestone or dolomite with a low content of insoluble residue, as well as in karst terrains.

Soils on sandy materials (Arenosols) are distinguished by these materials which may be of different origin and composition.

In the case of reclaimed soils (Technosols), some have developed on deposited materials transported from other places.

Vegetation
Soils manifesting slight development are formed under different vegetation and in almost all types of environments.

In the high mountains soils have a cover of lichens, mosses, and sparse mixed grasses. With decreasing altitude in the mountains, soils are under coniferous and deciduous forests and shrubs.

The subalpine meadows (over 2,000 m altitude) consist of perfect tall grasses that gradually give way to alpine short grasses (over 2,300–2,500 m altitude). The type of vegetation is in compliance with the moisture and temperature regimes, as follows: mat grass (*Nardus.* sp.), fescue (*Festuca* sp.), sesleria (*Sesleria comosa*), meadowgrass species (*Poa* sp.), clover (*Trifolium montanum*), reed grass (*Calamagrostis* sp.), bur reed (*Sparganium* L.); also with the presence of blueberry (*Vaccinium myrtillus*), whortleberry (*Vaccinium vitis idaea*), primrose (*Primula hinensis*), squat (), juniper (*Juniperus* L.), and dwarf mountain pine (*Pinus mugo*). In the high mountain lands the forest is of spruce species (*Picea excelsa*). Soils are characterized by short seasonal biological activity.

In the coniferous belt at the altitude of 1,500–2,100 m on silicate soils the vegetation is boreal, comprising *Pineta sylvestris, Piceeta abies,* and *Pineta peucis*. In areas that

Fig. 13.2 Sands and sandy soils at the Black Sea coast

were forest are now found *Junipereta sibiricae, Chamaecytiseta absinthioides, Festuceta penzesii,* and *Festuceta hyrtovaginatae*. At the altitude of 600–1,200 m deciduous trees of hornbeam and oak woods are located.

In the valleys, vegetation may comprise coniferous or broadleaf forest as well as mixed grass associations. In many places the vegetation has been planted by man for a specific purpose. Forests of *Pinis nigra, Pinus sylvestris, Robinia pseudoacacia* have been planted in some places.

Sandy soils may be bare sand or covered with grass, trees, and even agricultural crops. On the seacoast sand psamophytic vegetation is of the grass formations *Leymeta racemosi, Ammophyleta arenariae, Galileeta mucronatae, Centaureeta arenariae, Trachomiteta veneti,* and bush *Cionureta.*

Time

The chronological time of soils manifesting slight development is fairly young on anthropogenic landforms, but this is not always the case on natural landforms. Their slow development, or the continuos exposure of the surface is due to erosion.

In high gradient mountains where geological erosion prevails, soils on hard rocks are young. Differentiation of the rocks appears very slowly and soil can remain at the initial stage for thousands of years.

In natural soils low thickness of profile is determined by the short time of evolution (absolute age).

In the case of severely eroded soils on fine unconsolidated parent materials (Regosols), the age of fine earth materials corresponds exactly to the age of original basic soils with advanced development.

Terra rossa is relict soil that can be of the Quaternary or Pleistocene Age but may be much older.

In the case of the group of reclaimed soils (Technosols) almost all were developed in the second half of the last century.

Human influence
Lands with shallow soils are suitable for grazing. Overgrazing can provoke erosion.

Slightly developed soils on sandy materials (Arenosols) are used for urbanization and resorts.

In the case of anthropogenic degradation the restoration of the soil profile will be in relation to the degree of profile destruction and is very costly.

13.2.2 Genesis

Soils manifesting slight development can be of natural origin or the result of anthropogenic activities that provoke soil degradation. There are no criteria for the diagnostic, thus these soils show an early stage of development.

Five soil groups can be defined that are linked to the parent rock materials, as follows: *Leptosols* formed on continuous rock; *Regosols* formed on fine earth materials; *Arenosoils* formed on sands; *Terra rossa* formed on hard limestone; and reclaimed varieties of Technosols: Revised Legend of the World Soils (FAO-UNESCO-ISRIC 1990).

Shallow soils developed on hard rock: Leptosols
In mountain areas a fine layer of silty material appears below the lichens cover mixed with sparse grasses. The material is on the exposed hard rock and is a result of weathering due to hydrolysis. Initial soil formation is localized in a small area between the exposed rocks and cliffs. There is lack of physical translocation of clay, carbonates, sulfates, etc. This fact determines comparatively shallow soil development and high rates of skeletal performance. Both physical and chemical weathering is important in the transformation of parent rock materials. Soils show slight weathering and the formation of iron oxide coatings. Primary minerals prevail in the soil mass. The amount of clay minerals is low. The soil cover is ruptured by rocks and boulders. Shallow soil varieties develop on consolidated rock (Fig. 13.3).

Eroded soils formed on the fine earth materials: regosols
The group of eroded soils with natural or accelerated erosion initially on the unconsolidated parent materials (Regosols) usually manifest a severe degree of erosion. Generally regosols are formed on fine earth materials, which have in fact inherited from the original soil. The European Commission (2002), in the Communication of 16 April 2002 regarding the Thematic Strategy for Soil Protection defines the threat of accelerated erosion as the most dangerous to soil functions.

The group of Regosols is with secondary origin from the original soils. Soils are developed in different landscapes and environments, in hilly mountainous regions as well as plains.

Sandy soils and sands: arenosols
The group of sands and sandy soils (Arenosols) have a poor assortment of soil mass, and are of different composition (mainly of quartz grains), different size, and different origin—generally inherited. The group of Arensols usually refers to soils with sandy and sandy loam texture (Fig. 13.4). This group of soils is with secondary origin as a result of water and wind activity. Soils are poorly developed in different landscapes: sand plains, ridgy sands, fixed sands, unfixed sands, chamois sands, cumulose sands, coastal or beach sands, and dune sands.

Sandy soils may differ greatly in terms of origin and composition as follows: bimineral sand, three mineral sand, monomineral sand, top sand, water sand, volcanic sand, green sand, afgillaceous sand, semigravel, coarse sand, fine sand, iron sand, loamy sand, calcareous sand, limestone sand, quartz sand, concretionary sand, copper sand, soft sand, tight sand, porous sand, productive sand, shell sand, rusty sand, loose sand, micaceous sand, firn, entrained sand, dry sand, sea sand, lacustrine sand, residual sand, cover sand, river sand, creek sand, fluvio-glacial sand, and eolian sand.

The earlier stage of sandy soils development is initiated by plants, but despite this they have poor nutrient composition and remain in this state for a long time.

In Bulgaria sandy soils are distributed in small areas compared to western Europe, where they occupy vast compact territories.

Clayey soils: terra rossa
Terra rossa is distributed in many regions of the world—in the south European Mediterranean region, the Crimea peninsula, and California. Terra rossa is not included in the Bulgarian soil classification, it is recognized among other reddish colored soils in some regions of the country's territory (Fig. 13.5).

Fig. 13.3 Shallow soil on hard rock

Terra rossa is a soil that can be between a few centimeters and 1 m in thickness. The shallow soil varieties generally prevail. Usually a thin layer is recognized in gaping joints and filled fissures of calcareous rocks.

The soil may or may not have a thin humus horizon on the surface and has no other genetic horizons. Soil is deeply weathered with the release of iron oxides; however, many other soils have the same specifics.

The soil has a deep red color, and a clayey texture mix with skeletal part which derives from limestone rock. Terra rossa soil shows evidence of high drainage with regular leaching. Clay content prevails and may increase at the depth in the case of deeper soils. Despite this, there is no differentiation of clay and iron oxides. Soil is characterized by neutral pH values and is saturated with Ca^{2+} and Mg^{2+}. Parent materials are insoluble material of limestone and dolomite.

The soil is widely recognized in the Mediterranean climatic region and is old enough to be relict. It is generally considered that this soil was formed in previous hot and humid subtropical climatic conditions.The Holocene period is negligible in terms of rock weathering compared to that which took place during the Tertiary period.

There is some debate as to whether limestone rock is a primary parent rock with old weathering crust, or whether the reddish material is secondary drift clay or the old sedimented clays that have been deposited without any link to weathering rock. In pedology, the theory of relict reworked weathering crust dominates.

The concept of the polygenetic origin of terra rossa considers that soil develops as a result of a combination of many factors, including aeolian dust and the debris deposited over limestones (Durn 2003).

Dominant mineral phases in the clay fraction in terra rossa are kaolinite, illitic material, Fe-oxides, and XRD amorphous inorganic compounds. Kaolinite is the dominant mineral phase in fine clay; however, the dominant mineral phase of the clay fraction of the insoluble residues is illitic material (Durn et al. 1999). Both haematite and goethite are iron oxides but haematite is more typical for terra rossa and prevails over goethite.

However, in Bulgaria fericalcareous soil is recognized which is characterized by non-differentiated profile ACk-CRk-R, depth between 10–50 cm and 2.5YR4/6 and 5YR6/6 Hues. The low percentage of humus of calcic mull type, the carbonates content of more than 20–45 %, the pH (lightly basic) and the saturation of 100 % do not correspond to the terra rossa. As a result of permanent secondary carbonatization of the profile it is hard to refer this soil as developed from relict terra rossa. There is no correlation of the red soil color and the total content of Fe_2O_3 either in the soil mass or in the clay. Obviously, this is a contemporary developed soil where the red color is determined not by total content of Fe_2O_3, but by its high content of free forms

Fig. 13.4 Uprooted tree on sandy soil

Fig. 13.5 Terra rossa soil in forest area of the Strandzha mountain

Fig. 13.6 Reclaimed soil from the coal basin of East Maritsa composed of excavated and deposited yellowish-green clays

(amorphous and crystalline—Fe_o, determined by the method of Tamm and Fe_d, determined by the method of Mehra–Jackson) precipitated as coatings (Shishkov and Jokova 2005).

Reclaimed soils

Most of the group of so-called reclaimed soils (Technosols) are the result of coal production in Bulgaria and originate from open cast mines and underground mining. Open cast mining releases a huge amount of mining waste into the upper part of the soil surface as overburden dump material. The overburden dumps are located at different mining areas in Bulgaria. The largest open cast mine is the basin of East Maritsa, with a capacity of over 35 million tons per annum. Since the beginning of operations up to 2011 the overburden was 4,115,819,135 m^3. These overburden dumps change the natural landscape and affect the drainage system. Open cast mining in huge areas results in the destruction of the natural ecosystem, and the complete excavation of soil with subsequent loss of biodiversity and cumulate spoil heaps. On the top surface of the overburden dump materials can be used for the reclamation of degraded open cast mines in coalfields. Some layers composed of organic substances can be used for reclamation, if the amount is sufficient and readily available. All of this is an argument for the use of the term Technosols, which originates from the Greek *technikos,* meaning skillfully made (Fig. 13.6).

13.3 Soil Diagnostic and Classification

13.3.1 Morphology

In the case of shallow soils on consolidated rock (Leptosols), the horizon sequence is (AC)-CR-R with thickness not greater than 25 cm; in Lithic Leptosols the thickness is less than 10 cm with a cover of lichens or thin sod comprising grass roots. The common diagnostic feature in the group of shallow soils on consolidated rock (Leptosols) is lack of diagnostic horizons except *ochric* AC horizon which can be discontinuous or fragmented. Soils are always shallow and highly skeletal, often with large rock fragments. Beneath the surface, lying on the hard rock, there is a very thin mineral layer mixed with a low amount of organic compounds. There are no signs of hydromorphism.

AC—a mineral horizon, less than 25 cm thick, the fine earth is of grayish-brown color, with loose structure or not coherent single grains (<1 mm diameter). There are a few fine (0.5–2.0 mm) size grass roots; beneath there is clear transition to the lower

CR—a mineral horizon, shows some change from parent rock, lighter colored compared to the upper horizon, structure less (no structure), there is abundance of rock fragments with different size; beneath is sharp change to ***R***—bedrock.

In the case of eroded soils on fine earth unconsolidated parent materials (Regosols), the horizon sequence is (AC)(k)–C(k) with unspecified depth; in the case of Leptic Regesols, unless the depth of rock is more than 25 cm they are classified as Leptosols. No diagnostic feature, there is lack of diagnostic horizons. Soils may be shallow. Beneath is an unconsolidated layer.

AC(k)—a mineral horizon, of brighter color, with loose structure or not coherent single grains (<1 mm diameter). There may be many fine (0.5–2.0 mm) size grass roots; beneath there is gradual transition to the lower

C(k)1—a mineral horizon of parent materials, shows little change, lighter colored compared to the upper horizon, structure less; beneath is gradual change to

C(k)2—parent materials.

In the case of the group of sands and sandy soils (Arenosols), the horizon sequence is 1(AC) or 1(C)(k)-2C(k)-3C(k) with unspecified depth, no diagnostic feature except sandy texture, no diagnostic horizons, sometimes only layers. Beneath is sand.

1AC(k) or ***1C(k)***—a mineral horizon, of brighter color, with loose structure or not coherent single grains (<1 mm diameter). There may be many fine (0.5–2.0 mm) size grass roots; beneath there is gradual transition to the lower

2C(k)—a mineral horizon of parent materials, shows little change, lighter colored compared to the upper horizon, structure less; beneath is gradual change to

3C(k)—parent materials.

In the case of terra rossa soil there is no diagnostic feature except reddish colored clayey material enriched with iron oxides.

In the case of reclaimed soils on the overburdened dump materials, the layer sequence is of great spatial heterogeneity and cannot be of specified options. No diagnostic feature except variegation.2

13.3.2 Classification

Shallow soil on hard bedrock: leptosols

The Bulgarian systematic does not recognize the single group of soils manifesting slight development.

Bulgarian classification of Koynov et al. (1964) defined rankers (light and dark) and lithosols (plain and mountainous).

In the classification scheme of Yolevski and Hadzhiyanakiev (1976) shallow soils are grouped on the level subtype in any soil type. The idea is that soils can evolve in any other basic soil group in response to the environment.

According to Bulgarian soil classification (Yolevski and Hadzhiyanakiev 1976) the shallow soil varieties on consolidated rock are distinguished on the level subtype within a soil type, for example: *mountainous forest soil* (type) *shallow on the hard rocks* (subtype); or *mountainous meadow soil* (type) *shallow on the hard rocks* (subtype); or *calcareous chernozem, shallow on the rocks* (subtype) may be *(slightly,* moderately, severely *eroded).*

A recent scheme for the diagnostic and classification of soils manifested slight development was proposed by Teoharov (2003), and Boyadzhiev and Teocharov (2005).

There was classification in 1992 (Penkov et al.) which defined types of Lithosols and Rankers which has not so far been applied.

Since 1986, the soil group of Leptosols (after FAO) has included the former soil units lithosols, rendzinas, and rankers (Legend of the World Soil Map 1977). In WRB (2006) the requirement for this soil group to be recognized as Leptosols was the diagnostic of continuous rock at the depth of 25 cm. In the case of thickness less than 10 cm the term lithic is applied instead of former ranker. The initial stage of development is marked in this way.

Soil Taxonomy (2010) requires *Ochric* or *Histic* epipedon and non-diagnostic features for Entisols order classification. Most subsurface horizons have none. Usually this is indicated with lithic.

Eroded soils formed on the fine earth materials: regosols

In Bulgarian classification there is no single type for the group of shallow soils formed on fine earth materials (Fig. 13.7) similar to the group of Regosols after WRB (2006) and FAO-UNESCO-ISRIC (1990). Usually, the group of Regesols refers to eroded soils without diagnostic horizons: FAO-UNESCO-ISRIC (1990) and WRB (2006). Soil Taxonomy (2010) requires *Ochric* epipedon and non-diagnostic features for Entisols order classification. Most subsurface horizons have none.

In many cases, active erosion is responsible for the shallow depth: Biolchev (1972), Lazarov et al. (2002), Rousseva and Stefanova (2006), Rousseva et al. (2010a, b).

According to the Bulgarian classification (Yolevski and Hadzhiyanakiev 1976) severely eroded soil segments of genetic horizon BC(k) of original soils may be recognized in the profile (Fig. 13.6). Subordination is on the level subtype in any soil type, fixed by the local environment and designated as slightly, moderately, and severely eroded. Thus soils manifesting slight development formed on fine earth materials do not form a single group and are subordinated within different soil types, for example: *calcareous chernozem, shallow on the rocks* (subtype) may be *(slightly,* moderately and severely *eroded)*—Leptic Regosol Calcaric after WRB (2006); or may be *calcareous chernozem, eroded* (subtype) *(slightly,* moderately and severely *eroded)*—Haplic Regosol Calcaric after WRB (2006).

Fig. 13.7 Severely eroded soils with ravines

There was classification in 1992 (Penkov et al.) which defined types of Regosols (eutric, dystric, calcaric, skeletic) which has not so far been applied.

Sandy soils and sands
Also, there is no certain diagnostic for the group of sandy soils based on sound principles except for sandy texture. According to the Bulgarian classification (Yolevski and Hadzhiyanakiev 1976), within the soil type of alluvial and deluvial soils are recognized subtypes of sands and cultivated sands (Fig. 13.8). For example: *alluvial and deluvial soils* (type) *sands and gravels* (subtype); or *alluvial soils* (type) *cultural sands* (subtype).

Soil Taxonomy (2010) requires diagnostic features for Arents order classification.

Clayey soils: terra rossa
Because of the similarity between terra rossa and other soils of the world, the soil group of red Mediterranean soil (terra rossa) is not included in the Revised Legend of the World Soils FAO-UNESCO-ISRIC (1990).

Usually terra rossa is classified as Chromic Luvisol or Cambisol (FAO-UNESCO-ISRIC 1990) or Alfisols (Soil Taxonomy 2010).

Reclaimed soils
The classification of the group of reclaimed soils was formed by Christov (2000). Technogenic soils are grouped in the class of anthropogenic soils that are characterized on the level type by origin as reclaimed soils; subordinated on the level subtype depending on whether the technology of reclamation is humic, non-humic, or biological; present basic processes on the level unit with differences in features related to humus horizon, chemical properties, carbonate content, base saturation, etc., and on the level of variety present in the texture of applied layers.

13.3.3 Slightly Developed Soils in Bulgaria

13.3.3.1 Shallow Soil on Hard Bedrock (Leptosols)

Shallow soil on hard rock (*subtype*) of basic developed soil *(type)* after the Bulgarian classification Yolevski and Hadzhiyanakiev (1976), (Fig. 13.9), (Tables 13.1, 13.2, 13.3, 13.4, 13.5, 13.6, 13.7 and 13.8).

Leptosol according to WRB (2006) and *Entisol* are according to Soil Taxonomy (2010).

Leptosol, the Revised Legend of the World Soils (FAO-UNESCO-ISRIC 1990).

Fig. 13.8 Cultural sands, Bulgarian classification Yolevski and Hadzhiyanakiev (1976), the alluvial valley of Kamchia river

Fig. 13.9 Shallow soil on bedrock

Table 13.1 Variables of the particle-size distribution in 35 referenced profiles of Haplic Leptosols (Eutric)

Profile count 35	Horizon	Upper depth	Lower depth	Loss HCl	Particle-size distribution (mm), % (according to method of Kachinskiy revised)							
					>1 mm	1–0.25	0.25–0.05	0.05–0.01	0.01–0.005	0.005–0.001	<0.001	<0.01[a]
Mean	AC	0	19	1.3	2.3	15.4	29.7	17.3	7.8	9.5	17.0	31.9
	R	–	–	–	–	–	–	–	–	–	–	–
Stdev	AC	0	4	0.7	4.0	12.9	11.1	6.0	3.4	4.1	10.9	15.1
	R	–	–	–	–	–	–	–	–	–	–	–

[a] Physical clay, the parameter characterized sum of the particle-size distribution, according to Kachinskiy method (mm) $\sum < 0.01 = <0.001 + (0.005\text{–}0.001) + (0.01\text{–}0.005)$

Table 13.2 Variables of the particle-size distribution in 17 referenced profiles of Haplic Leptosols (Eutric)

Profile count 17	Horizon	Upper depth	Lower depth	Loss HCl	Particle-size distribution (mm), % (according method of Kachinskiy revised)							
					>1 mm	1–0.25	0.25–0.05	0.05–0.01	0.01–0.005	0.005–0.001	<0.001	<0.01[a]
Mean	AC	0	17	2	11	25	23	13	6	7	13	26
	Rk	–	–	–	–	–	–	–	–	–	–	–
Stdev	AC	0	5	2.2	11.2	15.9	7.6	5.2	3.4	3.4	11.4	16.7
	Rk	–	–	–	–	–	–	–	–	–	–	–

[a] Refer Table 13.1

Table 13.3 Variables of the particle-size distribution in 11 referenced profiles of Haplic Leptosols (Calcaric)

Profile count 11	Horizon	Upper depth	Lower depth	Loss HCl	Particle-size distribution (mm), % (according to method of Kachinskiy revised)							
					>1 mm	1–0.25	0.25–0.05	0.05–0.01	0.01–0.005	0.005–0.001	<0.001	<0.01[a]
Mean	ACk	0	16	15.7	30.2	2.0	11.1	10.7	7.5	8.9	13.9	30.2
	Rk	–	–	–	–	–	–	–	–	–	–	–
Stdev	ACk	0	4	7.8	16.3	1.6	8.5	4.6	2.9	3.2	5.2	8.4
	Rk	–	–	–	–	–	–	–	–	–	–	–

[a] Refer Table 13.1

Table 13.4 Variables of the particle-size distribution in 14 referenced profiles of Lithic Leptosols (Dystric)

Profile count 14	Horizon	Upper depth	Lower depth	Loss HCl	Particle-size distribution (mm), % (according to method of Kachinskiy revised)							
					>1 mm	1–0.25	0.25–0.05	0.05–0.01	0.01–0.005	0.005–0.001	<0.001	<0.01[a]
Mean	AC	0	19	1.1	20.3	20.5	22.1	15.7	6.8	5.3	8.2	20.4
	R	–	–	–	–	–	–	–	–	–	–	–
Stdev	AC	0	4	0.8	19.1	7.7	12.9	5.5	2.1	2.1	4.0	6.5
	R	–	–	–	–	–	–	–	–	–	–	–

[a] Refer Table 13.1

Table 13.5 Variables of humus, pH, carbonates, and particle-size distribution (mm) according to USDA in 35 referenced profiles of Haplic Leptosols (Eutric)

Profile count 35	Horizon	Upper depth	Lower depth	Humus (%)	pH_{KCl}	Bulk density (g.cm^3 calc)	Clay <0.002	Silt 0.002–0.05	Fine sand 0.05–0.2	Total sand 0.05–2.0
Mean	AC	0	19	2.5	5.2	1.3	22.7	30.8	30.2	46.9
	R	–	–	–	–	–	–	–	–	–
Stdev	AC	0	4	1.2	0.7	0.1	12.1	9.9	11.2	18.1
	R	–	–	–	–	–	–	–	–	–

Table 13.6 Variables of humus, pH, carbonates, and particle-size distribution (mm) according to USDA in 17 referenced profiles of Haplic Leptosols (Eutric)

Profile count 17	Horizon	Upper depth	Lower depth	Humus (%)	pH_{KCl}	Bulk density (g.cm^3 calc)	Clay <0.002	Silt 0.002–0.05	Fine sand 0.05–0.2	Total sand 0.05–2.0
Mean	AC	0	17	3	5	1	19	25	25	57
	Rk	–	–	–	–	–	–	–	–	–
Stdev	AC	0	5	2.16	0.5	0.23	13.6	8.6	7.5	20.3
	Rk	–	–	–	–	–	–	–	–	–

Table 13.7 Variables of humus, pH, carbonates, and particle-size distribution (mm) according to USDA in 11 referenced profiles of Haplic Leptosols (Calcaric)

Profile count 11	Horizon	Upper depth	Lower depth	Humus (%)	pH_{KCl}	Bulk density (g.cm^3 calc)	Clay <0.002	Silt 0.002–0.05	Fine sand 0.05–0.2	Total sand 0.05–2.0
Mean	ACk	0	16	6.20	7.0	1.23	32.2	34.4	14.9	33.4
	Rk	–	–	–	–	–	–	–	–	–
Stdev	ACk	0	4	3.12	0.2	0.18	8.2	7.8	9.8	11.7
	Rk	–	–	–	–	–	–	–	–	–

Table 13.8 Variables of humus, pH, carbonates, and particle-size distribution (mm) according to USDA in 14 referenced profiles of Haplib Leptosols (dystric)

Profile count 14	Horizon	Upper depth	Lower depth	Humus (%)	pH_{KCl}	Bulk density (g.cm^3 calc)	Clay <0.002	Silt 0.002–0.05	Fine sand 0.05–0.2	Total sand 0.05–2.0
Mean	AC	0	19	5.7	3.9	–	13.0	29.2	24.2	57.8
	R	–	–	–	–	–	–	–	–	–
Stdev	AC	0	4	2.7	0.2	–	5.6	7.1	11.6	10.4
	R	–	–	–	–	–	–	–	–	–

Fig. 13.10 Cinnamonic forest soil severely eroded, Bulgarian classification Yolevski and Hadzhiyanakiev (1976), *Haplic Regosol Eutric Siltic*, WRB (2006), *Typic Xerorthents*, Soil Taxonomy (2010), *Eutric Regosol*, Revised Legend FAO-UNESCO-ISRIC (1990)

The aeration of Leptosols is high; soils are well drained. Characterized by high water infiltration and very low water-holding capacity. Soils are characterized by seasonable biological activity.

The natural spatial relationship is with Rendzinas, mountainous meadow soils, and some mountainous forest soils (Cambisols).

13.3.3.2 Eroded Soil Varieties (Regosols)

Moderate and severely eroded (*subtype*) of basic developed soil *(type)* after the Bulgarian classification Yolevski and Hadzhiyanakiev (1976), (Figs. 13.10, 13.11).

Regosols according to WRB (2006) and *Entisols* according to Soil Taxonomy (2010).

Fig. 13.11 Cinnamonic forest soil, shallow on sandstone, severely eroded, Bulgarian classification Yolevski and Hadzhiyanakiev (1976), *Leptic Regosol Arenic,* WRB (2006), *Lithic Xerorthents*, Soil Taxonomy (2010), *Eutric Regosols* Revised Legend FAO-UNESCO-ISRIC (1990)

Table 13.9 Variables of the particle-size distribution in 17 referenced profiles of Haplic Regosol (Eutric)

Profile count 17	Horizon	Upper depth	Lower depth	Loss HCl	Particle-size distribution (mm), % (according to method of Kachinskiy revised)							
					>1 mm	1–0.25	0.25–0.05	0.05–0.01	0.01–0.005	0.005–0.001	<0.001	<0.01[a]
Mean	ABp	0	25	1.8	8.3	27.2	22.5	12.8	4.3	5.3	17.7	27.4
	BC	26	45	2.2	7.1	27.7	22.2	11.8	5.0	5.6	18.4	29.0
	C	46	67	4.8	10.4	31.6	19.6	10.9	4.0	4.2	14.5	22.7
Stdev	ABp	0	3	1.4	7.5	13.0	7.7	3.6	1.4	2.8	13.5	14.3
	BC	4	7	2.0	10.4	13.5	10.9	4.2	2.0	3.3	12.5	13.5
	C	7	10	10.1	12.4	13.4	10.1	4.7	1.4	1.9	9.5	10.6

[a] Physical clay, the parameter characterized sum of the particle-size distribution, according to Kachinskiy method (mm) $\sum < 0.01 = <0.001 + (0.005–0.001) + (0.01–0.005)$

Regosols according to Revised Legend of the World Soils (FAO-UNESCO-ISRIC 1990).

The particle-size distribution shows that coarse and fine sand fractions (1.0–0.25 and 0.25–0.05 mm) dominate (Table 13.9 and Fig. 13.12). The clay content is uniformly distributed. Texture is homogeneous at the depth (Table 13.10 and Fig. 13.13).

Humus content is <1.5 %; base saturation and cation exchange capacity are the same as in the basic soil.

13.3.3.3 Sands and Sandy Soil Varieties (Arenosols)

Sands and gravels or **Cultural sands** (*subtype*) of **Alluvial soil**(*type*) after the Bulgarian classification Yolevski and Hadzhiyanakiev (1976), (Fig. 13.14). *Arenosols* according to WRB (2006) and *Quartzipsamments* according to Soil Taxonomy (2010), *Arenosols* according to Revised Legend of the World Soils (FAO-UNESCO-ISRIC 1990).

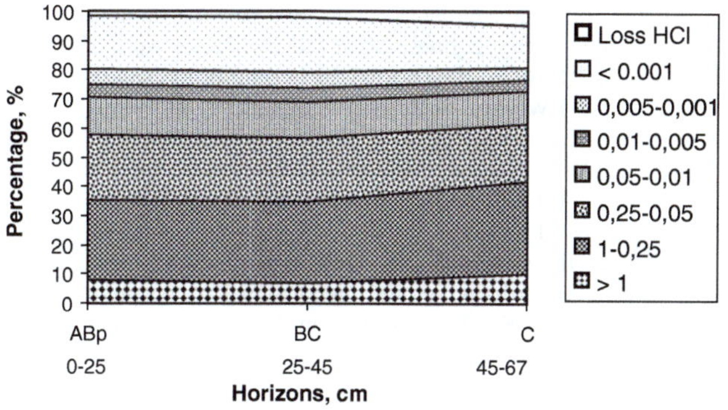

Fig. 13.12 Pattern of the particle-size distribution (mm) in Haplic Regosol (Eutric) (according to revised Kachinskiy method)

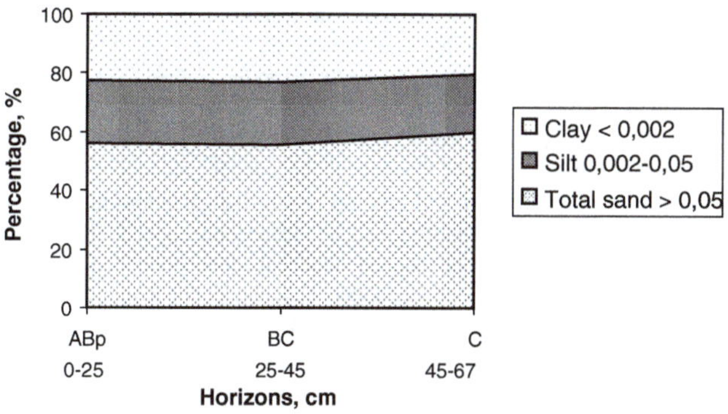

Fig. 13.13 Pattern of the particle-size distribution (mm) in Regosol (Eutric) (USDA)

Fig. 13.14 Alluvial cultural sands, Bulgarian classification Yolevski and Hadzhiyanakiev (1976), *Protic Arenosol Calcaric,* WRB (2006), *Typic Quartzipsamments,* Soil Taxonomy (2010), *Calcaric Arenosols,* Revised Legend FAO-UNESCO-ISRIC (1990)

Table 13.10 Variables of humus, pH, and particle-size distribution (mm) according to USDA in 17 referenced profiles of Haplic Regosol (Eutric)

Profile count 17	Horizon	Upper depth	Lower depth	Humus (%)	pH_{KCl}	Bulk density (g.cm^3 calc)	Clay <0.002	Silt 0.002–0.05	Fine sand 0.05–0.2	Total sand 0.05–2.0
Mean	ABp	0	25	1.3	5.2	1.2	22.3	21.6	23.7	56.2
	BC	26	45	0.7	5.2	1.5	23.3	21.2	23.1	55.6
	C	46	67	0.3	5.6	–	20.2	19.8	21.1	60.0
Stdev	ABp	0	3	0.5	0.9	0.2	13.4	4.9	7.6	14.2
	BC	4	7	0.4	0.9	0.1	12.8	6.1	10.5	14.2
	C	7	10	0.2	1.0	–	14.1	6.2	9.6	17.2

Texture is homogeneous at the depth. The particle-size distribution shows that coarse and fine sand fractions (1.0–0.25 and 0.25–0.05 mm) dominate.

13.4 Soil Properties and Soil Management

The group of soils manifesting slight development have diverse parent materials, climate, vegetation, topography, and properties. The properties are inherited from parent rocks.

Due to the limitation of topography, limited depth and considerable skeletal part, the slightly developed soils on consolidated rock are not suitable for agriculture. The shallow soil varieties on consolidated rock should be left to develop the natural vegetation and wildlife.

Severely eroded soils show degradation and following the United Nations Convention to Combat Desertification (1994) are given protective care. The hilly mountainous soils can be used for pastures and for afforestation, where possible. Moderately eroded soils formed on unconsolidated fine earth materials may be involved in land use after amelioration and with strict erosion control. In Bulgaria, partly severely eroded areas have been planted with *Pinus nigra, Robinia pseudoacacia,* and *Pinus sylvestris*. Some Regosols can be suitable for growing potatoes, tobacco, raspberries, and black currants.

In the case of sandy soils and reclaimed soils, grassing and afforestation benefit the surface.

References

Biolchev A (1972) Erosion control in Bulgaria. In: First national soil science congress, Bulgarian Academy of Sciences Press, Sofia, 23–25 Sept 1969, pp 53–63 (In Bulgarian)

Boyadzhiev T, Teocharov M (2005) Diagnostic and classification of slightly developed soils in Bulgaria. Soil Sci Chem Ecol J 2:3–1. Center of Agricultural Sciences, Pony Press, Sofia (In Bulgarian)

Christov B (2000) Contribution to the classification of reclaimed soils in Bulgaria. Soil Sci Agrochemistry Ecol J 35(1):2–7. Center of Agricultural Sciences. (in Bulgarian)

Durn G, Ottner F, Slovenec D (1999) Mineralogical and geochemical indicators of the polygenetic nature of terra rossa in Istria, Croatia. Geoderma 91:125–150

Durn G (2003) Terra rossa in the Mediterranean region: parent materials, composition and origin. Geol Croat 56(1):83–100

European Commission (2002) Thematic strategy for soil protection. In: Communication from the commission to the council, the European Parliament, the Economic and Social Committee and the Committee of the Regions, COM (2002):179, 16 April 2002

FAO-UNESCO-ISRIC (1990) Revised Legend of the soil map of the world. FAO World Soil Resources Report No 60, Rome

Keys to Soil Taxonomy (2010) United States Department of Agriculture (USDA). Natural Recourses Conservation Service (NRCS) 11th edn. Soil Survey Staff, Washington, DC, p 338

Koynov V et al (1964) Classification and systematic of soils in Bulgaria. In: Gerasimov IP et al. (eds) Soils of south–eastern Europe. Proceedings of the international symposium of soil science, June 1963, Sofia. Bulgarian Academy of Sciences Press, Sofia: 231–246 (In Russian)

Lazarov A, Rousseva SS, Stefanova V et al (2002) Geographic database and evaluation of different soil erosion prediction models for the purposes of the soil information system. Final report of research project contract No 1108–2556. Ministry of Environment and Water, Sofia (In Bulgarian)

Penkov M et al (1992) Soil classification and diagnostic in Bulgaria in relation to the land reform fragmentation. Zemizat Press, Sofia (In Bulgarian)

Rousseva S, Stefanova V (2006) Assessment and mapping of soil erodibility and rainfall erosivity in Bulgaria. In: Morel M et al. (eds) BALWOIS 2006 conference on water observation and information system for decision support, Ohrid, 23–26 May 2006. Book of abstracts (ISBN 9989-9594-1-2), pp 28–29 (Full text at: http://balwois.com/balwois/administration/full_paper/ffp-607.pdf

Rousseva S et al (2010a) Soil erosion risk in Bulgaria and recommendations for soil protective use of agricultural lands. In: Part I Northern Bulgaria. Agricultural Academy, Institute of Soil Science Nikola Poushkarov, PublishSaySet-Eco, Sofia (In Bulgarian)

Rousseva S et al (2010b) Soil erosion risk in Bulgaria and recommendations for soil protective use of agricultural lands. In: Part II Southern Bulgaria. Agricultural Academy, Institute of Soil Science Nikola Poushkarov, PublishSaySet-Eco, Sofia (In Bulgarian)

Shishkov T, Jokova M (2005) Characteristics of some soils from the plateau of Kamen Bryag and the fault-block terrace of the reserve Iailata. Ecol Ind 7(1):154–156. BALKANERECO, Sofia8–10 June 2005

Teoharov M (2003) Pedogenetic characteristic of Leptosols and Regosols in Bulgaria. Ecol Future II(3–4):86–87

United Nations Convention to Combat Desertification (1994) In those countries experiencing serious drought and/or desertification, particularly in Africa. Annex V. Regional implementation annex for Central and Eastern Europe

World Reference Base for Soil Resources (2006) A framework for international classification, correlation and communication. FAO-ISRIC, Rome

Yolevski M, Hadzhiyanakiev A (1976) Agro-grouping of soils in Bulgaria, extended systematic list. In: Code of groups and soil varieties. Institute of Soil Science Nikola Poushkarov, Sofia (In Bulgarian)

Yolevski M, Ninov N, Krastanov S et al (1983) Brief diagnostic and criterion of the main taxonomic soil divisions in Bulgaria. In: Precise classification scheme, Institute of Soil Science Nikola Poushkarov, Sofia (In Bulgarian)

14 Rendzinas (Humus-Calcareous) Soils

The term rendzic is derived from the Polish "*ren-dzic rze-dzic*" and is a composed syllable meaning the sound of an implement being scraped across a stone.

Rendzinas are shallow dark colored soils developed on consolidated calcareous rock.

14.1 Concept of the Type

Main concepts

1. Soil has a *mollic* horizon mixed with skeletal part overlapped by calcareous rock;
2. Well-drained mineral soil that develops in many environments.

Rendzinas (or humus-calcareous) soils show evidence of regular organic substances accumulation at the bottom limited by calcareous bedrock.

14.2 Ecology of Soil Formation

14.2.1 Distribution

Mostly during the Tertiary Period, tectonic movements uplifted the terrains dominated by basic and ultra basic rocks. Anticline and syncline activity lead to the formation of intermediate valleys and coastal plains. Rendzinas (humus-calcareous) soils are distributed in many environments throughout the country (Fig. 14.1). Rendzinas occupy about 265,841 ha, or 2.4 % of the entire territory. Shallow soil varieties prevail.

Climatic conditions
The term intrazonal soil means a formal independence from climatic factor. Rendzinas are intrazonal soils that occur in different landscapes with various types of climate. Usually there is a yearly drought period.

Topography
Relief determines conditions for intensive lateral and subsoil geochemical flood migration. Common are old landscapes that vary from flat to sloping. Karsts are typical phenomenon in the areas of redzinas occurrence.

Parent materials
Parent materials always contain high rates of calcium and/or magnesium carbonate. Consolidated rocks such as limestone and chalk are common. Consolidation of parent materials has a marked influence on the soil development.

In the southern part of the country, rendzinas have developed on the Precambrian Age marbles, dolomitic marbles, and schists.

In the western part, rendzinas have developed on the Upper Triassic Age variously colored terrigenous and terrigenous-carbonate rocks; and on the Lower-Upper Triassic Age limestones, dolomitic limestones, dolomites, and less shales.

In the northern part, rendzinas have developed on the Lower-Upper Triassic Age limestones, dolomitic limestones, and dolomites; and on the Cretaceous Age (Coniacian-Maastrichtian) terrigenous-carbonate rocks in different combinations in different localities.

In the Balkan Mountains, rendzinas have developed on the Lower Cretaceous (Berriasian-Barremian) Age limestones, clayey limestones and marls; and on the Upper Cretaceous (Maastrichtian) Age limestones.

Vegetation
Rendzinas (humus-calcareous) soils have formed under different vegetation and in almost all types of environment. Vegetation can be coniferous or broadleaf forest as well as grassy or mixed associations belonging to the different landforms. The type of vegetation is in compliance with the moisture and temperature regimes.

The shallow depth and shortage of moisture is a limiting factor for plant growth.

T. Shishkov and N. Kolev, *The Soils of Bulgaria*, World Soils Book Series, DOI: 10.1007/978-94-007-7784-2_14,

Fig. 14.1 Stoniness is common on the surface of cultivated land

Time
Soils are of Quaternary or Pleistocene Age but some may be much older.

Human influence
The hilly lands are suitable for grazing. The mat of dense roots at the surface limits the pronouncement of erosion. The removal of woodland exposed a loose litter layer and provoked rapid soil degradation. Due to agricultural use, the physical properties of these soils have deteriorated.

14.2.2 Genesis

Due to the high humus and carbonate content these soils are called humus-calcareous or rendzinas. The high carbonate content slows the decomposition of organic matter which means that mineralization is slow and they accumulate a large amount of humic substances. During soil evolution a dark clayey layer is formed. The mineral part has inherited insoluble residues of primary rock minerals. Rendzinas formed on dolomites can have an enrichment of Mg.

The main soil forming process is the continuous solution and removal of inherited bedrock carbonate in the drainage water.

The degree of humification depends on many factors. Rendzinas (or humus-calcareous) soils have formed "mull" type of humus. Dominated humic acids are bound with calcium, thus forming a strong granular structure.

Soils are very well drained. The high value of total porosity results in rapid moisture percolation.

14.3 Soil Diagnostic and Classification

14.3.1 Morphology

Most rendzinas have AC(k)/Rk horizon sequence (Fig. 14.2). Soils are shallow, and vary in thickness from 10 to 60 cm. The soil profile is uniform in texture, very skeletal, and often with large rock fragments. The classes of percentage of surface rock outcrops and stoniness according to the FAO (1990) are as follows: very few 0–2 %, few 2–5 %, common 5–15 %, many 15–40 %, abundant 40–80 %, dominant ≥80 %.

Rendzinas have a mollic horizon and no other diagnostic horizons. The thickness of the mollic horizon is not less than 20 cm. The enrichment with clay is fixed from the surface.

Below, lying on the calcerous rock, there is a thin mineral layer penetrated by mixed organic matter and partly weathered in situ.

Rendzinas show the following morphological features.

Ah, (***Ak***)—a thin sod or loose litter layer, the fine earth is of very dark color, between the roots there are strong

Fig. 14.2 Rendzina or humus-calcareous soil, Bulgarian classification Yolevski and Hadzhiyanakiev (1976), Rendzic Leptosol, WRB (2006), Lithic Haprendolls, Keys to Soil Taxonomy (2010), Rendzic Leptosol, Revised Legend FAO-UNESCO-ISRIC (1990)

structural aggregates of fine granular structure (spheroids or polyhedrons), with size of 1–2 mm diameter; beneath there is gradual transition to the lower

ACk—a mineral humus accumulative horizon, the fine earth is homogeneous, color can vary from dark brown or black to lighter. Structure is granular (spheroids or polyhedrons), with medium size (2–5 mm) diameter, strong pedality grade, and consistence is firm (moist) and slightly hard (dry). Abundance of fine roots with very fine (<0.5 mm) and fine size (0.5–2.0 mm). Carbonate is present throughout. Many parent rock fragments of different sizes are in the soil mass; beneath is clear change to

CRk—a thin half-weathered mineral horizon, transitional to the parent rock, light colored compared to the upper horizon. Structure is massive (structureless), many rock fragments of different size occur. There is sharp transition into

Rk—calcareous continuous hard rock.

14.3.2 Classification

Since 1986 the soil group of Rendzic Leptosols after the Revised Legend of the World Soils (FAO-UNESCO-ISRIC 1990) has included the former soil unit of rendzinas. A defining characteristic is carbonate materials with equivalent content over 40 % calcium carbonate. In WRB (2006) this is the sub-group of Rendzic Leptosols based on the diagnostic mollic horizon.

The Bulgarian soil classification (Koynov et al. 1964) defined calcareous and typical rendzinas (humus-calcareous) soils. The classification of Yolevski et al. (1976) subdivided the type of rendzina into non-eroded and eroded subtypes. The classification of Yolevski et al. 1983 renamed subtypes as haplic rendzina and pararendzinas (formed on unconsolidated calcareous materials).

The principal variation within the type is soil thickness (<20 cm, 20–40 and >40 cm depth) and degree of erosion (non-eroded, slightly, moderately, and severely eroded).

Classification after WRB (2006) required *Mollic* diagnostic horizon in the reference group of Rendzic Leptosols. The *Mollic* horizon has base saturation over 50 %.

Keys to Soil Taxonomy (2010) required *Mollic* epipedon for Rendolls sub-order classification in the order of Mollisols. Soil Taxonomy requires subsurface horizons: *calcic* is allowed.

14.3.3 Rendzina (Humus-Calcareous) Soil Type in Bulgaria

14.3.3.1 Rendzina (Humus-Calcareous) Soil Non-eroded and Eroded

Non-eroded and eroded (*subtype*) **rendzina** (*type*) is named after the Bulgarian classification of Yolevski and Hadzhiyanakiev (1976).

Rendzic Leptosol is named according to WRB (2006) and *Typic Haprendolls* according to Keys to Soil Taxonomy (2010).

Rendzic Leptosol: Revised Legend of the World Soils (FAO-UNESCO-ISRIC 1990).

The thickness of the soil is defined by topography, where the bottom is hard calcareous rock.

Texture is homogeneous at the depth. The clay content is uniformly distributed.

Particle-size distribution shows that the clay fraction (<0.001 mm) dominates (Tables 14.1 and 14.2).

The average particle density is 2.50–2.55 g/cm^3 at the surface and 2.65 g/cm^3 downward. Bulk density is about 1.2 g/cm^3 at the surface and 1.25–1.30 downward.

Table 14.1 Variables of the particle-size distribution in referenced profiles of rendzinas in Bulgaria

Profile count 11	Horizon	Upper depth	Lower depth	Loss HCl	Particle-size distribution (mm), % (according method of Kachinskiy revised)							
					>1 mm	1–0.25	0.25–0.05	0.05–0.01	0.01–0.005	0.005–0.001	<0.001	<0.01[a]
Mean	ACk	0	22	22.6	10.7	3.9	7.0	13.2	8.0	10.8	23.7	40.5
	Rk	–	–	–	–	–	–	–	–	–	–	–
Stdev	ACk	0	8	9.0	10.6	2.5	6.2	4.1	2.5	3.4	7.3	8.5
	Rk	–	–	–	–	–	–	–	–	–	–	–
9												
Mean	ACkp	0	29	16.3	18.0	6.6	11.2	9.8	5.3	7.9	24.9	38.0
	Rk	–	–	–	–	–	–	–	–	–	–	–
Stdev	ACkp	0	4	10.6	18.2	5.5	11.5	3.5	2.7	3.2	3.3	5.8
	Rk	–	–	–	–	–	–	–	–	–	–	–

[a] Physical clay, the parameter characterized sum of the particle-size distribution, according to Kachinskiy method (mm) $\sum <0.01 = <0.001 + (0.005 - 0.001) + (0.01 - 0.005)$

Table 14.2 Variables of humus, pH, carbonates, and particle-size distribution (mm) according to USDA in referenced profiles of rendzina in Bulgaria

Profile count 11	Horizon	Upper depth	Lower depth	Humus (%)	pH_{KCl}	Bulk density $g.cm^3$ calc	Clay <0.002	Silt 0.002–0.05	Fine sand 0.05–0.2	Total sand 0.05–2.0
Mean	ACk	0	22	4.05	7.1	1.08	44.2	37.5	9.2	18.3
	Rk	–	–	–	–	–	–	–	–	–
Stdev	ACk	0	8	1.3	0.2	0.3	5.7	6.7	5.9	6.3
	Rk	–	–	–	–	–	–	–	–	–
9										
Mean	ACkp	0	29	6.07	7.1	0.96	42.4	27.8	13.4	29.8
	Rk	–	–	–	–	–	–	–	–	–
Stdev	ACkp	0	4	3.18	0.2	0.29	8.5	6.4	11.1	10.8
	Rk	–	–	–	–	–	–	–	–	–

Water stable aggregates with size >1 mm are about 55–65 %. Prevailing water stable aggregates with size over 0.25 mm are 80–85 %.

In rendzinas nearly 15–20 % of pore space is represented by macropores or the pores of aeration (with diameter >300 μm); and 20–25 % is mesopores or pores with rapid moisture drainage (>30 μm diameter). Micropore space or water-filled is about 15–20 %, represented by slow drainage pores (diameter 0.2–30.0 μm) and 15 % of micropores is with none remaining available for roots moisture (diameter <0.2 μm) (Dilkova 1985; Dilkova et al. 1998).

Total porosity is 55 % (at field capacity) at the surface and gradually decreases with depth. Soils have good water permeability.

14.4 Soil Properties and Soil Management

14.4.1 Soil Properties

The organic matter content and composition shows an advanced stage of humification. The humus content is 4–10 % in pristine land and is less in arable land. Humus is entirely composed of humic acids bounded with calcium. The C/N ratio is about 8–12 and indicates humus enrichment with nitrogen.

Soil reaction is most often neutral to alkaline with pH_{KCl} values 7.0–8.5. The amount of calcium carbonate is very high at 35–40 % and the base saturation is 100 %. Calcite crystals are present throughout the soil mass.

Cation exchange capacity is 15–25 meq/100 g soil. Among the exchangeable cations Ca^{2+} and Mg^{2+} dominate. The amount of mobile phosphorous and potassium is low.

Soils are characterized by favorable physical properties. Porosity is high, structural aggregates are well formed, and of crumb and granular structure.

Aeration of soil (when dry) affects the bottom part. Soils are characterized by high water infiltration rates (when wet) and have high water-holding capacity.

Rendzinas are characterized by high biological activity.

14.4.2 Soil Management

Rendzinas have high natural fertility, but shallow depth limits agricultural use. The other drawbacks are the chemical composition, abundance of stones, and low moisture retention.

The high content of calcium can induce microelement deficiencies. Iron is bounded with siderite and is not available. Chlorosis may develop due to the high value of active calcium that often inhibits plant growth. Acid phosphorous fertilizers can be applied to improve fertility.

Rendzina can be used for pasture and for afforestation where possible. Wood can be of poor quality, however. Soils are suitable for growing cereals and establishing vineyards.

References

Dilkova R (1985) Structure and aeration in main soil varieties in Bulgaria. Theses, Institute of Soil Science Nikola Poushkarov, Sofia (in Bulgarian)

Dilkova R, Kerchev Zh, Kercheva M (1998) Evaluating and grouping of soils according to their susceptibility to anthropogenic degradation, Advances in GeoEcology, 31. CATENA VERLAG, Reiskirchen, pp 125–131

FAO-UNESCO-ISRIC (1990) Revised legend of the soil map of the world. FAO World soil resources report no 60, Rome

Kachinskiy NA (1965) Soil physics, part I and II. High School Press, Moscow

Keys to Soil Taxonomy (2010) USDA United States Department of Agriculture, NRCS Natural Recourses Conservation Service, 11th edn. Soil Survey Staff, Washington, DC, p 338

Koynov V et al (1964) Classification and systematic of soils in Bulgaria. In: Gerasimov IP et al. (eds) Soils of south–eastern Europe. Proceedings of the international symposium of soil science, June 1963, Sofia. Bulgarian Academy of Sciences Press, Sofia: 231–246 (In Russian)

World Reference Base for Soil Resources (2006) A framework for international classification, correlation and communication. FAO-ISRIC, Rome

Yolevski M, Hadzhiyanakiev A (1976) Agro-grouping of soils in Bulgaria, extended systematic list. Code of groups and soil varieties. Institute of Soil Science Nikola Poushkarov, Sofia

Yolevski M, Ninov N, Krastanov S et al (1983) Brief diagnostic and criterion of the main taxonomic soil divisions in Bulgaria. Precise classification scheme, Institute of Soil Science Nikola Poushkarov, Sofia)

15 Land Use/Cover in Bulgaria

15.1 General Soil Functions

In Bulgarian, "*soil*" has another meaning which is: *beginning, start, commence, lead off, start up, initiate, jump off, prototype, creation.*

Soils have environmental functions and are linked to human activities that foster cultural, social, and economic development (Figs. 15.1, 15.2, 15.3, 15.4, 15.5 and 15.6).

Protection of soils and policies for sustainable use should, at all stages of development, be based on knowledge and reliable information. In cases where there is significant reduction in the capacity of soil functions, precautionary measures have to be taken to minimize threats such as erosion, decline in soil organic matter, compaction, contamination, sealing, salinization, decline in soil biodiversity, desertification, floods, and landslides (Huber et al. 2009; Jones et al. 2012).

Assessment, adopted on 5 June 2002 (COM (2002) 276 final), identified the positive and negative, direct and indirect, short-term and long-term social, economic and environmental impacts of proposed policy actions.

The European Council in Goteborg (June 2002) endorsed the European Union strategy for sustainable development, adding the Agricultural Council conclusions on environmental integration and sustainable development in the Common Agricultural Policy (CAP). In Communication (Com (2006) 508 final), integration of environmental concerns into the Common Agricultural Policy (CAP) was focused on the set of agro-environmental indicators which among others identified "land use, soil erosion, soil quality and landscape."

The most serious threat to soils in Bulgaria is erosion. Erosion can be natural/geological or accelerated by human activity (Figs. 15.7, 15.8 and 15.9). As a result of erosion, there is loss of the upper detached soil particles, followed by the degradation of soil structure.

The average annual erosion was assessed as approximately about 32 Mt, of which more than 70 % occurs in arable lands. Since 1989, the area of land used for agriculture in Bulgaria has decreased and been superseded by abandoned lands. The significant reduction of the arable land area has caused a substantial reduction in average annual soil loss as well as in total annual soil loss. However, the increasing area of abandoned lands has provoked and/or accelerated processes such as linear erosion which reduces the infiltration capacity of soils (Rousseva 2006).

The model developed for the evaluation risk of area water erosion based on the forecasting of factors and intensity of area water erosion that have been validated for the conditions in Bulgaria shows that the potential risk of area water erosion of soil exceeds 100 t/ha per year for 26.4 % of the country's territory; 16.2 % of the soils are at risk at a rate of 40–100 t/ha per year; 21.0 % of soils are at risk at a rate between 10 or 40 t/ha per year; and 33.6 % of the soils are at risk at a rate of less than 20 t/ha per year (Rousseva et al. 2010a, b).

For agricultural lands, the highest risk of area water erosion of soils exists in the regions of the cities of Burgas, Razgrad, and Russe, with average forecast intensity of 12–15 t/ha y, followed by regions of the cities of Dobrich, Silistra, Kardzhali, Gabrovo, Lovech, and Sofia (10–12 t/ha y); regions of the cities of Sliven, Haskovo, Targovishte, Velico Tarnovo, and Varna (7–10 t/ha y), and regions of the cities of Blagoevgrad, Pazardzhik, Smoyan, Pleven and Yambol (5–7 t/ha y), (Rousseva and Stefanova 2006; Rousseva et al. 2006).

Farmlands areas of about 534,953 ha are at risk of deflation at a rate of over 0.5 t/ha per year.

The main sources of soil contamination in Bulgaria are ore and coal mining, industry, transport, and others. 44,900 ha of farmland have been polluted by industrial

T. Shishkov and N. Kolev, *The Soils of Bulgaria*, World Soils Book Series, DOI: 10.1007/978-94-007-7784-2_15,

Fig. 15.1 Soils are an element of the landscape and a nonrenewable resource for agricultural production of food and other biomass production, including forestry

activities including heavy metals and metalloids. Natural radioactive elements from uranium mining contaminated 1,000 ha. 130 ha have been polluted by petroleum products (Atanasov 2006).

Sustainable land use is based on socioeconomic factors. The problem upon which sustainable land use policy focuses is how to preserve the soils and their specific functions.

Agriculture and forestry are the main uses of land; they play a key role in the management of natural resources in rural areas and determine the rural landscape.

Rural development is a priority for the agricultural sector. Since Agenda 2000, the Common Agricultural Policy (CAP) has focused on sustainable development, encouraging innovation, equal opportunities in rural areas, and support for diversification within and outside farms. However, it is difficult to evaluate the effect of changes of land use in isolation from others such as urbanization and changes in infrastructure, which also occur in rural areas. Changes in the infrastructure, leading to a change of land use, reflect on sustainable use of arable lands, and may lead to updated crop rotation.

The restoration and maintenance of the traditional rural landscape and its characteristics, which have cultural, natural, or environmental value is particularly crucial in terms of biodiversity. Support for agricultural lands was included at the ecological network Nature 2000.

15.2 Facts Related to Land Use/Cover in Bulgaria

> Article 21 (1) *Land is a major national wealth, which is under the special protection of the state and society.*
>
> (2) *Arable land is used for agricultural purposes. Change of land use shall be permitted in exceptional circumstances proved and under conditions and procedures specified by law.*
>
> ***Constitution of the Republic of Bulgaria***

According to the Cadastrial Areas Classification, land use in Bulgaria should meet the needs of agriculture, forestry, industry, mining, and waste disposal; it should accommodate and support settlements and infrastructure as well as parks and resorts used for recreation; and its areas of natural value including gullies, ravines, barren lands, bodies of water, underground springs and mineral waters, rocks, and sands should be protected.

Fig. 15.2 Soils have filtering, storing, and transformation capacity, strongly influencing the water cycle at the land surface

Fig. 15.3 Soils are a biological habitat and gene reserve with a large variety of organisms

Fig. 15.4 Soils are the physical basis for socioeconomic structures and their development, including industrialization and urbanization

Fig. 15.5 Soils are a source of raw materials, providing clay, sand, gravel, minerals, and peat

Fig. 15.6 Soils are an element of cultural heritage

Fig. 15.7 Ravine landscape of gully erosion

Fig. 15.8 High-gradient slope affected by severe sheet water erosion

Fig. 15.9 Natural landmark Stob's Rock Pyramids is a result of geologic erosion

A minimum spatial set of data requires the digitalization of existing maps on land use/land cover, soils, topography, drainage, geomorphology and hydrogeomorphology, and the incorporation of settlements including villages, transport networks including canals, administrative boundaries, irrigation areas, and sectors of industry. Such reference data should be provided by authorized institutions.

The first European Earth Observation program, "Global Monitoring for the Environment and Security" (GMES) provides real-time and long-term data about the use of land and natural resources.

Sometimes the term land cover may be confused with the term land use. Generally, land cover is associated with the physical, chemical, ecological, and biological categories of the Earth's surface, while land use refers to categories of territory based on planned socioeconomic purpose. This means that land-use planning is able to regulate the use of land and land consumption.

Initially the primary data of land use/land cover can be visually interpreted on the surface of Bulgarian territory in terms of forest boundaries and bodies of water, and areas of croplands, fallow lands, grasslands, wet lands, and urban lands (Fig. 15.10).

Data about the territory of the Republic of Bulgaria were obtained from the Statistical Yearbook 2010 issued by the National Statistical Institute (Yancheva et al. 2011). The total area of the territory of Bulgaria is 111,001.9 km^2, where agricultural area is 637,64.8 km^2, forestry is 371,57.5 km^2, settlements, and other urbanized areas are 460,3.4 km^2, water flows and water areas are 201,0.4 km^2, territory of mining and quarrying raw materials is 271,0.9 km^2, transport and infrastructure territory is 754.9 km^2.

In 2009, total protected natural areas amounted to 582,076.3 ha, including reserves 770,21.9 ha, natural landmarks 168,76.8 ha, protected areas 768,44.5 ha, National Parks 150,362.3 ha, Nature Parks 256,455.7 ha, and maintained reserves 451,5.1 ha.

The population of Bulgaria was 7,563,700 in 2009. Population density (per km^2) was 45 km^2 at the end of 1910, and 68.2 km^2 at the end of 2009.

In 2008–2010 the Bulgarian agrarian sector was strongly hit by the financial and economic crises. Gross value added (GVA) generated by the agrarian sector (agriculture, forestry, hunting and fishing) has maintained the downward trend of previous years; the reduction of its relative share in the Bulgarian economy is, as follows: 11.7 % in 2003, 10.0 % in 2004, 9.4 % in 2005, 8.5 % in 2006, 6.2 % in 2007, 7.3 % in 2008, 5.6 % in 2009, and 5.4 % in 2010 (according to the National Institute of Statistics included in the National program for statistical surveys).

In Bulgaria, areas with agricultural use amounted to 5,492,891 ha or nearly 50 % of the total territory in 2010; 5,490,113 ha in 2009; 5,648,206 ha or 50.9 % of the total territory in 2008; 5,666,336 ha or 51.04 % of the total territory in 2007; 5,709,733 ha or 51.4 % of the total territory

Fig. 15.10 A landsat satellite image of Bulgarian territory with 30 m split facility. *Source* Landsat ETM, 1998–2002

in 2006; 5,725,663 ha in 2005; 5,672,000 ha in 1960; and 5,721,000 ha in 1950 (Agrarian Report 2006, 2007, 2008, 2009, 2010, 2011).

The utilized agricultural area, which includes arable lands, land with perennial crops, permanent grasslands, family gardens, and greenhouses was 5,051,886 ha or 45.5 % of the total territory in 2010; 5,029,585 ha or 43.5 % of the total territory in 2009; 5,100,825 ha or 46.0 % of the total territory in 2008; 5,116,220 ha or 46.1 % of the total territory in 2007; 5,190,053 ha or 46.8 % of the total territory in 2006; and 5,264,521 ha or 47.4 % in 2005 (Table 15.1).

Arable land (which includes areas with annual crop rotation, temporary legume crops, and fallows) was 3,162,526 ha or 62.6 % of the total utilized agricultural area of the country in 2010; 3,122,516 ha or 62.1 % of the total utilized agricultural area of the country in 2009; 3,060,543 ha or 60 % in 2008; 3,057,740 ha or 59.8 % in 2007; 3,089,531 ha or 59.5 % in 2006; and 3,128,210 ha or 59.4 % of the total utilized agricultural area of the country

Table 15.1 Arable land and utilized agricultural area in ha; data were obtained from the Annual Agrarian Reports compiled at the department of Agro-statistics of the ministry of agriculture and food in Bulgaria, as well as the statistical yearbooks issued by the National Statistical Institute

Year after the Agrarian report and statistical yearbook	Total area of arable land occupied with crops during the years, ha					
	2005	2006	2007	2008	2009	2010
Utilized agricultural area	5,264,521	5,190,053	5,116,220	5,100,825	5,029,585	5,051,886
Total area under cereal crops	1,815,000	1,586,000	1,701,000	1,726,000	1,857,000	1,810,820
Total area under industrial crops	758,000	871,000	818,000	897,000	896,000	1,071,470
Wheat	1,134,354	979,925	1,120,510	114,427	1,254,151	1,095,703
Barley	276,472	192,539	193,840	223,004	264,689	250,640
Rye and triticale	24,158	17,285	12,030	15,296	17,034	16,116
Oats	36,172	20,707	25,412	40,230	28,894	24,627
Maize	340,847	386,772	408,880	348,402	303,881	360,046
Rice	4,683	5,082	6,454	5,042	6,521	5,2,77
Other cereals	15,943	6,779	5,224	8,175	3,288	3,982
Sunflower	653,371	785,064	686,692	723,962	687,209	734,314
Tobacco	43,005	27,932	31,144	26,742	27,865	31,652
Potatoes	25,441	23,825	21,890	21,648	14,068	13,824
Industrial oleaginous crops	13,094	22,012	59,389	102,899	115,013	209,347
Other industrial crops	44,217	35,325	39,954	48,824	60,629	74,536
Beans, peas, broad beans	4,785	4,379	5,401	5,868	3,803	3,670
Fresh vegetables	41,584	39,899	41,088	30,001	28,715	29,420
Fodder crops	13,028	4,885	6,108	7,352	6,349	7,366
Grassland under legumes	92,772	91,162	92,213	99,362	94,226	87,718
Grassland under cereal grasses	3,205	2,492	3,353	2,468	1,183	1,191
Fallow land	348,118	436,508	291,751	229,471	196,336	207,616
Arable land	***3,128,210***	***3,089,531***	***3,057,740***	***3,060,543***	***3,122,516***	***3,162,526***
Family gardens	42,315	40,388	25,790	25,763	21,411	21,629
Orchards	71,457	71,084	75,035	69,893	71,995	72,913
Vineyard pure culture	106,635	100,564	103,949	100,873	84,438	82,675
Greenhouses	1,998	1,912	2,024	2,129	2,094	2,113
Permanent pro-ductive grasslands	536,259	524,106	515,429	500,008	453,274	–
Rough grazing	1,204,185	1,189,930	1,165,191	1,169,823	1,118,872	–
Meadow-orchards	26,639	24,443	23,883	21,809	19,057	–
Permanent pro-ductive grasslands	1,904,016	1,876,392	1,842,141	1,828,865	1,719,028	1,701,990
Mixed perennials crops	9,892	10,182	9,541	12,759	16,996	–
Alpine pastures	136,933	137,913	137,638	137,225	127,825	–

in 2005. The tendency was increased in 2010 compared to 2009 due to larger areas occupied by industrial oilseed crops (maize, sunflower, tobacco); and was increased in 2009 by 2.1 % compared to 2008 due to larger areas occupied by cereals (wheat, barley, rye, and oilseed rape crops).

Fallows (which are arable lands that be cultivated or not, and remain in this state for no more than 2 years) were about 207,616 ha or 4.1 % of the total utilized agricultural area in 2010; 196,336 ha, or 3.9 % of the total utilized agricultural area in 2009; 229,471 ha or 7.5 % in 2008; 291,751 ha or 9.5 % in 2007; 436,508 ha or 14.1 % in 2006; and 348,118 ha in 2005.

The area of orchards was 1.4 % of the total utilized agricultural area in 2010, the same as in 2009.

The area of vineyards was 82,675 ha or 1.6 % of the total utilized agricultural area in 2009; 84,438 ha or 1.4 % of the total utilized agricultural area in 2009; 100,873 ha or 2.0 % of the total utilized agricultural area in 2008; 103,949 or 2.03 % of the total utilized agricultural area in 2007; 100,564 ha or 1.9 % of the total utilized agricultural area in 2006; and 106,663 ha in 2005.

Areas of permanent pastures and meadows were 1,701,990 ha or 33.7 % of the total utilized agricultural area in 2010; 1,719,028 ha or 34.2 % of the total utilized agricultural area in 2009; 1,828,865 ha or 35.9 % of the total utilized agricultural area in 2008; 1,842,141 ha or 36.0 % of the total utilized agricultural area in 2007; 1,876,392 ha in 2006; and 1,904,016 ha in 2005.

Areas of family gardens were 21,629 ha or 0.4 % of the total utilized agricultural area in 2010; 21,411 ha or 0.4 % of the total utilized agricultural area in 2009; 0.5 % of the total utilized agricultural area in 2008; 40,388 ha in 2006; and 42,315 ha in 2005.

Uncultivated lands (which are areas not included in the crop rotation and not used for agricultural production for more than 2 years) were 4.0 % of the total country's territory in 2010; 4.1 % of the total country's territory in 2009; and 547,381 ha or 4.9 % in 2008.

Areas referred to as under irrigation were 744,100 ha in 2010 and 744,200 ha in 2009.

Engineering and melioration constructions like protective dikes, river correction, and drainage systems and channels that prevent flooding and inundation were more than 166,200 ha of the country's land.

Nitrogen (N) fertilizers were applied on 1,957,800 ha in 2010; 2,190,500 ha in 2009; 1,972,600 ha in 2007; phosphorous fertilizer (P_2O_5) were applied on 327,500 ha in 2010; 318,400 ha in 2009; 174,480 ha in 2007; and potassium fertilizers (K_2O) were applied on 235,900 ha in 2010; 124,900 ha in 2009; and 95,340 ha in 2007.

The situation of the land market was that demand for land was greater than supply.

Rental regulations are mostly applied in the planning regions with cereal crops: in the north-central region—103,377 ha and in the northeast region—834,33 ha. Despite the fact that during recent years the subsidies scheme requires a single area payment, the rental rates for soil use are low.

Private state owned land was 224,734 ha in 2010; 229,290 ha in 2009; 235,059 ha in 2008; 235,163 ha in 2007; and 237,240 ha in 2006. These areas can partly be used for investment purposes.

The development of construction and tourism in recent years has led to increased demand for property near sites of good communication, shoreline, water bodies and coastal sea, and cultural centers, etc.

In Bulgaria the identification system for agricultural land parcels is GIS, developed as digital orthophoto objects of physical and agricultural land parcels. It reflects any restructuring of land, crop rotation, infrastructure, farm land parcels, roads, rivers, and forests.

Areas of cereal cultivation were mostly in the region of the city of Dobrich—10.5 % of the total area in the country occupied by cereals; followed by the region of the city of Pleven with 9.9 % and the region of the city of Veliko Tarnovo with 5.7 %.

Area occupied by cereals was 56 % of the total utilized agricultural area in 2007.

In 2009, the areas occupied by cultivated wheat were 40.2 % of the total arable land in the country and barley was 8.5 %.

15.2.1 Some Facts Related to Land Use/Cover in Bulgaria in the Last Century

All types of farm agricultural land occupied a total of 5,718,300 ha in 1939; 5,721,400 ha in 1950; 6,021,700 ha in 1969; and 6,206,400 ha in 1979.

Total arable land was 4,788,300 ha in 1939; 4,859,900 ha in 1950; 4,799,300 ha in 1969; and 4,725,300 ha in 1979.

Areas occupied by fields were 4,296,700 ha in 1939; 4,343,800 ha in 1950; 4,158,200 ha in 1969; and 3,897,200 ha in 1979.

Cereals occupied areas of 2,929,400 ha in 1948; 2,536,500 ha in 1960; 2,256,500 ha in 1969; and 2,349,800 ha in 1979.

Irrigated areas occupied a total of 1,185,222.4 ha in 1979, where the amount with gravity irrigation was 722,921.1 ha and the amount with furrowed rainwater was

458,287.5 ha. The total area of irrigated cereals was 483,213 ha in 1979.

15.2.2 Facts Related to Land Use/Cover in Bulgaria of the Forest Fund

The total forest area comprises all wooded, non-wooded, and non-timber areas. Deciduous high-stem forest includes oaks (*Quercus* L.), beech (*Fagus* L.), hornbeam (*Carpinus* L.), elm (*Ulmus* L.), ash (*Fraxinus* L.), chestnut (*Castanea* L.), cerris (*Quercus* L.), linden (*Tilia* L.), and others that have been grown from seed and planted. Coniferous forests include pine (*Pinus* L.), spruce (*Picea* L.), fir (*Abies* L.) and others that have been grown from seed. Non-coniferous short-stem forests include the same kind of trees, but grown from suckers.

Areas not afforested include glades, burnt-down areas, wasteland, forest paths, rocks, arable land and meadows, marshlands, and others.

Reserve forest areas are around historical and natural monuments, areas that preserve rare flora and fauna, and areas with natural centuries-old forests.

The total area of forest fund was 4,138,147 ha in 2010; 4,130,896 ha in 2009; 4,108,494 ha in 2007; 4,090,000 ha in 2006; and 4,077,000 ha in 2005 (Agrarian Report 2006, 2007, 2008, 2009, 2010, 2011). Of this, forested land was 3,737,542 ha in 2010; 3,725,494 ha in 2009; and 3,680,384 ha in 2007. Dwarf pines occupied 23,757 ha in 2010; 23,635 ha in 2009; and 23,631 ha in 2007.

In 2010, the unforested land was 70,758 ha (including burnt-down areas 2,818 ha, felled areas 57,644 ha, bare land 10,296 ha); land that did not produce wood was 306,090 ha (including cultivated land 4,513 ha, meadows 3,805 ha, glades 111,162 ha; roads and clearings occupied 42,665 ha, rocks, rivers, taluses, etc., 306,090 ha).

In 2009, the unforested land was 73,959 ha (including burnt-down areas 3,302 ha, felled areas 11,005 ha, bare land 59,652 ha; land that did not produce wood was 307,808 ha (including cultivated land 4,759 ha, meadows 4,026 ha, glades 110,969 ha; roads and clearings occupied 43,323 ha, rocks, rivers, taluses, etc., 143,046 ha).

In 2007, the unforested land was 93,081 ha (including burnt-down areas 5,364 ha, felled areas 16,105 ha, bare land 71,612 ha); land that did not produce wood was 310,889 ha (including cultivated land 4,239 ha, meadows 3,934 ha, glades 110,508 ha; roads and clearings occupied 43,442 ha, rocks, rivers, taluses, etc., 143,739 ha).

Total area occupied by deciduous forests was 2,615,125 ha or 69.5 % in 2010; 2,602,666 ha or 69.4 % in 2009; and 2,565,571 ha or 69.7 % in 2007.

Total area occupied by coniferous forests was 1,145,781 ha or 30.5 % in 2010; 1,146,463 ha or 30.6 % in 2009; and 1,114,813 ha or 30.3 % in 2007.

The area of natural habitats was 2,925,125 ha, of which 476,441 ha or 16.3 % was coniferous in 2009; and 2,844,668 ha, of which 433,913 ha or 15.25 % was coniferous in 2007.

According to the functions the forests can be wooded and forest land for wood production was 2,554,564 ha or 61.7 % in 2010; 2,831,644 ha or 68.5 % in 2009; and 2,381,132 ha or 68.1 % in 2007.

Protected and recreational lands occupied 920,247 ha or 22.3 % in 2010; 963,871 ha or 23.4 % in 2009; and 963,871 ha or 23.7 % in 2007.

Protected natural areas occupied 663,336 ha or 16.0 % in 2010; 335,381 ha or 8.1 % in 2009; and 335,381 ha or 8.2 % in 2007.

State forests were 3,066,771 ha or 74.1 % in 2010; 3,072,976 ha or 74.4 % in 2009; and 3,099,890 ha in 2007.

Municipal forests occupied 503,694 ha or 12.2 % of the total forest fund in 2010; 500,754 ha or 12.1 % in 2009; and 478,624 ha in 2007.

Private individual ownership was 421,885 ha or 10.2 % in 2010; 421,165 ha or 10.2 % in 2009; and 408,619 ha in 2007.

Legal entities owned 29,945 ha or 0.7 % in 2010; 29,467 ha or 0.7 % in 2009; and 24,813 ha in 2007.

Religious communities owned 23,243 ha or 0.6 % in 2010; 23,460 ha or 0.6 % in 2009; and 23,484 ha in 2007.

The area of agricultural land within the forest fund was 92,609 ha or 2.2 % in 2010; 83,074 ha or 2.0 % in 2009; and 73,064 ha in 2007.

The area of forested territory affected by fires was 6,526 ha in 2010 and 43,000 ha in 2007. The main causes of forest fires were the burning of stubble, which accounted for 53.5 % in 2010 and 49 % in 2007, and negligence, which accounted for about 30.0 %.

15.2.3 Facts Related to Land Use/Cover in Bulgaria of Forest Fund in the Last Century

Total forest area was 3,671,000 ha in 1955; 3,635,000 ha in 1960; 3,709,000 ha in 1970; and 3,839,000 ha in 1979.

Total wooded area was 3,153,000 ha in 1955; 3,190,000 ha in 1960; 3,162,000 ha in 1970; and 3,281,000 ha in 1979.

Coniferous type forest was 558,000 ha in 1955; 636,000 ha in 1960; 953,000 ha in 1970; and 1,231,000 ha in 1979.

Deciduous high type forest was 1,709,000 ha in 1955; 1,676,000 ha in 1960; 1,830,000 ha in 1970; and 1,661,000 ha in 1979. Deciduous low-type forest was 1,404,000 ha in 1955; 1,323,000 ha in 1960; ,926,000 ha in 1970; and 947,000 ha in 1979.

References

Agrarian Report (2006, 2007, 2008, 2009, 2010, 2011) Department of Agro-statistics at the Ministry of Agriculture and Food of Bulgaria, Sofia

Atanasov I (2006) Soil degradation processes and options for sustainable land management in Bulgaria, Project "Capacity Building for Sustainable Land Management in Bulgaria", Baseline Report, Sofia, p 303

Huber S et al (2009) Environmental assessment of soil for monitoring. Volume I: Indicators and Criteria. EUR 23490 EN/1. JRC 47184, Office for official publication of the European Communities, Luxembourg, p 339

Jones A et al (2012) The state of soil in europe. A contribution of the JRC to the european environment agency's environment state and outlook report—SOER 2010, JRC reference reports, european commission, EUR 25186 EN-N, Joint research centre 68418, Institute for environment and sustainability. Publication office of the european union, Luxembourg. 76 pp

Rousseva S (2006) Stocktaking of land degradation in agriculture. Discussion report. Project 00043507 "Capacity Building for Sustainable Land Management in Bulgaria". http://www.unccd-slm.org/files/LD_Agri_ENG_Svetla%20Rousseva.pdf

Rousseva S, Stefanova V (2006) Assessment and mapping of soil erodibility and rainfall erosivity in Bulgaria. In: Morel M et al. (eds) BALWOIS 2006 conference on water observation and information system for decision support, 23–26 May 2006, Ohrid, Republic of Macedonia. Book of abstracts (ISBN 9989-9594-1-2), pp 28–29. http://balwois.com/balwois/administration/full_paper/ffp-607.pdf

Rousseva S, Lazarov A, Stefanova V et al (2006) Soil erosion assessments in Bulgaria. In: Morel M et al. (eds) BALWOIS 2006 conference on water observation and information system for decision support, 23–26 May 2006, Ohrid, Republic of Macedonia. Book of abstracts (ISBN 9989-9594-1-2): 29–30. http://balwois.com/balwois/administration/full_paper/ffp-608.pdf

Rousseva S et al (2010a) Soil erosion risk in Bulgaria and recommendations for soil protective use of agricultural lands. Part I Northern Bulgaria, Agricultural Academy. Institute of Soil Science Nikola Poushkarov, PublishSaySet-Eco, Sofia

Rousseva S et al (2010b) Soil erosion risk in Bulgaria and recommendations for soil protective use of agricultural lands. Part II Southern Bulgaria, Agricultural Academy. Institute of Soil Science Nikola Poushkarov, PublishSaySet-Eco, Sofia

Yancheva D et al (2011) Statistical yearbook 2010. In: Blatski L et al. (eds) National Statistical Institute. Education and Science inc. Co.

16 Soils and Humans

16.1 Formulation of the Issue

In the context of globalization of the planet, the relation of soils and humans is a decisive matter directly related to the quality of life of mankind. Our relation to soil is connected to all the components of the environment, including land, air, and water, that are vital conditions for human existence. The surface of the dry land on earth is 148,920,000 km^2 or 29.2 %, and only 13.13 % of land is fit for cultivation.

The presence of chlorophyll and protein, both of which are closely related to soil, may make the Earth a unique planet: soil, as a natural earthly reality of intrinsic functions, perhaps distinguishes Earth from all the other planets in the Universe. Until scientifically proven otherwise, this hypothesis remains valid.

During the early Pliocene Period 4.4 million years ago, in the jungles of what is now Ethiopia, a hominine named *Ardipithecus ramidus* evolved that was able to walk upright. In conditions of climate change, with the onset of the Ice Age 3 million years ago, a pre-human species named *Australopithecus* evolved that lived a nomadic existence in order to survive. From the middle of the Pleistocene Period around 2 million years ago, in East Africa, *Australopithecus* evolved into *Homo erectus,* a species that developed brain capacity, used fire, cooked meat, and lived in temporary dwellings. *Homo sapiens* originated as primates 200,000 years ago in East Africa.

The relationship between soils and humans is an intransitive global issue that is gaining greater critical importance over time, that is, since the appearance of the modern *Homo sapiens* 100,000 years ago, through migration to a Europe inhabited by *Neanderthals* 40,000 years ago, up to the present when there are 7 billion people on the planet. During the Neolithic Period, humans began to tame the environment and use what nature had provided. *Homo sapiens* culture began to develop at a faster rate around 30,000 years ago. Modern human behavior involves component skills—abstract thinking, planning, innovation, and communication. Since the dawn of human civilization, humans have organized themselves into bigger and bigger communities and aimed to gain control over natural resources. About 12,000 years ago, humans passed a significant milestone in the development of civilization when they started to build monuments, thereby changing the landscape. Nearly 20,000 years ago, humans developed agriculture, which 10,000 years ago revolutionized ancient society as a new technology, and 7,000 years ago spread to western Europe. About 3,000 years ago, the script appeared in Mesopotamia known as cuneiform not as symbols look like but as sound like. There was a time when fertility was deified. It was a story about genres of art. It will continue to exist for centuries as long as there are soils and humans.

The relationship between human beings and soils is primarily a relationship between man and earth. This fact permeates the development of mankind in all civilizations, ages and historical social formations.

Human thoughts are the manner how he perceives the world around him. In ancient Rome, a classical maximum of the land was formulated. It is *locus standi* "place of existence and medium for action." On this basis, related to the category of "land ownership," classical Roman law formulated fundamental notions, namely *nuda propretas*—property nude, *res nullius*—nobody's property, and *res communis*—common property, nowadays updated in modern regulation as land ownership, and possession and use of land, including management of land by man. In general, the management of the soil is targeted to meet man's need for agricultural products.

Thus the nature of the relationship between soils and human beings is complex. Soil is a natural phenomenon *sui generis* to humans; it is fundamental to mankind, a necessary, nonrenewable, objective condition for the existence, development, and prosperity of the human race. Soil is fundamental to the reproduction of biota including human beings. Despite advances in technology, man has so far been unable to produce or reproduce soil as an industrial product.

T. Shishkov and N. Kolev, *The Soils of Bulgaria*, World Soils Book Series, DOI: 10.1007/978-94-007-7784-2_16,

The concept of soil under discussion focuses mainly on the soil as a source of goods vital to man and society. Because of the nature of the relationship between soils and humans, this is a global issue. Developments in marine science may present us with the option to live on areas of the seabed; and developments in space science may also allow people to settle on other planets. These possibilities are not utopian—they are related to the growing imbalance between population growth on the planet and quality of life, and the advance of desertification, erosion, and aggression to the land caused by oceans and due to climate change.

Human activity alters the factors responsible for soil functions and ambient soil conditions. Awareness of the challenges created by global climate change and an increasing human population should lead us as a global society to seek solutions by making changes, albeit difficult changes, in our value systems. Modern civilization is characterized by the use of electricity and fossil fuel consumption. In the course of one day, humans burn as much carbon dioxide as was stored by nature for 11,000 years.

Effective strategies are required with regard to forest clearing and bioenergy production. A significant part of the problematic aspects regarding human attitudes toward soils can be covered by legal regulation.

The scientific knowledge of soils expands our capacity to find answers to problems and formulate creative solutions. Attempts to utilize existing facts to make mathematical models for the purposes of prediction have become important and are possible when quantitative databases become available.

16.2 National and Global Issue

The issue concerning the soils and humans is common to all countries around the world; it does not exclude particular regions or presuppose special treatment for others.

Nature generously provides various natural and ready-to-use resources, which are crucial not only to strengthen the economy and raise the standard of material well-being, but also to develop culture, science, education, and health care. The range and importance of such resources are constantly expanding along with scientific and technical progress. All these natural gifts have to be preserved, so that they to continue to bring maximum benefit in the future.

At the turn of nineteenth and twentieth centuries, the science and practice based on the belief that there was an inexhaustible wealth of natural renewable resources was finally discarded. It was this powerful realization that led to the idea that we need to be lean in the use of all natural resources. The protection of nonrenewable resources (soils, minerals, and mineral raw materials) consists primarily in the highest efficiency of their use with the lowest losses.

This was one of the motivations for President Kennedy's special Message on Conservation to the US Congress on 1 March, 1962, where he appealed for national thrift with regard to all the country's natural resources, to prevent them from loss and depredation. The speech of the President of the USA at the XVIII session of the General Assembly of the United Nations in 1963 strongly encouraged the adoption of the Global Program on conservation, not providing any difference in the legal status of natural resources.

The XVII session of the General Assembly of the United Nations of 18 December 1962, approved the resolution on "Economic Development and Conservation." The progressive idea was aired that conservation and productiveness are natural partners, not contradictory forces. The recommendation concerning the Safeguarding of the Beauty and Character of Landscapes and Sites, was adopted by the General Conference at its 12th Session, UNESCO, Paris, 11 December 1962. Mutual cooperation in the area of rational nature management is a progressive and inevitable factor of contemporary international relations. This entails an objective knowledge about the laws of nature and certain international legal provisions regarding a number of natural resources (including soils), and the socioeconomic and cultural interests of the people.

Sovereignty of a state extends to the airspace of its territory, but not to the air itself. State boundaries cannot stop the air circulation, flow of rivers, and disposal of the areal with soils, flora, and fauna. Hence, the coordination of country's programs on the rational use of national natural resources is an absolute necessity. The boundaries of biogeocenosis (origin from the Greek *bios*—life, *ge*—earth, *kainos*—common) for the most part do not correspond with state boundaries, but spread over the territory of more than one country.

Many international organizations are involved in the issue of soil conservation and protection. The Food and Agricultural Organization (FAO) was founded in 1945. In 2011, the framework of the Global Partnership for the Soil (GPS) FAO highlighted the importance of productive soil resources as key to maintaining both the food supply and the ecosystem. Recognizing this crucial role, FAO recommended that soils be placed at the top of the agenda of development programs worldwide. The International Union of Soil Sciences (IUSS) was originally founded as the International Society of Soil Science (ISSS) in Amsterdam in 1924. In 1998 it was reorganized, and its activity supports a strong relationship with FAO and UNESCO. The Bulgarian Soil Science Society is a member of the IUSS. Statutes of the International Union for Conservation of Nature and Natural Resources (IUCN) were approved on 5

October 1948 at the Assembly in Fontainebleau, France, when soils are included in Chap. 1 (2).

In 1992, an international environmental treaty was signed by 154 nations at the United Nations Conference on Environment and Development known as the Earth Summit held in Rio de Janeiro.

In 2002, when the United Nations adopted the Millennium Development Goals toward ending world poverty, it was noted that while they included environmental sustainability, agriculture was not mentioned.

In the sixth Community Environment Action Program OJ L242, decision 1600/2002/EC of the European Parliament and at the Council of 22 July 2002, soil was included alongside water and air, and sustainable land use was promoted.

On 16 April 2002, the Commission adopted a Communication (COM (2002) 179) toward a Thematic Strategy for Soil Protection. The European Soil Framework Directive and Impact Assessment that ensued were debatable in terms of ease of use and relevance of soil data. At the present time, long-term soil protection policies are legally enforced by the Waste and Water Framework Directives, CAP, and Flood Directive. Spatial soil information is regulated by the INSPIRE Directive Annex III, which came into force in 2007 to make spatial or geographical information more accessible and addresses 34 spatial data themes and between these is soil.

The present time is known as a period of global climate change, which raises particular concerns about consequences for soil quality in the future. Under the auspices of the United Nations, there is an annual conference at which world leaders aim to reach a common position to deal with the growing climate crisis and to limit the rise of global temperature to no more than 2 °C.

16.3 Relation Between Soils and Humans

The problem in the relationship between soils and humans is the given primacy of the soil, which exists as an objective natural phenomenon outside and independently of man. The historical development of mankind is related to the progressive development of man's activity on the land and management of the impact he has on its management for targeted impact of man on the ground.

The natural resource of land and its component soils have historically been the source of constant enmity and wars of conquest, distribution, and redistribution. Man becomes a productive force when he influences soils in a conscious, systematic, and effective way for rational and productive use. Increase in the quantity and the quality of demand for agricultural products, both fresh and processed, is determined by many factors.

The scientific and theoretical study of practical problems in the relation of soils and humans is based on their linkage and dependance—soil as the granted universal object and man as a subject of ownership, possession, and use of land. The relation of soils and humans prompts the current requirements for the concrete establishment and evaluation of factors that determine soil fertility.

The natural fertility of soils is the result of chemical, physical, biological, and ecological characteristics which are usually determined by natural conditions. Natural fertility occurs without human intervention and investment, while artificial fertility is the result of human efforts, according to the agroecological approach to a specific area. Man's use of the land for science and technology purposes, and standards of national and international legislation do not in themselves reduce fertility; on the contrary, they increase it in line with the parameters of soil fertility. As owners, possessors, and users of land, humans own rights to the land in commercial terms such as leases, rents, and concessions, among others; everything is subject to market interests.

Farmers determine the directions of soil management, taking into account a set of informative sources related to demand, market, health, climate, and other agroecological conditions. Through enter into a contract, farmers orient directions of agricultural production—for direct marketing or for processing in the food, textile, pharmaceutical, cosmetic, diet, nutrition, and other industries. At domestic and international markets, fairs, and auctions competitive farmers are identified by a logo (brand) and a certificate to confirm the ecological purity of the product.

Farmers develop economical soil fertility for economic purposes, as facilitated by scientific, and technological innovations; and the development of new varieties and other innovations continuously enhances their professional skills. Farmers have legal clout with a wide range of powers in production and trade, and they are a party to a wide range of contractual arrangements and relationships. The landowner is defined as the person or entity (farmer company) who possesses competence and the requisite legal capacity, rights, and obligations, and holds legal responsibility—administrative or penal—for the infringement of law or contract.

The relationship between soils and humans defines man mainly as a manager, farmer, and producer who implements agricultural production on a competitive basis on the stock exchange market.

Man is a legal entity with wide-ranging powers in production and trade, with a wide range of contractual and regulatory arrangements with partners across the world. In addition to national legislation, contemporary man is bound by the provisions of the European Union.

Investing in professionalism, knowledge, and culture determines the optimum efficiency of soil fertility and functions. This investment is not absolute as a source market, but requires a reasonable agrarian policy in order to synchronize market and regulatory provisions.

The role played by differential ground rent becomes increasingly important as an economic category that reflects production and the relationships between people and between market and regulatory provisions.

16.4 The Contract for Land Rent

In general, the landowner is a land user. However, the common practice is that the renter of land is a land user who manages soil. In Bulgaria, in the last 20 years the rent contract has been the primary regulator of the relationship between the lessor and the renter of land, and is associated with omissions and contradictions in favor of the land renter. Before such contracts came into being in 1944, there was a metayage in the country.

Historically, land rent agreements arose with the emergence of private land ownership and developed in all countries of the world. For example, in France, they covered 58 % of the land in 1892, 42 % in the United States in 1940, and 75 % in England in 1932. The rent is a contract between the lessor and the renter to provide land for temporary use for income against an agreed fee.

However in the case of Bulgaria, this is not the typical renting arrangement. More often than not, the purchase of land is controlled by legal persons who are not involved in farming and who aim to lease the land or often the unregulated purchase of land is made by legal persons who are not engaged in farming with the aim of leasing the land. The analysis of mass tenancy practice in Bulgaria shows that it is associated with poor management of renting land, practices detrimental to the lessor, systematic reduction of soil fertility, accelerated soil erosion, and negative ecological impact. The contract and agrarian law do not guarantee equal relations to the detriment of the lessor, as follows: intensive land use, poor environmental protection of soils, conquest by the renter of land, non-payment, or late payment of the agreed equivalent in cash or in kind, and with European Union subsidies to create benefits for monopoly (cereal) and discrimination of crucial products. It is imperative that the contents of the land rent agreement are structured in such a way that it emphasizes the obligations of the renter as follows: responsibility for guarding against violations, maintenance of quality, increase of soil fertility, effective soil protection and generational transmission of this national wealth. In this regard, the rent agreement serves the functional economic relationship between man and land.

16.5 Moral and Ethical Aspects

Historically, along with the legal regulation of the relation between soils and humans, there have been public regulators in the form of morals, ethics, traditions, and customs which acquired crucial importance in civil society. Law, morals, and religion as legal and social systems appeared after the emergence of human society.

The term morals originates from the Latin *mores,* meaning a value system of principles dealing with our manner of behavior and attitude toward the world. Morals and the law can be applied only within society. Moral behavior is expected from individuals within society in the interests of the community. After the law, morals are the second source of law in countries with legal systems. Morals are closely related to human ethics.

Ethics contain rules for human behavior, principles for human action, priorities and values which are the basis for the preservation, and continuation of organized social life.

Ethics as a philosophy developed in ancient Greece in the fifth to sixth century BC, and has the original meaning of temper, custom, and character. It is concerned with the fundamental relationship between the individual and the social world. Ethical principles should be respected along with human rights—civil, political, economic, social, and cultural—in accordance with the constitution, laws, international agreements, programs, and strategies. In addition to human rights there are human obligations. Ethical arguments are involved in the decisions and long-term plans made by leaders and experts.

In 1933, during Franklin Roosevelt's (1933–1945) presidency of the United States, governmental programs were implemented that were designed to conserve soil and restore a sustainable environment. The Soil Conservation Service was established in 1935. Also, in 1935 President Roosevelt announced the most successful program of afforestation in the United States (Sauer et al. 2011). In 1937, the federal government began a campaign to educate farmers on soil conservation practices through paying a dollar an acre for stimulating new methods.

Today, in the technologically advanced era, the demands of human society in terms of industry, energy, and food are determined on global markets. There is a common human feeling that one man is entitled to have what another man has. In a world characterized by have and have-nots, i.e., overproduction of agriculture in some countries and the existence of millions of hungry people in others, there has been an unprecedented conflict in the consumption of land. Areas suitable for agriculture are gradually decreasing as they are converted into other uses such as energy production.

Individual efforts have had little effect on soil protection, but collaboration among people or companies is made difficult by competing interests. This raises the question whether the public incentive for long-term policies to protect soil fertility will diminish. It is essential to reassess the investment priorities of traditional land management practices.

The diversity of land users needs to be borne in mind. Actions in society that increase man's satisfaction and sense of well-being can only be viewed as positive. Humans act according to procedures established in society and since society is subject to change, so the nature of these acts is subject to change. In the past, poor practices and behaviors have been accepted by society. People should come to realize that moral, ethical values, and priorities are ultimately in their own interests and act in consciousness of this realization. Human feeling and behavior is cultivated. Our children need to be cultivated in such way that they come to love and to value the earth.

References

IUCN (1948) Statutes of the International Union for Conservation of Nature and Natural Resources (IUCN) approved in October 5th, 1948 at the Assembly in Fontainebleau, France. Amendments to IUCN Statutes. In: Proceedings of the IUCN 8ts general assembly, Nairobi, Kenya, Sept 1963, Morges, 1964, p 126

Sauer TJ, Norman JM, Mannava Sivakumar VK (2011) Sustaining soil productivity in response to global climate change: science, policy, and ethics. Wiley, Oxford, p 264

Author Biography

Toma Shishkov obtained a degree in Agronomy (1983) at the Agricultural University in city of Plovdiv, Bulgaria. The first professional years covered the field of large-scale soil survey. He received his postgraduate education at the Institute of Soil Science "V.V. Dokuchaev" in Moscow, Russia, where he obtained a Ph.D. degree in the subject of Soil Sciences (1993). Nowadays, he is an Assistant Professor at the Department of Soil Genesis, Geography and Classification of the Institute of Soil Science, Agrotechnologies and Plant Protection "Nikola Poushkarov," Sofia, where he conducts research on soil pedology, classification, and soil conservation.

Nikola Kolev obtained a degree in Electronic Engineering (1968) at the Technikal University in Sofia, Bulgaria. He obtained a PhD degree in the subject of application of electronic in agriculture (1978). He was a Professor at the Department of Soil Physics, Institute of Soil Science "Nikola Poushkarov" since 1998. He was a Director of the Institute of Soil Science "Nikola Poushkarov" (2003–2008), and a Chief Scientific Secretary of the Agricultural Academy until his retirement in 2011. Nowadays, he teaches mainly on soil science, monitoring of soil resources, application of remote sensing in agriculture and the spatial analisys of data.

T. Shishkov and N. Kolev, *The Soils of Bulgaria*, World Soils Book Series, DOI: 10.1007/978-94-007-7784-2,

Index

T. Shishkov and N. Kolev, *The Soils of Bulgaria*, World Soils Book Series, DOI: 10.1007/978-94-007-7784-2,

MIX
Papier aus verantwortungsvollen Quellen
Paper from responsible sources
FSC® C105338
FSC
www.fsc.org

If you have any concerns about our products,
you can contact us on
ProductSafety@springernature.com

In case Publisher is established outside the EU,
the EU authorized representative is:
Springer Nature Customer Service Center GmbH
Europaplatz 3, 69115 Heidelberg, Germany

Printed by Libri Plureos GmbH
in Hamburg, Germany